Wetting Experiments

Wetting: Theory and Experiments
Two-Volume Set

Eli Ruckenstein
Gersh Berim

Volumes in the Set:

Wetting Theory (ISBN: 9781138393301)

Wetting Experiments (ISBN: 9781138393332)

Wetting Experiments

Eli Ruckenstein

Gersh Berim

CRC Press is an imprint of the
Taylor & Francis Group, an **informa** business

CRC Press
Taylor & Francis Group
6000 Broken Sound Parkway NW, Suite 300
Boca Raton, FL 33487-2742

© 2019 by Taylor & Francis Group, LLC
CRC Press is an imprint of Taylor & Francis Group, an Informa business

No claim to original U.S. Government works

Printed on acid-free paper

International Standard Book Number-13: 978-1-1383-9333-2 (Hardback)

This book contains information obtained from authentic and highly regarded sources. Reasonable efforts have been made to publish reliable data and information, but the author and publisher cannot assume responsibility for the validity of all materials or the consequences of their use. The authors and publishers have attempted to trace the copyright holders of all material reproduced in this publication and apologize to copyright holders if permission to publish in this form has not been obtained. If any copyright material has not been acknowledged please write and let us know so we may rectify in any future reprint.

Except as permitted under U.S. Copyright Law, no part of this book may be reprinted, reproduced, transmitted, or utilized in any form by any electronic, mechanical, or other means, now known or hereafter invented, including photocopying, microfilming, and recording, or in any information storage or retrieval system, without written permission from the publishers.

For permission to photocopy or use material electronically from this work, please access www.copyright.com (http://www.copyright.com/) or contact the Copyright Clearance Center, Inc. (CCC), 222 Rosewood Drive, Danvers, MA 01923, 978-750-8400. CCC is a not-for-profit organization that provides licenses and registration for a variety of users. For organizations that have been granted a photocopy license by the CCC, a separate system of payment has been arranged.

Trademark Notice: Product or corporate names may be trademarks or registered trademarks, and are used only for identification and explanation without intent to infringe.

Library of Congress Cataloging-in-Publication Data

Names: Ruckenstein, Eli, 1925- author. | Berim, G. O. (Gersh Osievich), author.
Title: Wetting experiments / Eli Ruckenstein and Gersh Berim.
Description: Boca Raton : Taylor & Francis, a CRC title, part of the Taylor & Francis imprint, a member of the Taylor & Francis Group, the academic division of T&F Informa, plc, 2018. | Includes bibliographical references and index.
Identifiers: LCCN 2018034126 | ISBN 9781138393332 (hardback : acid-free paper) | ISBN 9780429401824 (ebook)
Subjects: LCSH: Wetting. | Surfaces (Technology)
Classification: LCC TP156.W45 R83 2018 | DDC 668/.14--dc23
LC record available at https://lccn.loc.gov/2018034126

Visit the Taylor & Francis Web site at
http://www.taylorandfrancis.com

and the CRC Press Web site at
http://www.crcpress.com

Contents

Preface vii
Authors ix

Chapter 1 Biology Related Experiments 1

Eli Ruckenstein and Gersh Berim

1.1 A Nondestructive Approach to Characterize Deposits on Various Surfaces 2
Eli Ruckenstein, Sathyamurthy V. Gourisankar, and R. E. Baier

1.2 A Surface Energetic Criterion of Blood Compatibility of Foreign Surfaces 8
Eli Ruckenstein and Sathyamurthy V. Gourisankar

1.3 Surface Characterization of Solids in the Aqueous Environment 25
Sathyamurthy V. Gourisankar and Eli Ruckenstein

1.4 Preparation and Characterization of Thin Film Surface Coatings for
Biological Environments 28
Eli Ruckenstein and Sathyamurthy V. Gourisankar

Chapter 2 Experiments on Wetting of Polymers 61

Eli Ruckenstein and Gersh Berim

2.1 Environmentally Induced Restructuring of Polymer Surfaces and Its
Influence on Their Wetting Characteristics in an Aqueous Environment 62
Eli Ruckenstein and Sathyamurthy V. Gourisankar

2.2 Surface Restructuring of Polymeric Solids and Its Effect on the Stability
of the Polymer-Water Interface 77
Eli Ruckenstein and Sathyamurthy V. Gourisankar

2.3 Stability of Polymeric Surfaces Subjected to Ultraviolet Irradiation 87
Sang Hwan Lee and Eli Ruckenstein

2.4 Estimation of the Equilibrium Surface Free Energy Components of
Restructuring Solid Surfaces 94
Eli Ruckenstein and Sang Hwan Lee

2.5 Surface Restructuring of Polymers 104
Sang Hwan Lee and Eli Ruckenstein

2.6 Adsorption of Proteins onto Polymeric Surfaces of Different
Hydrophilicities—A Case Study with Bovine Serum Albumin 113
Sang Hwan Lee and Eli Ruckenstein

vi Contents

Chapter 3 Experiments on Wetting by Catalysts 129

Eli Ruckenstein and Gersh Berim

3.1 Redispersion of Platinum Crystallites Supported on Alumina–Role of Wetting.. 130
Eli Ruckenstein and Y. F. Chu

3.2 Redispersion of Pt/Alumina via Film Formation 143
I. Sushumna and Eli Ruckenstein

3.3 Events Observed and Evidence for Crystallite Migration in Pt/Al$_2$O$_3$ Catalysts.. 162
I. Sushumna and Eli Ruckenstein

3.4 Role of Physical and Chemical Interactions in the Behavior of Supported Metal Catalysts: Iron on Alumina—A Case Study 184
I. Sushumna and Eli Ruckenstein

3.5 Wetting Phenomena during Alternating Heating in O$_2$ and H$_2$ of Supported Metal Crystallites... 232
Eli Ruckenstein and J. J. Chen

3.6 The Behavior of Model Ag/Al$_2$O$_3$ Catalysts in Various Chemical Environments.. 244
Eli Ruckenstein and Sung H. Lee

3.7 Role of Wetting in Sintering and Redispersion of Supported Metal Crystallites.. 263
Eli Ruckenstein

3.8 Optimum Design of Zeolite/Silica–Alumina Catalysts 269
Sung H. Lee and Eli Ruckenstein

3.9 Effect of the Strong Metal-Support Interactions on the Behavior of Model Nickel/Titania Catalysts... 292
Eli Ruckenstein and Sung H. Lee

3.10 Simulation of the Behavior of Supported Metal Catalysts in Real Reaction Atmospheres by Means of Model Catalysts................................... 312
Sung H. Lee and Eli Ruckenstein

Index... 371

Preface

This book contains papers published by Professor Ruckenstein and his coworkers on the theoretical and experimental investigation of wetting of solid surfaces. It is one of two standalone books that comprise *Wetting: Theory and Experiments, Two-Volume Set*, which contains six chapters, each of which is preceded by a short introduction. Each volume making up the set is available to be read and understood on its own. Reading both volumes together provides the reader with a comprehensive view of the subject. The papers of the chapters are selected according to the specific features being considered and they are arranged in logical rather than chronological order. The main attention is given to the wetting on the nanoscale (nanodrops on solid surfaces, liquid in the nanoslit) considered on the basis of microscopic density functional theory, and to dynamics of fluid on the solid surface considered on the basis of hydrodynamic equations. Along with this, experimental studies of wetting related to various applications are presented. A description of the contents of each volume within the Set follows.

Wetting Theory (ISBN: 9781138393301):

In Chapter 1, various microscopic processes (static and dynamic) in a liquid in contact with a solid are considered. They are the flow of liquid along horizontal and inclined surfaces, slipping of the contact line of a liquid on a solid, etc.

Chapter 2 is about the symmetry breaking of fluid density distribution in nanoslit between parallel solid walls. One component classical and quantum fluids and binary mixtures are considered and conditions for symmetry breaking to occur are examined.

In Chapter 3, the microscopic approach is applied to the treatment of macroscopic drops on smooth or rough, planar or curved, solid surfaces. It is based on fluid–fluid and fluid–solid interaction potentials and considers the drop equilibrium state as that having the minimum of total potential energy. The concept of microscopic contact angle is introduced and both macroscopic (classical) contact and microscopic angles are calculated.

In the next chapter, Chapter 4, a liquid drop on a smooth or rough planar solid surface is examined on the basis of a microscopic nonlocal density functional theory in canonical ensemble. The variety of characteristic features are examined including nonuniform fluid density distribution inside the drop, drop profile, microscopic contact angle, sticking force, etc. The results are compared with predictions of classical theories for macroscopic drops and similarities and differences between them are analyzed.

In the last chapter, Chapter 5, the theory of the rupture of liquid and solid films is developed, first in the linear approximation, and then extended to the case of perturbations of finite amplitude. The theory provides, in particular, the conditions for the instability, the dominant wavelength of the disturbances, and the time of rupture of the films. Along with this, the rupture of liquid films supported on a solid surface is examined on the basis of a thermodynamic approach which considers the change of the free energy of the film after the formation of a hole in it. The theory is applied to the practically important problem of tear film stability and rupture.

Wetting Experiments (ISBN: 9781138393332):

This volume focuses on experimental studies of wetting that are related to biological problems, polymers, and catalysts. The biology-related studies are devoted to the problem of selecting synthetic materials for use in biological media. The polymers are examined to estimate experimentally various surface characteristics such as the ability of polymeric solids to alter their surface structures between different environments in order to minimize their interfacial free energy. The investigation of catalysts concentrates on their physical and chemical changes, formed of small crystallites of Pt, Pd, Ni, Co, Fe, or Ag supported on alumina.

Authors

Eli Ruckenstein, National Academy of Engineering and National Academy of Art and Science member, National Medal of Science winner, SUNY Distinguished Professor of CBE Eli Ruckenstein's copious and pioneering contributions to chemical engineering have been rewarded with numerous distinctions, including the 2004 National Academy of Engineering Founders Award, the 2002 American Institute of Chemical Engineers (AIChE) Founder's Award, AIChE's 1988 Walker Award, the 1977 AIChE's Alpha Chi Sigma Award, the 1986 American Chemical Society's (ACS) Kendall Award, the 1994 ACS Langmuir Lecture Award, the 1996 ACS E.V. Murphree Award, the 1985 the Alexander von Humboldt Foundation's Senior Humboldt Award, and the 1985 NSF Creativity Award. Ruckenstein joined the School's faculty in 1973, and was the first full-time SUNY system professor elected to the NAE. A leading influential chemical engineer, he has made numerous contributions to modernizing research and development in key areas of chemical engineering. He is a fellow of AIChE which with the occasion of its 100th anniversary designated him as one of 50 eminent chemical engineers of the Foundation age. Dr. Ruckenstein published about 900 papers in various areas of Chemical and Biological Engineering.

Gersh Berim earned a Ph.D. degree in Physics in 1978 from Kazan State University, Russia. Till 2001, his research was focused on nonequilibrium properties of low dimensional spin system. In 2001 he joined the group of Professor Eli Ruckenstein at SUNY at Buffalo where have been studying various topics of Chemical Physics especially related to nanosystem. He has authored or co-authored more than 70 papers.

1 Biology Related Experiments

Eli Ruckenstein and Gersh Berim

INTRODUCTION TO CHAPTER 1

Chapter 1 consists of four sections and contains biology-related experimental studies of wetting. These studies are devoted to the problem of selecting synthetic materials for use in biological media. In Sec. 1.1, the suitable methods to deposit thin films on substrates of interest (such as the apatitic tooth surface or stainless-steel heat exchange surfaces) are developed. In Sec. 1.2, a surface energetic criterion of biocompatibility of foreign surfaces is suggested, which is based on an analysis of the surface interactions between a typical biological fluid (i.e. blood) and synthetic surfaces. In the second part of this investigation (Sec. 1.3), it is shown that the experimental approach involving the radio frequency sputter deposition of thin solid films of tightly adhering polymeric compounds on materials with the desired bulk characteristics is a promising method of tailoring the surface properties of many types of synthetic materials for use in biological environments. In the final part of this investigation (Sec. 1.4), the possibility of affecting a drastic reduction in the solid-water interfacial free energy of the sputtered polymer surfaces by physical and/or chemical modification of their surfaces and thereby improving their biocompatibility is illustrated.

1.1 A Nondestructive Approach to Characterize Deposits on Various Surfaces*

Eli Ruckenstein[a†], Sathyamurthy V. Gourisankar,[a] and R. E. Baier[b]

[a] Department of Chemical Engineering, State University of New York at Buffalo, Amherst, New York 14260

[b] Calspan Corporation, Advanced Technology Center, Buffalo, New York 14225

Corresponding Author

[†] To whom correspondence should be addressed.

Received February 4, 1983; accepted May 2, 1983

1. INTRODUCTION

Surface interactions play an important role in several situations such as remineralization of teeth (1), thrombus formation (2–4), metastasis (5), fouling of heat transfer surfaces by food fluids (6), and the deposition of Brownian particles or cells on collector surfaces (7–9). In all these cases, deposit formation on various types of substrates is involved.

Multiple internal reflection spectroscopy (10) is a very sensitive approach to nondestructively characterize either *in vivo* or *in vitro* films deposited on artificial substrates. In the application of this technique, the choice of a suitable internal reflection element is of great importance. The internal reflection element must not only simulate the surface of interest, but must also be transparent to the appropriate radiation (infrared, ultraviolet, or visible) over a wide range of frequencies. Frequently, however, it is found that many surfaces of interest (such as the apatitic tooth surface or stainless steel heat exchange surfaces) are unsuitable as internal reflection elements. To overcome this problem, germanium internal reflection elements have been widely used to characterize the films formed in various environments. Recently, oral *in vivo* films were also characterized on germanium internal reflection elements of varying surface energies (11). Though germanium is an ideal substrate for studies of this kind, it still does not accurately mirror the properties of the real surfaces of interest, such as the tooth surface in the oral environment.

---

* *Journal of Colloid and Interface Science.* Vol. 96. No. I, p. 245, November 1983. Republished with permission.

In this paper, we suggest an approach by which it is possible to study nondestructively, the deposits formed on real surfaces of interest, by using multiple internal reflection spectroscopy. This involves the deposition of thin solid films of the substrata of interest on suitable internal reflection elements. For illustrative purposes, we provide details concerning the preparation and characterization of a hydroxyapatite film on a germanium internal reflection element and the use of multiple internal reflection infrared spectroscopy to detect the molecular structure of salivary components adsorbed on the apatite surface.

2. DEPOSITION OF HYDROXYAPATITE ON GERMANIUM SURFACES

Hydroxyapatite ($Ca_{10}(PO_4)_6(OH)_2$) forms the major constituent ($\sim 95\%$) of the outermost part of the human tooth enamel (surface enamel). The actual tooth contains 1% of organic material which also play an important role in the salivary adsorption processes. However, here we illustrate our procedure with pure hydroxyapatite. To identify the adsorbed species on hydroxyapatite coated germanium internal reflection elements by multiple internal reflection infrared spectroscopy, the apatite coating must be considerably thinner than the depth of penetration of the infrared beam. The depth of penetration (d_p) of the beam can be calculated from (10)

$$d_p = \frac{\lambda_1}{2\pi \left(\sin^2\theta - n_{21}^2 \right)^{1/2}},$$

where λ_1 is the wavelength of the radiation in the denser medium, θ is the angle of incidence, and n_{21} ($=n_2/n_1$) is the ratio of the refractive indices of the rarer and denser media. Germanium is transparent to infrared radiation in the wavelength range of 2 to 12 μm. At the lowest wavelength (2 μm) the value of d_p for germanium surfaces coated with hydroxyapatite is of the order of 10^3 Å, for $\theta = 45°$. Therefore, to detect the adsorption of salivary components on hydroxyapatite, the apatite coating must be at least less than 1000 Å thick.

The deposition of multicomponent compounds such as hydroxyapatite, poses one major problem, namely, that of preserving the chemical composition of the bulk material when it is coated as a thin solid film on a substrate. During vacuum evaporation, these materials tend to fractionate, giving rise to chemically dissimilar films. To overcome these limitations, we tried sputtering, a technique which is widely recommended for the deposition of multicomponent materials. Sputtering also often gives rise to films which are more tightly adherent than those obtained by evaporation. In this technique, energetic particles bombard the surface of any material and eject surface atoms from it. Under sufficiently high vacuum, the sputtered atoms travel until they strike another surface and deposit there. For conducting materials, dc sputtering can be used, while for insulators, such as hydroxyapatite, radiofrequency sputtering is necessary.

To sputter the powder form of hydroxyapatite, we used a sputter-up configuration, in which, the target (hydroxyapatite, Bio-Gel HTP, Bio-Rad Labs, Richmond, Calif.) was pressed well on an aluminum plate and the plate was placed below the substrate (germanium internal reflection element) in the sputtering chamber, as shown in Fig. 1. Argon was used as the sputtering gas. Typical conditions adopted for sputtering were (a) rf power of 500 W, (b) pressure of 10 to 15 μm in the chamber, (c) target to substrate spacing of 2.5 cm, and (d) sputtering time of 2 hr.

The above conditions led to the reproducible deposition of tightly adherent films of this compound.

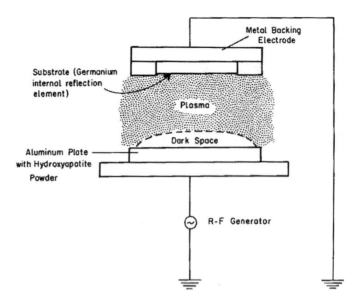

FIG. 1. Essential features of the radiofrequency sputtering setup.

3. CHARACTERIZATION OF HYDROXYAPATITE FILMS ON GERMANIUM SURFACES

The sputtered films of hydroxyapatite were characterized for their molecular structure, calcium to phosphorus ratio, surface texture and thickness, by multiple internal reflection infrared spectroscopy, energy dispersive X-ray analysis, scanning electron microscopy and ellipsometry, respectively. The infrared spectrum in Fig. 2 reveals the presence of the phosphate group (characteristic absorption band at 1070 cm^{-1}) in the sputtered film. A comparison of this

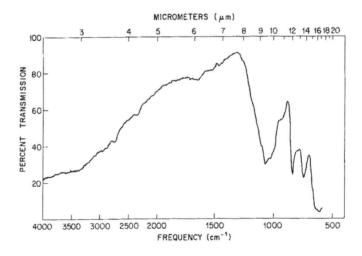

FIG. 2. Multiple internal reflection infrared spectrum of a film of hydroxyapatite, deposited on a Ge internal reflection element, under the following conditions: (a) rf power of 500 W, (b) pressure of 10 to 15 μm in the chamber, (c) target to substrate spacing of 2.5 cm, and (d) sputtering time of 2 hr.

TABLE I
Results of Energy Dispersive X-Ray Analysis

Material	(Ca/P) Ratio 1	2	3	4	$(Ca/P)_{avg}$	Remarks
Hydroxyapatite film on germanium	1.977	1.981	1.99	1.999	1.987	Count time was 192 sec for all readings
Hydroxyapatite powder (Bio-Gel HTP, Bio-Rad)	2.006	1.982	2.021	1.979	1.997	Count time was 20 sec for all readings

spectrum with that of pure hydroxyapatite powder placed on a germanium internal reflection element showed that the molecular structure of the solid film was very close to that of the bulk material. The calcium to phosphorus ratios of a thick film formed in a 3.5 hr sputtering run, are listed in Table I, along with the ratios of the bulk powder. It can be seen that the average (Ca/P) ratios of the film formed by sputtering, differed only by 0.5% from that of the bulk powder. This provides satisfactory evidence that the chemical composition of the solid film is very close to that of hydroxyapatite. By scanning electron microscopy, it was found that the films were deposited uniformly over the exposed region of the substrate, as shown in Fig. 3.

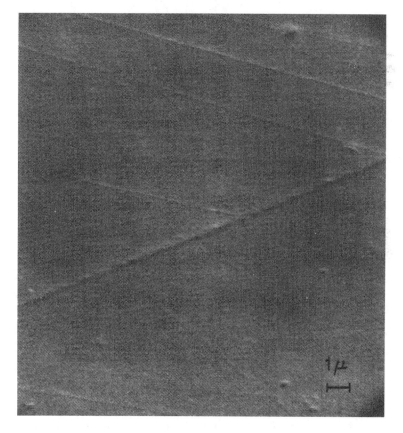

FIG. 3. SEM picture of the hydroxyapatite film of Fig. 2.

Film thicknesses measured by ellipsometry showed them to be considerably less than 1000 Å thick (about 200 Å for films formed in 2 hr sputtering experiments).

4. DISCUSSION

To test the usefulness of the sputter coated germanium internal reflection elements, an experiment was performed, in which, saliva flowed for 15 min between two parallel surfaces of apatite coated germanium internal reflection elements. The flow conditions of saliva resembled those encountered in the oral cavity. The test plates were then rinsed with distilled water for 7.5 min (to remove loosely adhering deposits) and air dried. An infrared spectrum of one of these dried surfaces is shown in Fig. 4, where protein adsorption is indicated (characteristic absorption bands at 3300, 1650, and 1550 cm^{-1}) on the hydroxyapatite surface. This shows the applicability of multiple internal reflection spectroscopy to characterize deposits formed on surfaces which closely resemble the real ones.

This approach, namely, to use appropriate procedures to deposit thin films of the substrata of interest on internal reflection elements and then to characterize the deposits formed on these surfaces by nondestructive techniques, promises to be a very useful method of studying the surface interactions in many environments. Studies on the mechanisms involved in the biofouling of heat exchangers in the dairy industry and marine environment and the adsorption of surfactants on minerals in surfactant enhanced oil recovery, seem particularly well suited to this approach.

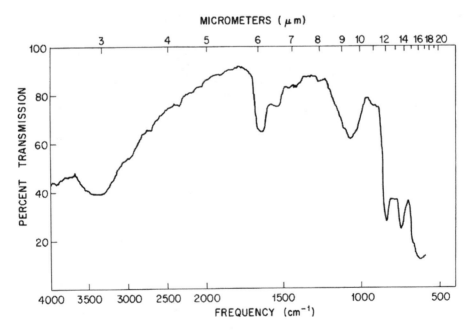

FIG. 4. Multiple internal reflection infrared spectrum of an apatite coated germanium internal reflection element, after it was exposed to saliva flowing at 1.2 ml/min for 15 min, rinsed with distilled water for 7.5 min, and air dried.

ACKNOWLEDGMENT

The authors are indebted to Dr. Cecilia Bengtsson for her help in conducting the saliva experiments, to Mr. Mark Fornalik for his help in the SEM analyses, and to Mr. E. Rastefano for his help with the sputtering equipment.

REFERENCES

1. Tencate, J. M., *in* "Dental Plaque and Surface Interactions in the Oral Cavity" (S. A. Leach, Ed.), p. 273. Proceedings of a workshop on Dental Plaque and Surface Interactions in the Oral Cavity, ICI Ltd, Macclesfield, Cheshire, England, November 7–9, 1979.
2. Marmur, A., and Ruckenstein, E., *in* "Advances in Biomedical Engineering" (D. O. Cooney, Ed.), part p. 341. Dekker, New York, 1980.
3. Gordon, J. L., and Mitner, A. J., *in* "Platelets in Biology and Pathology" (J. L. Gordon, Ed.), p. 1. North-Holland, Amsterdam, 1976.
4. Berger, S., Saltzman, E. W., Merrill, E. W., and Wong, P. S. L., *in* "Platelets: Production, Function, Transfusion and Storage" (M. Baldini and S. Ebbe, Eds.), p. 299. Grune & Stratton, New York, 1974.
5. Weiss, L., *in* "Chemotherapy of Cancer Dissemination and Metastasis" (S. Garattani and G. Franchi, Eds.). Raven Press, New York, 1973.
6. Sandu, C., and Lund, D. B., *in* "Proceedings of the Food, Pharmaceutical, and Bioengineering Division Symposium," 89th National meeting of the American Institute of Chemical Engineers, Portland, Oreg., in press.
7. Ruckenstein, E., and Kalthod, D. G., *in* "Fundamentals and Applications of Surface Phenomena Associated with Fouling and Cleaning in Food Processing," (B. Hallström, D. B. Lund, and Ch. Trägärdh, Eds.), p. 115. Proceedings of an international workshop arranged by the division of Food Engineering, Lund University, Sweden, April 69, 1981.
8. Ruckenstein, E., and Prieve, D. C., *in* "Testing and Characterization of Powders and Fine Particles," (J. K. Beddow and T. Meloy, Eds.), Heydon, London, 1980.
9. Ruckenstein, E., and Prieve, D. C., *J. Chem. Soc. Faraday Trans. 2* **69**, 1522 (1973).
10. Harrick, N. J., "Internal Reflection Spectroscopy." Interscience, New York, 1967.
11. Baier, R. E., and Glantz, P-O., *Acta Odontol. Scand.* **36**, 289 (1978).

1.2 A Surface Energetic Criterion of Blood Compatibility of Foreign Surfaces*

Eli Ruckenstein and Sathyamurthy V. Gourisankar
Department of Chemical Engineering, State University
of New York at Buffalo, Amherst, New York 14260
Received December 28, 1983; accepted April 2, 1984

1. INTRODUCTION

In the selection of biomaterials for use in long-term biomedical applications, two important criteria are involved, namely (i) the blood compatibility of the foreign surface and (ii) suitable mechanical properties for the specific blood contact application. By blood compatibility, it is meant that, when a foreign surface is placed in contact with blood, it should not provoke adverse responses, such as thrombosis, destruction of the cellular components of blood, alteration of plasma proteins, damage to adjacent tissue, and toxic and allergic reactions.

Several materials have been tried to date in the continuing search for a suitable biomaterial and although many of them exhibited some desirable virtues (like a satisfactory degree of antithrombogenicity or good mechanical properties or short-term blood compatibility), none of them has emerged as truly outstanding in terms of fulfilling the rather rigorous twin requirements of long-term blood compatibility as well as good mechanical strength. Some examples of biomaterials which have shown promise for use as blood-contacting devices, but are still far from being biocompatible on a long-term basis, include the following: (a) hydrogels or grafted hydrogels, such as those based on poly (hydroxyethyl methacrylate) and polyacrylamide, (b) LTI (low-temperature isotropic) carbons, (c) silicone–urethane copolymers (such as "Avcothane"), and (d) segmented polyurethanes (such as "Biomer").

A central feature of the search for new and better biomaterials has been the tendency to empirically correlate the performance of materials which have already been tested as blood-contacting devices (in either *in vivo, ex vivo,* or *in vitro* tests), with some characteristic properties of these materials. Many such correlations have been based on the surface properties of biomaterials, which is not surprising, in view of the fact that adsorptive events at the blood–biomaterial interfaces undoubtedly play a major part in determining the blood compatibilities of foreign surfaces. One of the earliest of such correlations was proposed by Baier (1), who suggested that, materials which possess a critical surface tension (γ_c) in the range of 20–30 dyn/cm would be more blood compatible than those with either lower or higher critical surface tensions. Nyilas *et al.* (2) have suggested that the polar component fraction of a solid's surface free energy $\left(\gamma_S^p / \gamma_S \right)$ is primarily responsible for its performance as a biomaterial. Kaelble and Moacanin (3) conclude that materials with high dispersion and low polar surface free energies will be more blood compatible than those with low dispersion and high polar surface free energies. Akers *el al.* (4) have defined a zone of biocompatibility, comprising a range of values of polar and dispersion components of solid surface free energies. Ratner *et al.* (5)

---

* *Journal of Colloid and Interface Science.* Vol. 101, No. 2. p. 436, October 1984. Republished with permission.

A Surface Energetic Criterion of Blood Compatibility of Foreign Surfaces

have suggested that a balance of polar and apolar sites on a surface may be important for its blood compatibility. Though each of the above empirical correlations may explain the observed behavior of some biomaterials, none of them can satisfactorily account for the performance of a wide variety of biomaterials tested to date. This fact led Bruck (6) to conclude that the surface energetic properties of materials cannot be correlated with their blood compatibilities. However, the above mentioned correlations [with the exception of ref. (3)] emphasize only the surface properties of the biomaterials, namely, either their total surface free energies, or the fractional contributions of their component surface free energies, as parameters with which their blood compatibilities could be related. But, in fact, all biomaterials interact with blood, which is a complex fluid and whose stability under normal conditions is governed by a delicately maintained equilibrium between its various components. Therefore, when a foreign surface is placed in contact with this fluid, the responses of its components to the presence of the "stranger" in their midst will be determined not only by the surface properties of the biomaterial but also by those of blood. Thus, it becomes necessary to consider the surface properties of both the biomaterial and blood as important determinants of the performance of the former as a blood contacting device.

It is the objective of this paper to suggest a surface energetic criterion of blood compatibility of foreign surfaces, which is based on an analysis of the surface interactions between the two media (blood and biomaterial). More specifically, it will be shown that, in order to achieve satisfactory blood compatibility, the magnitudes of the individual surface free energy components of a biomaterial, namely, the polar and dispersion components of the solid surface free energy, must separately be sufficiently near to their respective surface free energy counterparts of blood, so as to cause a low (but not very low) blood–biomaterial interfacial tension. Based on an analogy with the highly biocompatible cellular elements of blood, it will be shown that a spectrum of combinations of polar and dispersion components of solid surface free energies can result in both satisfactory blood compatibility of biomaterials as well as mechanical stability of the blood-biomaterial interface.

Second, the possibility of improving the surface properties of polymeric materials, in order to enhance their blood compatibilities, will be discussed.

2. SURFACE INTERACTIONS BETWEEN BLOOD AND BIOMATERIAL

In order to analyze the surface interactions between blood and a biomaterial, it will be necessary to briefly review the adsorptive events which take place when a foreign surface is exposed to blood. Almost instantaneously upon encountering a foreign surface, specific blood proteins, mainly fibrinogen, adsorb onto the solid surface (7). Continued adsorption of proteinaceous components takes place until a film of proteins has covered the surface of the biomaterial. Adhesion of the cellular components of blood begins only after this protein film has built up to a critical thickness, which varies for different substrates, depending on the surface properties of the materials. Platelets are known to be the first cellular components to adsorb onto the proteinated solid surface. These adherent platelets can flatten and become activated, which then leads to their irreversible aggregation (with other platelets from blood) and a chain of highly unpredictable events involving several components of blood, and culminating in the formation of either a red thrombus (interwoven mesh of red blood cells in fibrin) or a white thrombus (interwoven mesh of white blood cells in fibrin).

From this highly simplified picture of the encounter between a thrombogenic foreign surface and blood, it is evident that the cellular components of blood, which are chiefly responsible for thrombus formation, do not interact directly with the surface of the biomaterial. Therefore, the influence of the substrate is conveyed to the cells only through the preadsorbed protein film. This then poses the following questions, in connection with the pursuit of relating the surface properties of biomaterials with their blood compatibilities: (i) How do proteins interact with solid surfaces? (ii) What are the consequences of the surface-induced conformational changes of initially adsorbing blood proteins on the events of cellular adhesion and thrombosis? In an attempt to answer the above questions, let us consider more specifically, the surface interactions of a biomaterial with the initially adsorbing plasma proteins.

When a foreign surface is placed in contact with blood, a new interface is created in the environment of the latter. Therefore, there is a thermodynamic driving force for minimizing the blood-biomaterial interfacial tension. As a result of this, the most surface active components of blood, namely, proteins (mainly fibrinogen), adsorb at the solid–liquid interface almost instantaneously following the contact of the foreign surface with blood. Upon adsorption, the first layer of proteins may denature on the solid surface, depending on the magnitude of the blood-biomaterial interfacial tension. If this interfacial tension is high, the adsorbed proteins will anchor at multiple sites on the solid surface, in order to interact strongly with the solid surface and thereby decrease the solid-liquid interfacial tension. Therefore the adsorbed proteins will be considerably denatured in this case. On the contrary, if the blood-biomaterial interfacial tension is low, the adsorbed proteins will undergo minimum distortion of their native configuration.

Let us first consider the case in which the initially adsorbed proteins are considerably denatured on the solid surface, as a result of a high solid-blood interfacial tension. These denatured proteins will be rigidly attached to the solid (at multiple adsorption sites on the solid surface) and the proteinated solid surface will now form a new interface with blood plasma. The interfacial tension at this new interface will also be quite high, though it will be smaller than the initial blood-biomaterial interfacial tension, for the following reason: The solution state configuration of proteins in blood plasma must be optimal in terms of increasing their interaction with the solvent and thereby minimizing the free energy of the system. This means that, in their solution state configuration, the hydrophylic portions of the blood proteins will be oriented toward the aqueous phase (subject to steric constraints). However, when these proteins adsorb at multiple sites on solid surfaces, their configuration is altered and, therefore, a large part of their hydrophylic portions may not be exposed to the aqueous phase. As a result of this, the adsorbed (denatured) proteins will no longer interact as strongly with blood plasma and therefore the interfacial tension between the denatured protein covered solid surface and the surrounding medium will not be small. Due to this interfacial tension, there will be a further driving force for the adsorption (followed by denaturation) of a second layer of proteins, resulting in the creation of another interface with the medium and likewise, a succession of new interfaces will thus be presented to the components of blood within the first few minutes of their encounter with a foreign surface. Consequently, a sequence of hierarchical adsorption of blood components takes place, with the most surface active components preferentially adsorbing at each stage, until, finally, platelets begin to adsorb. Since the proteinated surface-blood interfacial tension is still high, it is likely that the adsorbed platelets will flatten and thereby become activated, thus releasing agents such as ADP. Due to this release, the platelets will interact more strongly with the proteinated solid surface, which causes their further activation. Platelet adsorption and activation generally herald the beginning of a chain of highly unpredictable and poorly understood events, involving several components of blood and culminating in thrombosis.

Contrary to the above example, let us consider the case in which the blood–biomaterial interfacial tension is relatively low. In this situation, the driving force for adsorption will be smaller because the interactions in the adsorbed state (protein-plasma and proteinsolid) are not too different from the interactions which exist between proteins and plasma in the bulk. Moreover, the conformation of the initially adsorbed proteins will not be very different from that of their solution state, because the latter configuration offers a greater entropic freedom. Therefore, the adsorbed proteins will retain much of their interactions with blood plasma in this case (because only a small part of their surface will be attached to the solid). As a result of this, the interfacial tension between the protein covered solid surface and blood plasma will be relatively small and so there will not be a significant driving force for the further adsorption of blood components. This provides a much higher blood compatibility to a foreign surface than in the previous case. However, if the blood-biomaterial interfacial tension is very low, the plasma proteins may not even adsorb to any significant extent on the surface of the biomaterial. As a result of this, one will not witness the formation of a cascade of interfaces or ultimate thrombosis on such surfaces.

From the above discussion, it is clear that the surface properties of a biomaterial must be selected with the view of minimizing the blood–biomaterial interfacial tension and thereby minimizing the denaturation of the initially adsorbing plasma proteins as well. Though an interfacial tension of zero will be ideal for blood compatibility of a foreign surface, it will be undesirable from the point of view of the mechanical stability of the blood-biomaterial interface. This is because, at very low interfacial tensions, any interface will be highly susceptible to perturbations (see Appendix 1). In other words, mechanically or thermally induced corrugations of such an interface will tend to grow in time. Due to this, some of the surface components of the solid can dissolve into blood and/or blood can be absorbed into the solid, events which can trigger the thrombotic sequence and thereby render the foreign surface incompatible with blood. Therefore, for satisfactory biocompatibility of a foreign surface, the blood biomaterial interfacial tension must be sufficiently low (but not very low) in order to comply with the dual requirements of a low driving force for the adsorption of blood components and a mechanically stable blood-biomaterial interface. In order to estimate a suitable non-thrombogenic value of the blood-biomaterial interfacial tension, one can note the high compatibility as well as the interfacial stability of the cellular elements, with blood plasma. The cell–medium interfacial tension is generally considered to be of the order of 1-3 dyn/cm, though for certain cells, it can be as low as 0.1 dyn/cm (8,9). How well these measured interfacial tensions portray the actual values is a matter that needs confirmation. Therefore, it seems reasonable to suggest that the blood-biomaterial interfacial tension should also be maintained close to this range ($\gamma_{SL} \approx 1 - 3$ dyn/cm) in order to ensure the compatibility of a foreign surface as well as the mechanical stability of its interface, with blood plasma.

3. BLOOD-BIOMATERIAL INTERFACIAL TENSION

Now, an expression which relates the surface energetic properties of a biomaterial to its blood compatibility, will be derived. Assuming that the principle of additivity of different interactions is valid, the total surface free energy of the biomaterial, denoted by γ_S, and that of blood, denoted by γ_L, can each be represented as the sum of several independent contributions, such as those arising from dispersion forces, dipole interactions, hydrogen bonding, etc. (10). Therefore,

$$\gamma_S = \gamma_S^p + \gamma_S^d + \gamma_S^h + \cdots \qquad [1]$$

and

$$\gamma_L = \gamma_L^p + \gamma_L^d + \gamma_L^h + \cdots, \qquad [2]$$

where the superscripts p, d, and h denote the polar, dispersion, and hydrogen bonding contributions, respectively, to the surface free energies. Similarly, the expression for the work of adhesion between the solid and liquid phases, W_{SL}, can also be expressed as the sum of several contributions, as follows:

$$\begin{aligned} W_{SL} &= \gamma_S + \gamma_L - \gamma_{SL} \\ &= W_{SL}^p + W_{SL}^d + W_{SL}^h + \cdots, \end{aligned} \qquad [3]$$

where γ_{SL} is the solid–liquid interfacial tension. The dispersion component of the work of adhesion could be related by a geometric mean expression of the form, $W_{SL}^d = 2\left(\gamma_L^d \gamma_S^d\right)^{1/2}$ (10). Kloubek (11) has shown that a geometric mean expression could be used for the polar component of the work of adhesion as well, i.e., $W_{SL}^p = 2\left(\gamma_L^p \gamma_S^p\right)^{1/2}$. Fowkes (12) has emphasized that hydrogen bonding interactions are only a subset of acid-base interactions and that the latter cannot be predicted by a geometric mean expression. Instead, the acid-base interactions could be predicted from the equation of Drago (13, 14). In the systems considered by Fowkes, the contribution of polar interactions was negligible in comparison to those of acid-base interactions and, therefore, he regarded the total work of adhesion as arising only from that due to dispersion and acid-base interactions. However, he noted that polar interactions can be considerable in solvents of high dipole moment. Since blood is a good example of such a system, one can expect a significant contribution from polar interactions between blood and a biomaterial, if the latter has sufficient polarity. Though acid-base interactions may also contribute to the work of adhesion between blood and a biomaterial, it is currently not possible to estimate the magnitude of such interactions when they are present alongside with dispersion and polar interactions. In addition, the following contributions to the work of adhesion between blood and a biomaterial are neglected: (1) the contribution of the double layer, which is always present in the vicinity of a solid surface in contact with blood (15, 16) and (2) that of the polarization layer generated in the liquid, by the dipoles present on the solid surface (17). Therefore, this analysis will be restricted to only those types of biomaterials which interact with blood largely by means of nonspecific interactions, such as dispersion and polar forces. For these cases, the work of adhesion between blood and a biomaterial can be approximated by

$$W_{SL} = W_{SL}^p + W_{SL}^d$$
$$= 2\left\{\left(\gamma_L^p \gamma_S^p\right)^{1/2} + \left(\gamma_L^d \gamma_S^d\right)^{1/2}\right\}. \quad [4]$$

From Eqs. [3] and [4], an expression for the blood-biomaterial interfacial tension follows as

$$\gamma_{SL} = \gamma_S + \gamma_L$$
$$-2\left\{\left(\gamma_L^p \gamma_S^p\right)^{1/2} + \left(\gamma_L^d \gamma_S^d\right)^{1/2}\right\}. \quad [5]$$

This equation can be further transformed by observing that $\gamma_S = \gamma_S^p + \gamma_S^d$ and $\gamma_L = \gamma_L^p + \gamma_L^d$ (see Appendix 2), to obtain (18)

$$\gamma_{SL} = \left\{\left(\gamma_L^p\right)^{1/2} - \left(\gamma_S^p\right)^{1/2}\right\}^2$$
$$+ \left\{\left(\gamma_L^d\right)^{1/2} - \left(\gamma_S^d\right)^{1/2}\right\}^2. \quad [6]$$

This equation is quite revealing from the point of view of blood-biomaterial interfacial tension considerations. It shows that, γ_{SL} can have its minimum, which is zero, when $\gamma_S^p = \gamma_L^p$ and $\gamma_S^d = \gamma_L^d$. When $\gamma_{SL} = 0$, a biomaterial can be theoretically expected to remain in perfect compatibility with the components of blood, without promoting any adhesive events. However, as mentioned before, one must attempt to maintain a blood-biomaterial interfacial tension of the order of 1-3 dyn/cm, in order to ensure the mechanical stability of the blood–biomaterial interface as well as the compatibility of the foreign surface in the environment of blood.

Equation [6] shows that several combinations of polar and dispersion components of solid surface free energies can lead to a blood-biomaterial interfacial tension (γ_{SL}) of about 1–3 dyn/cm. To graphically illustrate this spectrum of combinations of blood compatible surface free energy components, it can be seen from Eq. [6] that a plot of $\left(\gamma_L^p\right)^{1/2} - \left(\gamma_S^p\right)^{1/2}$ vs $\left(\gamma_L^d\right)^{1/2} - \left(\gamma_S^d\right)^{1/2}$ describes the locus of all points on a circle of radius $(\gamma_{SL})^{1/2}$. Figure 1 shows such a plot. The shaded area in this figure represents the zone of suggested biocompatibility (i.e., blood compatibility of the biomaterial as well as mechanical stability of the blood-biomaterial interface). Table I shows a few typical combinations of surface free energy components, taken from that zone of Fig. 1. From this table it can be seen that, even solids whose total surface free energies differ considerably (for example, a surface such as C with a γ_S of 46.7 dyn/cm and a surface such as C' with a γ_S of 104.49 dyn/cm) can bear the same interfacial tension with blood and thereby remain equally compatible with blood. This happens because the individual surface free energy components (γ_S^p and γ_S^d) of the solids are

A Surface Energetic Criterion of Blood Compatibility of Foreign Surfaces

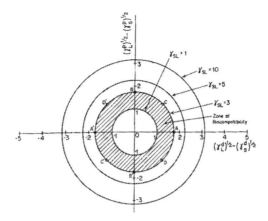

FIG. 1. Plot of Eq. [6]. The circles of this figure represent the locus of all points (different combinations of γ_S^p and γ_S^d values), which give rise to the same value of blood–biomaterial interfacial tension (i.e., $\gamma_{SL} = 1$, 3, 5, or 10).

in suitable conformity with those of their respective counterparts of blood (γ_L^p and γ_L^d). From this, it is clear that, neither the total surface free energy of the solid (γ_S), nor the fractional contributions of the solid's component surface free energies $\left(\gamma_S^p/\gamma_S, \gamma_S^d/\gamma_S\right)$ can be considered as indicators of its blood compatibility. Rather, one must separately consider the polar and dispersion components of a solid's surface free energy, in relation to their respective surface free energy counterparts of blood, in attempting to correlate the surface energetic properties of biomaterials with their blood compatibilities.

As early as 1973, Andrade (19) postulated that a minimum blood–biomaterial interfacial tension (preferably zero) can provide ideal blood compatibility to a biomaterial. He considered that this situation can be brought about by using materials such as hydrogels, whose high water contents can lead to low values of blood–biomaterial interfacial tensions. Andrade regarded the total surface free energy of the solid as the parameter to be related with the blood–material interfacial tension. However, Eq. [6] shows that, even if the total surface free energies of different solids are

TABLE I
Typical Combinations of Surface Free Energy Components and the Corresponding Solid–Water Interfacial Tensions from the Biocompatible Zone of Fig. 1

Surface (from Fig. 1)	γ_S^p (dyn/cm)	γ_S^d (dyn/cm)	γ_S (dyn/cm)	γ_{SW} (dyn/cm)
A	50.8	8.63	59.43	3
A'	50.8	40.97	91.77	3
B	29.11	21.8	50.91	3
B'	78.49	21.8	100.29	3
C	34.84	11.86	46.7	3
C'	69.76	34.74	104.49	3
D	69.76	11.86	81.62	3
D'	34.84	34.74	69.58	3

equal to that of water (72.6 dyn/cm), they can give rise to different interfacial tensions with water, depending on the magnitudes of their individual surface free energy components. For example, let us consider two hypothetical surfaces, A and B, whose total surface free energies are both equal to 72.6 dyn/cm, but whose polar and dispersion components differ considerably, as follows: $\gamma_A^p = 15$ dyn/cm, $\gamma_A^d = 57.6$ dyn/cm, $\gamma_B^p = 45$ dyn/cm, and $\gamma_B^d = 27.6$ dyn/cm. From Eq. [6], it can be seen that the interfacial tensions of these two surfaces with water are: 19 dyn/cm for surface A and 0.5 dyn/cm for surface B. This illustrates the fact that any attempt to relate the surface energetic properties of materials with their blood compatibilities on the basis of interfacial tension considerations must be based on an appropriate matching of the respective individual surface free energy components of both the biomaterial and blood.

The above considerations are valid for solid surfaces which interact with blood via dispersion as well as polar forces. We will consider below the case of a special class of materials, namely, nonpolar solids, which contain little or no polarity and, therefore, can interact with blood only via dispersion forces.

4. BLOOD COMPATIBILITY OF NONPOLAR SOLIDS

Many solids which are considered nonpolar are also usually low surface free energy materials. Since a number of these low surface energy, nonpolar materials (especially polymers) possess some desirable qualities for blood contact applications (like good mechanical strength, chemical inertness to blood components, nontoxicity, etc.), they have been widely tried as biomaterials, with the expectation that their low surface free energies would induce minimal adhesion of blood components on their surfaces. The results, however, have been contrary to such expectations. For example, Teflon is a low surface energy material, which is relatively nonpolar. Though this material has a high degree of inertness to many chemicals, its blood compatibility has been found to be poor (6). The reason for this can be explained by examining Eq. [6]. For nonpolar solids, $\gamma_S^p = 0$, $\gamma_S^d = \gamma_S$ and Eq. [6] reduces to

$$\gamma_{SL} = \left\{ \left(\gamma_L^d \right)^{1/2} - \left(\gamma_S \right)^{1/2} \right\}^2 + \gamma_L^p. \qquad [7]$$

This equation shows that, for the case of nonpolar surfaces, the blood–biomaterial interfacial tension attains a minimum value when $\gamma_S = \gamma_L^d$. However, the magnitude of this minimum is equal to $\gamma_L^p [= 50.8$ dyn/cm, considering the surface tension properties of water and blood plasma to be equal (3)]. This represents a considerable blood–biomaterial interfacial tension, well outside the range of 1–3 dyn/cm that is considered suitable for providing satisfactory blood compatibility to a foreign surface. From this, it is clear that, nonpolar surfaces (like Teflon) can never remain in long term compatibility with blood.

Schrader (20) has also used Eq. [7] to note that γ_{SL} has a minimum when $\gamma_S = \gamma_L^d$ and he has used this fact to explain the experimental observations of Baier (21) concerning the existence of minimal bioadhesion on solids whose critical surface tensions $\left(\gamma_c \right)$ range from 20 to 30 dyn/cm. Even though Eq. [7] was used by both of us, our conclusion concerning the blood compatibility of nonpolar solids is distinctly different from that of Schrader.

5. CHARACTERIZATION OF BIOMATERIAL SURFACES

One important consideration remains to be emphasized in order to ensure an unambiguous interpretation of any surface energetic criterion of blood compatibility of foreign surfaces. It has been noted that many solids possess the tendency to rearrange their surface structures in response to their local environments (22,23). This observation is particularly applicable to polymeric surfaces, many of which possess sufficient mobility to adopt different surface configurations in different environments. As a result of this, such materials may be hydrophobic when exposed to air (unhydrated state), but in the aqueous environment (hydrated state), they will display a reasonable hydrophilicity. This behavior is much more pronounced in the case of many hydratable and mobile polymers, like hydrogels (24) and polydimethyl siloxane (25), respectively. Even some relatively rigid polymers like polymethyl methacrylate seem to display this type of behavior (23). An important implication of this is the fact that the polar and dispersion components of the surface free energies of such solid surfaces can be grossly different in the aqueous environment (hydrated state) in comparison to the air environment (unhydrated state). A recent study by Ko *et al.* (26) provides a good illustration of the above consideration. These authors determined the individual surface free energy components of several grafted copolymers of hydroxy ethyl methacrylate (HEMA) and ethyl methacrylate (EMA) of varying compositions and water contents, in both the hydrated and unhydrated states. The values obtained for some solid surfaces in the two environments are listed in Table II. Unprimed quantities refer to the surface free energies in the hydrated state, while the primed symbols represent the surface free energies in the unhydrated state. It can be seen that, for all the solid surfaces, the polar component of the solid's surface free energy is increased and its dispersion component is decreased in the hydrated state, in comparison to the unhydrated state. However, in general, the change in the polar component is much more pronounced than that of the dispersion component. The solid–water interfacial tensions, γ_{SW}, of these surfaces, calculated from Eq. [6], for both the cases (hydrated as well as unhydrated solids), are also included in Table II and plotted in Fig. 2. From Fig. 2, it can be seen that the values of the individual surface free energy components in the unhydrated state can cause a misleading estimate of their blood compatibilities. For example, the experimental results of Ratner *et al.* (5) revealed that a grafted copolymer of poly(hydroxy ethyl methacrylate) and poly(ethyl methacrylate) which contained about 15% water had the highest blood compatibility in comparison to surfaces which contained either higher or lower water contents. These authors did not determine the individual surface free energy components of their materials. However, from the measurements of Ko *et al.* (26) for similar surfaces, the values obtained for a solid surface which contained about 17% water (which is near to Ratner *et al's.* 15% water content surface) are included in Table II. One can see from Table II and Fig. 2 that the interfacial tension of this surface with water is 0.92 dyn/cm, if it is based on the hydrated solid's surface free energy components (γ_S^p and γ_S^d), while it is as high as 21.57 dyn/cm, if it is based on the unhydrated solid's surface free energy components ($\gamma_S^{p'}$ and $\gamma_S^{d'}$). The former value, which is closer to the actual value, is certainly more indicative of the high blood compatibility of this material. From this example, it appears that the surface characterization of biomaterials in their hydrated state only can provide an accurate indication of their blood compatibilities.

TABLE II

Water Content, Surface Free Energies, and Solid–Water Interfacial Tensions of Mixtures of HEMA–EMA Grafted Polyethylene Films, Characterized in Air and under Water[a]

HEMA (%)	Water content (%)	$\gamma_s^{d'}$ (dyn/cm)	γ_s^d (dyn/cm)	$\gamma_s^{p'}$ (dyn/cm)	γ_s^p (dyn/cm)	$\gamma_s^{d}/\gamma_s^{d}$	$\gamma_s^{p}/\gamma_s^{p'}$	γ_{sw}' (dyn/cm)	γ_{sw} (dyn/cm)
0	1.7 ± 1.1	39.1	_	3.7	24.3	_	6.57	29.59	_
20	2.0 ± 0.5	42.6	7.5	3.8	44.6	5.68	11.74	30.26	3.93
40	3.5 ± 0.3	41	4.4	4.4	47	9.32	10.68	28.31	6.68
50	8.0 ± 1.6	53.1	8.3	1.8	48.4	6.40	26.89	40.33	3.23
60	8.6 ± 0.6	50.7	10.6	2.0	49.2	4.78	24.60	38.65	2.01
80	17.3 ± 0.3	42.4	13.8	8.2	49.8	3.07	6.07	21.57	0.92
100	25.6 ± 0.8	35.9	14.6	20.0	50.0	2.45	2.50	8.80	0.72

[a] From Ref. (26).

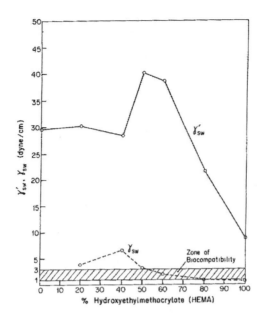

FIG. 2. Solid-water interfacial tension, calculated on the basis of the surface free energy components of the hydrated as well as unhydrated surfaces of HEMA-EMA grafted polyethylene films, which were measured by Ko *et al.* (26).

In this connection, it is necessary to mention some of the limitations of the techniques [like Hamilton's technique (27)] which are currently used to evaluate the polar and dispersion surface free energy components of biomaterials in the hydrated state. In all such techniques, the solid-fluid-water contact angles of only two fluids (i.e., either two water immiscible liquids or one water immiscible liquid and one vapor) are used for the evaluation of the polar and dispersion surface free energy components of the hydrated surface. It is well known that the magnitudes of the individual surface free energy components (polar and dispersion) of unhydrated solids are sensitive to the test liquids used for measuring the solid-liquid-air contact angles. For this reason, the contact angles of several pairs of test liquids are used to calculate the average polar and dispersion surface free energy components of solids in the unhydrated state (18, 28). In the light of this fact, it may perhaps be important to rely on the solid-fluid-water contact angles of several pairs of test fluids, in order to arrive at a more accurate estimate of the polar and dispersion surface free energy components of solids in the hydrated state also. In addition, it must be noted that, due to the lack of general validity of the expressions which are used to evaluate the individual surface free energy components, there may be some uncertainty in the calculated values of γ_S^p and γ_S^d.

6. DISCUSSION OF THE SURFACE ENERGETIC CRITERION OF BIOCOMPATIBILITY OF FOREIGN SURFACES

We will now discuss the performance of some prominent biomaterials, in relation to their surface energetic properties. For this discussion, we will consider the following materials, whose blood compatibilities range from excellent to poor: (i) hydrophilic, cross-linked polymeric gels (hydrogels), (ii) LTI (low temperature isotropic) carbons, (iii) ethyl cellulose perfluorobutyrate, (iv) "Avcothane," (v) "Biomer" polyurethane, (vi) polyalkylsulfone, (vii) Teflon, and (viii) glass. Hydrogels have shown good promise for blood contact applications since the early sixties (29). Their poor mechanical properties are also being overcome by covalently grafting them onto the surfaces of materials (like polymers) which possess suitable mechanical properties. Ratner *et al.* (5) recently evaluated the

A Surface Energetic Criterion of Blood Compatibility of Foreign Surfaces

performance of several grafted hydrogels, by conducting different blood compatibility tests on them. Relevant to this discussion is their results on a series of p (HEMA-EMA) copolymers of varying water contents. In this series of copolymers, they observed that the 1:1 p (HEMA–EMA) graft, which contained about 15% water, showed superior blood compatibility in comparison to surfaces which contained either higher or lower water contents. This result was observed in both the tests conducted on such materials, namely, the low shear rate vena cava ring test and the high shear rate A–V shunt test. This fact led to Ratner *et al*'s. hypothesis that a balance of hydrophylic and hydrophobic sites on a surface may be important for the blood compatibility of a biomaterial. As noted before, this superior blood compatible surface is likely to bear an interfacial tension of about 0.92 dyn/cm with water.

The experimental results of Ratner *et al.* (5) thus lend support to our point of view that, just as a high blood–biomaterial interfacial tension (such as that caused by the low water content (p (HEMA-EMA)) surfaces) is undesirable for the blood compatibility of the foreign surface, a very low blood-biomaterial tension (such as that caused by the high water content [p (HEMA–EMA) surfaces] is also equally detrimental to the performance of the biomaterial. A high thermodynamic driving force for the adsorption of blood components is responsible for the low blood compatibility of the p (HEMA–EMA) surface in the former case, while in the latter case, the poor mechanical stability of the blood-biomaterial interface may be the cause of adverse effects such as the dissolution or leaching of some of the surface components of the solid into blood and/or the absorption of blood into the solid, events which are capable of triggering the thrombotic sequence and thereby resulting in the poor blood compatibility of the foreign surface. From this, it appears that an optimum blood-biomaterial interfacial tension [which is sufficiently low (but not very low)] can provide a satisfactory blood compatibility to a foreign surface. Of course, the exact range of optimum interfacial tension values must be determined by suitable experiments (it may vary slightly from the range of 1–3 dyn/cm suggested here).

Let us next consider the case of LTI (low-temperature isotropic) carbons, whose blood compatibility is excellent (6). The surface of this material has been characterized only in the unhydrated state, for its polar and dispersion surface free energy components (3), which are as follows: $\gamma_S^{p'} = 4.16$ dyn/cm and $\gamma_S^{d'} = 54.46$ dyn/cm. As noted earlier, the polar surface free energy component of a number of solids is grossly underestimated and the dispersion component is usually overestimated in the unhydrated state (in comparison to the hydrated state). This fact is evident from the study of Ko *et al.* also (26). This is because many solids possess the ability to reorient their surface structures and maintain a minimum interfacial energy with the surrounding medium, so as to comply with the thermodynamic requirement of minimizing the free energy of the system. This behavior has been demonstrated for the case of hydrogels by Holly and Refujo (24). Subsequently, Yasuda *et al.* (22) have demonstrated (by means of contact angle hysteresis experiments) that the concept of orientation of surface molecules at an interface is also applicable to some other polymer surfaces (like oxygen-plasma treated polypropylene). More recently, Andrade *et al.* (23) have shown (by means of contact angle hysteresis experiments) that, even a rigid, hydrophobic polymer surface, such as polymethyl methacrylate, could display modest hydrophilicity under water due to interfacial restructuring in the aqueous environment. In the light of these observations, it appears plausible to suggest that the surface structure of LTI carbons may also be significantly altered in the blood environment (in comparison to the air environment) so as to maintain a sufficiently low interfacial tension with blood, which can provide this material with a high blood compatibility. Note that some specific interactions such as the acid-base interactions may also play a role. Based on the same reasoning, one may be able to account for the excellent blood compatibility of ethyl cellulose perfluorobutyrate (6), the good blood compatibilities of "Avcothane" and "Biomer" polyurethane (6) and the moderate–good blood compatibility of polyalkylsulfone (6).

The characterization of these biomaterial surfaces (for their polar and dispersion surface free energy components) in the aqueous environment can confirm the validity of the above explanation.

Let us now discuss the blood compatibilities of two materials which differ considerably, among other things, in their polar surface free energies. Teflon is a classic example of a low surface energy, nonpolar surface, whose blood compatibility is poor (6). The surface of this material remains nonpolar even in the aqueous environment (27). Therefore, as mentioned earlier, any nonpolar surface will bear a very high interfacial tension with blood plasma ($\gamma_{SL} \geqslant 50.8$ dyn/cm), as a result of which, its blood compatibility will be poor, as is indeed the case with Teflon. Another example of a thrombogenic material is glass, which has been characterized only in the unhydrated state, in which its surface free energy components are as follows (19): $\gamma_S^{p'} = 90$ dyn/cm and $\gamma_S^{d'} = 80$ dyn/cm. Since this material presents such a highly polar surface character even to an apolar phase like air, one can conclude that it has a very rigid surface structure and so it cannot undergo significant surface restructuring in the aqueous environment. Therefore, if it is assumed that the unhydrated surface free energy components of this material will be valid for its hydrated state also, then one can note that this material will bear a very high interfacial tension with blood ($\gamma_{SL} = 23.8$ dyn/cm), and consequently, it will be thrombogenic, as is the observed case.

7. THERMODYNAMIC AND KINETIC CONSIDERATIONS IN THE BLOOD COMPATIBILITY OF FOREIGN SURFACES

It is necessary to mention that the surface energetic criterion suggested in this paper is based on thermodynamic considerations and, therefore, it is valid for the long-term blood compatibility of foreign surfaces. In some instances, however, it is possible for a biomaterial to remain in compatibility with blood for a fairly long time, even when this thermodynamic criterion of biocompatibility is not satisfied. This may take place due to many factors like steric repulsion, double layer forces, hydration forces, etc., which can give rise to the "kinetic compatibility" of a biomaterial for a reasonable length of time, much like the well-known kinetic stability of thermodynamically unstable colloidal suspensions (30-32). The following observations illustrate this point: Bruck (6) has hypothesized that the electrical and semiconduction properties of some natural and synthetic polymers may bear a relationship to their blood compatibilities. This may be due to double layer repulsion. Ikada *et al.* (33) suggest that, if a material presents a diffuse interfacial structure (as opposed to a rigid one) when it is placed in the aqueous environment, then it may exhibit excellent antithrombogenicity. This may arise as a result of steric repulsion between the solid surface and the adsorbing components. Barenberg *et al.* (34) conclude that, if a surface appears as an ordered ionic array, it will invite a thrombogenic response, while if it appears as a disordered array, only limited thrombogenesis will take place on it. All these and many other surface properties of materials can result in their compatibility with blood for varying periods of time (sometimes sufficiently long for practical purposes), thereby providing a "kinetic compatibility" to a biomaterial. While a low blood-biomaterial interfacial tension can ensure the long-term blood compatibility of a foreign surface, "kinetic biocompatibility" may often be adequate for practical purposes.

8. SURFACE MODIFICATION OF POLYMERS BY ULTRAVIOLET IRRADIATION AND ITS IMPLICATIONS FOR BIOMATERIAL BLOOD COMPATIBILITY

In order to satisfy the criterion of blood compatibility suggested in this paper, it is necessary to either find materials whose individual surface free energy components belong to the biocompatible range ($\gamma_{SL} \simeq 1-3$ dyn/cm) or else, it is necessary to modify the surface properties of existing materials, in order to improve their blood compatibilities. The former alternative places a high premium on a trial and error search for suitable biomaterials. On the other hand, it is a more attractive proposition to attempt an improvement of the blood compatibilities of existing biomaterials which possess good mechanical properties.

A Surface Energetic Criterion of Blood Compatibility of Foreign Surfaces

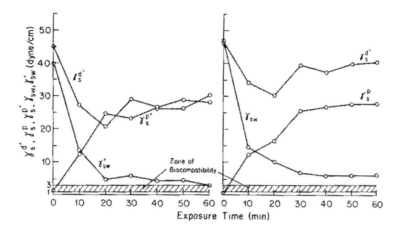

FIG. 3. Effect of ultraviolet irradiation (for various exposure times) on the surface free energy components of poly(diphenyl siloxane) (35). The solid–water interfacial tension is also plotted for different exposure times. The left-hand side shows the values of $\gamma_S^{p'}$ and $\gamma_S^{d'}$ (evaluated by Kaelble's method), while the right-hand side shows the values of γ_S^p and $\gamma_S^{d'}$ (evaluated by Fowkes–Hamilton method).

A number of polymers, like Teflon for example, possess some desirable properties for blood contact applications (such as nontoxicity, good mechanical properties, etc.), but they are incompatible with blood largely because of their poor surface properties. Moreover, most polymeric biomaterials have a very small value of γ_S^p, which results in their high interfacial tension with blood plasma, since the latter has a relatively high polar component of surface tension ($\gamma_L^p \approx 50.8\,\text{dyn/cm}$). Therefore, in order to improve the blood compatibilities of such polymeric biomaterials, it will be necessary to considerably increase their polar surface free energy components.

The recent study of Esumi et al. (35) shows that ultraviolet irradiation of polymer surfaces can result in considerable changes in their polar surface free energy components. Therefore, this technique seems particularly promising for altering the surface properties of several types of polymeric biomaterials, in order to enhance their blood compatibilities.

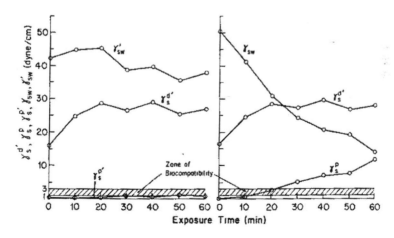

FIG. 4. Effect of ultraviolet irradiation (for various exposure times) on the surface free energy components of copolymer dimethyl siloxane PS255 (35). The solid–water interfacial tension is also plotted for different exposure times. The left-hand side shows the values of $\gamma_S^{p'}$ and $\gamma_S^{d'}$ (evaluated by Kaelble's method), while the right-side shows the values of γ_S^p and $\gamma_S^{d'}$ (evaluated by the Fowkes-Hamilton method).

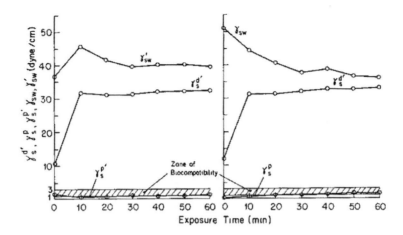

FIG. 5. Effect of ultraviolet irradiation (for various exposure times) on the surface free energy components of copolymer dimethyl siloxane PS054 (35). The solid-water interfacial tension is also plotted for the different exposure times. The left-hand side shows the values of $\gamma_S^{p'}$ and $\gamma_S^{d'}$ (evaluated by Kaelble's method), while the right-side shows the values of γ_S^p and $\gamma_S^{d'}$ (evaluated by Fowkes–Hamilton method).

Figures 3 to 6 and Table III show some of the results obtained by Esumi *et al.* (35) regarding the effects of ultraviolet irradiation on polymer surfaces. The surface free energy components, evaluated by two different techniques, namely, Kaelble's method (18, 28) and Fowkes-Hamilton method (27), are shown in these figures. While the former technique was used to calculate the solids' surface free energy components in the unhydrated state ($\gamma_S^{p'}$ and $\gamma_S^{d'}$), the latter method provided the value of the polar surface free energy component in component in the unhydrated state $\left(\gamma_S^{d'}\right)$. For the hydrated state $\left(\gamma_S^p\right)$ and the dispersion each combination of component surface free energies, the value of the solid-water interfacial tension (calculated from Eq. [6]), is also shown in Figs. 3 to 6 and Table III.

Figure 3 and Table III show that, exposure to UV radiation for 1 hr increases the polar surface free energy component of poly- (diphenyl siloxane) in the hydrated state, from an initial value of 0.37 dyn/cm to a value of 28 dyn/cm. The dispersion component changed far less dramatically

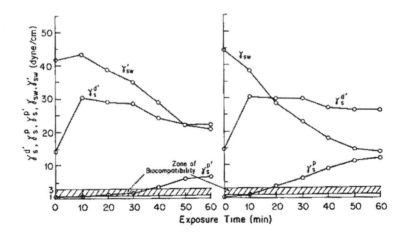

FIG. 6. Effect of ultraviolet irradiation (for various exposure times) on the surface free energy components of copolymer dimethyl siloxane PS264 (35). The solid-water interfacial tension is also plotted for different exposure times. The left-hand side shows the values of $\gamma_S^{p'}$ and $\gamma_S^{d'}$ (evaluated by Kaelble's method), while the right-side shows the values of γ_S^p and $\gamma_S^{d'}$ (evaluated by Fowkes-Hamilton method).

TABLE III

Changes in Surface Free Energy Components and Solid-Water Interfacial Tensions of Polymers as a Result of Exposure to UV Radiation[a]

No.	Polymer	Exposure time (min)	Kaelble surface energies (dyn/cm)			Fowkes–Hamilton surface energies (dyn/cm)		
			$\gamma_S^{d'}$	$\gamma_S^{p'}$	γ_{sw}'	$\gamma_S^{d'}$	γ_S^{p}	γ_{sw}
1	Poly(diphenyl	0	45.32	1.27	40.26	46.5	0.37	47.12
	siloxane)	10	27.36	12.19	13.54	34.27	12.26	14.55
		20	20.74	24.58	4.72	30.3	16.48	10.11
		30	29.03	23.40	5.76	39.61	25.70	6.87
		40	26.58	26.25	4.25	37.54	26.85	5.91
		50	28.81	26.16	4.54	40.11	28.00	6.14
		60	28.32	30.34	3.05	40.60	28.00	6.27
2	Copolymer	0	16.09	0.41	42.52	16.48	0.0012	50.68
	dimethyl	10	24.84	0.18	45.03	24.56	0.51	41.21
	siloxane	20	28.92	0.18	45.43	28.58	2.63	30.77
	(PS255)	30	26.62	0.84	38.82	27.42	4.97	24.31
		40	29.13	0.72	39.95	29.73	6.98	20.73
		50	25.47	1.36	35.68	26.85	7.63	19.32
		60	27.09	0.95	38.14	28.00	11.83	13.99
3	Copolymer	0	10.58	1.49	36.90	11.83	0.046	49.3
	dimethyl	10	31.85	0.18	45.88	31.45	0.29	44.30
	siloxane	20	31.15	0.53	41.78	31.45	0.70	40.46
	(PS054)	30	31.32	0.80	39.71	32.02	1.18	37.47
		40	32.01	0.73	40.33	32.59	1.00	38.63
		50	32.01	0.73	40.33	32.59	1.39	36.46
		60	32.37	0.85	39.55	33.15	1.50	36.03
4	Copolymer	0	14.00	0.54	41.72	14.53	0.26	44.53
	dimethyl	10	30.28	0.37	43.19	30.30	1.00	38.24
	siloxane	20	28.85	0.91	38.60	29.73	3.37	28.62
	(PS264)	30	28.17	1.59	34.82	29.73	5.78	22.92
		40	24.13	3.19	28.59	26.85	8.67	17.76
		50	22.21	6.00	21.88	26.28	10.99	14.74
		60	21.99	6.52	20.92	26.28	11.83	13.81

(from an initial value of 46.5 dyn/cm to a final value of 40.6 dyn/cm after 1 hr of exposure), though this component was estimated only for the unhydrated solid. As a result of these changes in the individual component surface free energies, the solid-water interfacial tension changed considerably, from an initial value of 47.12 dyn/cm to a value of 6.27 dyn/cm after only 1 hr of exposure to UV radiation. Though these values of γ_{sw} may not represent the actual values attained by such surfaces when they are placed in contact with blood (as a result of using the unhydrated solid's dispersion surface free energy component), they are expected to reasonably approximate the actual values of γ_{sw} (which are based on the hydrated solids' surface free energy components, γ_S^p and γ_S^d).

Figures 4 to 6 show the effect of ultraviolet irradiation on other types of polymers. Even though these materials (in Figs. 4 to 6) were relatively nonpolar initially, exposure to ultraviolet radiation caused an increase in their component surface free energies (both γ_S^p and $\gamma_S^{d'}$), so as to cause a marked decrease in their interfacial tensions with water. It must be noted that Esumi *et al.* have characterized the surfaces of the UV-irradiated polymers, after 2 days of dark aging (to enable the escape of volatile irradiation products). However, the long-term stability of these irradiated surfaces in an aqueous environment (such as that of blood) is not yet known.

It can be seen from Figs. 3 to 6 that the component surface free energies of various polymers are affected to different extents, by this surface modification technique. But this example is discussed only to illustrate the possibility that such a technique may be useful in improving the blood compatibilities of those polymers that can undergo significant surface modification, as a result of this treatment.

From Figs. 3 to 6 and Table III, it can be seen that such a surface treatment usually leads to a considerable decrease of the solid–water interfacial tension. This presents the promising possibility that the above technique may be useful in enhancing the blood compatibilities of a wide variety of polymeric biomaterials, whose mechanical properties are already suitable for blood contact applications.

9. CONCLUSIONS

A surface energetic criterion of biocompatibility of foreign surfaces which is based on two considerations, namely (i) a low blood–biomaterial interfacial tension and (ii) a mechanically stable blood-biomaterial interface, is suggested. To fulfill both these conditions, a sufficiently low (but not very low) blood- biomaterial interfacial tension is necessary. Since the cellular elements are compatible with blood and their interface with the medium (blood plasma) is also mechanically stable, it is considered that a blood-biomaterial interfacial tension of about the same magnitude as the cell-medium interfacial tension ($\gamma_{SL} \simeq 1\text{–}3$ dyn/cm) will provide a foreign surface with both long-term compatibility as well as a mechanically stable interface, with blood. In order to satisfy these conditions, a suitable correspondence between the respective individual surface free energy components of the biomaterial and blood is necessary. As a result of this condition, it is possible that even solids which differ appreciably in their total surface free energies can exhibit equal compatibility with blood. It is also noted that nonpolar surfaces $\left(\gamma_S^p = 0\right)$ cannot remain in long-term compatibility with blood.

On the basis of some examples, it is shown that the characterization of biomaterial surfaces (for their surface free energies) in their hydrated state (as opposed to their unhydrated state) is an important requirement for the successful interpretation of any surface energetic criterion of biocompatibility of foreign surfaces.

Finally, a surface modification technique, involving the irradiation of surfaces with ultraviolet radiation, is suggested as a promising method of improving the surface energetic properties of polymeric biomaterials, in order to enhance their blood compatibilities.

APPENDIX 1: INTERFACIAL INSTABILITY AT LOW INTERFACIAL TENSIONS

There are no theoretical treatments concerning the stability to large perturbations of solid-liquid interfaces, when their interfacial tensions are low. However, the information which is available, in particular, for the stability to small perturbations, provides support to our considerations on the stability of the blood-biomaterial interface. Investigations concerning the stability to small perturbations of fluid-liquid and fluid-solid interfaces of sufficiently thin films (<1000 Å) supported on a solid have shown that the time of rupture of the films becomes smaller as the interfacial tensions of these interfaces decrease. In fact, for liquid films, the available analytical expressions show that the time of rupture of sufficiently thin films becomes zero for a zero liquid-fluid interfacial tension (36–38). In the case of sufficiently thin solid films, analytical expressions which relate the time of rupture to the solid-fluid interfacial tension, are not available. Numerical calculations, however, appear to suggest that the time of rupture of thin solid films will also become very small when the solid-fluid interfacial tension attains very low values (39).

A Surface Energetic Criterion of Blood Compatibility of Foreign Surfaces

In the case of brittle solids, fracture mechanics demonstrates (40) that the critical stress σ_c needed to initiate crack propagation is given by the expression

$$\sigma_c = \left(\frac{2E\gamma_{SL}}{\pi r} \right)^{1/2},$$

where E is the Young elastic modulus, γ_{SL} is the interfacial tension, and r is a length characterizing the crack. This expression also shows (although it is limited to brittle solids only) that at very low values of γ_{SL}, any crack of the interface can grow easily (since σ_c is small).

In addition, let us note that a thermal perturbation of the interface can also generate large amplitudes of the interfacial fluctuations (41). The mean of the square of the amplitude of the interfacial fluctuations is proportional to kT/γ_{SL}. Therefore, the effect of the thermal perturbations is expected to become more pronounced when γ_{SL} is small.

APPENDIX 2: JUSTIFICATION OF THE EXPRESSION $\gamma_L = \gamma_L^p + \gamma_L^d$

The evaluations of Coulson (42) indicate that in water the covalent interactions (≈ -8.0 kcal/mol) and overlap repulsions ($\approx +8.4$ kcal/mol) nearly cancel each other. As a result of this, the orientation plus induction interactions (≈ -6 kcal/mol) and the dispersion interactions (≈ -3 kcal/mol) are mainly responsible for the overall interaction energy. For this reason, one can approximate γ_L by the sum $\gamma_L^p + \gamma_L^d$.

APPENDIX 3: NOMENCLATURE

γ_L Surface tension of the liquid, dyn/cm

γ_S Surface free energy of the solid, dyn/ cm

γ_{SL} Solid–liquid interfacial tension, dyn/ cm

γ_{SW} Solid–water interfacial tension, dyn/ cm

W_{SL} Work of adhesion between the solid and liquid phases, dyn/cm

Superscripts

′   Denotes solid–air interface

d   Denotes the dispersion contribution

h   Denotes the hydrogen bonding contribution

P   Denotes the polar contribution

REFERENCES

1. Baier, R. E., *in* "Adhesion in Biological Systems" (R. S. Manly, Ed.). Academic Press, New York, 1970.
2. Nyilas, E., Morton, W. A., Cumming, R. D., Lederman, D. M., Chiu, T. H., and Baier, R. E., *J. Biomed. Mater. Res. Symp.* **8**, 51 (1977).
3. Kaelble, D. H., and Moacanin, J., *Polymer* **18**, 475 (1977).
4. Akers, C. K., Dardik, I., Dardik, H., and Wodka, M., *J. Colloid Interface Sci* **59**, 461 (1977).
5. Ratner, B. D., Hoffman, A. S., Hanson, S. R., Harkar, L. A., and Whiffen, J. D., *J. Polym. Sci. Polym. Symp. No.* **66**, 363 (1979).
6. Bruck, S. D., *J. Polym. Sci., Polym. Symp. No.* **66**, 283 (1979).
7. Baier, R. E., and Dutton, R. C., *J Biomed. Mater. Res.* **3**, 191 (1969).
8. Troshin, A. S., "Problems of Cell Permeability." Pergamon Press, New York, 1966.
9. Weiss, L., "The Cell Periphery, Metastasis and Other Contact Phenomena." North Holland, Amsterdam, 1967.
10. Fowkes, F. M., *Ind. Eng. Chem.* **56**, 40 (1964).

11 Kloubek, J., *J. Adhesion* **6**, 293 (1974).

12 Fowkes, F. M., *in* Physicochemical Aspects of Polymer Surfaces (K. L. Mittal, Ed.), Vol. 2. Plenum Press, New York, 1983.

13 Drago, R. S., Vogel, G. C., and Needham, T. E., *J. Amer. Chem. Soc.* **93**, 6014 (1970).

14 Drago, R. S., Parr, L. B., and Chamberlain, C. S., *J Amer. Chem. Soc.* **99**, 3203 (1977).

15 Verwey, E. J. W., and Overbeek, J. M. G. "Theory of the Stability of Lyophobic Colloids." Elsevier, Amsterdam, 1948.

16 Ruckenstein, E., and Krishnan, R., *J. Colloid InterfaceSci.* **76**, 201 (1980).

17 Schiby, D., and Ruckenstein, E., *Chem. Phys. Lett.* **95**,' 435 (1983); **95**, 439 (1983); **100**, 277 (1983).

18 Kaelble, D. H., and Uy, K. C., *J. Adhesion* **2**, 50 (1970).

19 Andrade, J. D., *Med. Instrumentation* **7**, 110 (1973).

20 Schrader, M. E., *J. Colloid Interface Sci.* **88**, 296 (1982).

21 Baier, R. E,, *in* "Proceedings of Third International Congress on Marine Corrosion and Biofouling" (R. F. Acker, B. F. Brown, J. R. Depalma, and W. P. Iverson, Eds.). Northwestern University Press, Evanston, 111., 1973.

22 Yasuda, H., Sharma, A. K., and Yasuda, T., *J. Polym.Sci., Polym. Phys. Ed.* **19**, 1285 (1981).

23 Andrade, J. D., Gregonis, D. E., and Smith, L. M., *in* "Physicochemical Aspects of Polymer Surfaces" (K. L. Mittal, Ed.), Vol. 2. Plenum Press, New York, 1983.

24 Holly, F. J., and Refujo, M. F., *J. Biomed. Mat. Res.* **9** 315 (1975).

25 Owen, M. J., *I EC Prod. Res. Dev.* **19**, 97 (1980).

26 Ko, Y. C., Ratner, B. D., and Hoffman, A. S., *J.Colloid Interface Sci.* **82**, 25 (1981).

27 Hamilton, W. C., *J. Colloid Interface Sci.* **47**, 672 (1974).

28 Kaelble, D. H., *J. Adhesion* **2**, 66 (1970).

29 Wichterle, O., and Lim, D., *Nature (London)* **185**, 117 (1960).

30 Overbeek, J. Th. G., *J. Colloid Interface Sci.* **58**, 408 (1977).

31 Ottewill, R. H., *J. Colloid Interface Sci.* **58**, 57 (1977).

32 Marmur, A., and Ruckenstein, E., *in* "Advances in Biomedical Engineering (D. O. Cooney, Ed.), Vol. II. Marcel Dekker, New York, 1980.

33 Ikada, Y., Iwata, H., Horii, F., Matsunaga, T., Taniguchi, M., Suzuki, M., Taki, W., Yamagata, S., Yonekawa, Y., and Handa, H., *J. Biomed. Mater. Res.* **15**, 697 (1981).

34 Barenberg, S. A., and Mauritz, K. A., *in* "Biomaterials: Interfacial Phenomena and Applications," ACS, Adv. Chem. Series, no. 199, p. 195. Amer. Chem. Soc., Washington, D. C., 1982.

35 Esumi, K., Schwartz, A. M., and Zettlemoyer, A. C., *J. Colloid Interface Sci.* **95**, 102 (1983).

36 Ruckenstein, E., and Jain, R. K., *J. Chem. Soc. Far aday Trans.II* **70**, 132 (1974). (Section 5.1 of Volume I).

37 Maldarelli, C., Jain, R. K., Ivanov, I. B., and Ruckenstein, E., *J. Colloid Interface Sci.* **78**,118 (1980).

38 Williams, M. B., and Davis, S. H., *J. Colloid Interface Sci.* **90**, 220 (1982).

39 Ruckenstein, E., and Dunn, C. S., *Thin Solid Films* **51**, 43 (1978) (Section 5.2 of Volume I).

40 Griffith, A. A., *Phil. Trans. R. Soc., Ser. A* **221**, 163 (1920).

41 Mandelstam, L., *Ann. Phys.* **41**, 609 (1913).

42 Coulson, C. A., *in* "Hydrogen-Bonding" (Hazdi, D., and Thomson, H. W., Eds.), p. 339. Pergamon Press, London, 1959.

1.3 Surface Characterization of Solids in the Aqueous Environment*

Sathyamurthy V. Gourisankar and Eli Ruckenstein

A surface energetic criterion was recently suggested for the selection of blood compatible surfaces (1). It was based on the thermodynamic consideration that the solid–blood plasma (i.e., water) interfacial free energy must be sufficiently low (but not too low) in order to ensure the long-term compatibility of a foreign surface in the environment of blood. In that treatment, it was also shown that widely differing combinations of polar and dispersion surface free energy components of solids in the aqueous environment can give rise to the same value for the solid–water interfacial free energy that is considered suitable for endowing a foreign surface with blood compatibility (see Table I of Ref. (1)).

The polar and dispersion surface free energy components of solids in an aqueous environment are usually estimated by underwater surface characterization techniques such as Hamilton's method (2) and the two liquid methods (35). While the former technique relies on the measurement of the contact angle of octane on the solid specimen underwater to obtain an estimate of the polar surface free energy component of the solid, the latter techniques rely on the measurement of the contact angles of two water-immiscible fluids (such as octane and air, for example), to estimate both the dispersion and polar surface free energy components of the solid in the aqueous environment. As rightly pointed out by Jho (6), these underwater contact angle techniques suffer from the limitation that they do not permit the estimation of the polar surface free energy component of a solid that is greater than that of water, i.e., 50.8 erg/cm^2. In the case of Hamilton's technique for example, the polar surface free energy component of the solid under water is given by the expression

$$\gamma_s^p = \left(\gamma_w + \gamma_{ow}\cos\phi_o - \gamma_o\right)^2/4\gamma_w^P \tag{1}$$

where γ_w is the surface tension of water ($=72.6\,\text{erg/cm}^2$), γ_o is the surface tension of octane ($=21.8\,\text{erg/cm}^2$), γ_{ow} is the octane-water interfacial tension $\left(=50.8\ \text{erg/cm}^2\right)$, γ_w^p is the polar component of the surface tension of water ($=50.8\,\text{erg/cm}^2$), and ϕ_0 is the solid–octane–water contact angle (measured through the water phase). From this equation, it is clear that γ_s^p, will be a maximum (i.e., $50.8\,\text{erg/cm}^2$) when $\phi_0 = 0°$. This expression for γ_s^p, which is independent of the dispersion surface free energy component of the solid under water $\left(\gamma_s^d\right)$, arises as a result of the coincidence that the surface tension of octane is equal to the dispersion component of the surface tension of water. If alkanes other than octane are employed, then it is necessary to measure the solid–fluid–water contact angles of a pair of fluids and solve two simultaneous equations to estimate the polar and dispersion surface free energy components of the solid in the aqueous environment. It is to be noted, however, that even in the case of the other alkane probe fluids, the maximum value of γ_s^p that can be detected by underwater contact angle goniometry will not be significantly different from the value of 50.8 erg/cm^2. This is because all the alkane fluids are nonpolar and the values of their surface tensions and interfacial

---

* *Journal of Colloid and Interface Science*, Vol 109, No. 2, p. 591, February 1986. Republished with permission.

tensions with water do not differ much. This limitation does not, however, mean that the underwater surface characterization technique is incapable of detecting γ_s^p values greater than 50.8 erg/cm^2. In order to extend the range of values of γ_s^p that can be experimentally detected, the problem is to select a pair of appropriate probe fluids which will be water immiscible and possess considerably different interfacial tensions with water than the conventional probe fluids in current use (such as the alkanes and air). To illustrate this point, let us first inspect Eq. [1], Although this equation is applicable to the case of octane only, it is still suggestive of the main limiting factor involved in the underwater surface characterization of solids. This expression shows that the maximum detectable value of γ_s^p is limited by the octane–water interfacial tension. Therefore, if we select a probe fluid such as mercury, which bears a very high interfacial tension with water (γ_{Hg-w} = dyn/cm (7)), then it appears possible to estimate solid polar surface free energy components that are greater than 50.8 erg/cm^2. The same observation was recently made by Andrade (8). The dispersion component of the surface tension of mercury has been estimated as 200 erg/cm^2 and its total surface tension is 484 erg/cm^2 (7). It has also been suggested that mercury and water interact mainly by means of dispersion forces and that the dipole-metal image forces are relatively insignificant (7). Therefore it appears reasonable as a first approximation to consider that the interaction of mercury with most solids will also take place predominantly by dispersion forces. Based on these considerations, mercury seems to be a potentially useful probe fluid for the underwater surface characterization of solids. However, one more probe fluid is still necessary to permit the determination of both the dispersion and polar surface free energy components of the solid in the aqueous environment. In our search for a second probe fluid which will also be water-immiscible and maintain a high interfacial tension with water, we find that gallium (which is a liquid metal at room temperature) appears to be suitable for underwater contact angle characterization of solids. The surface tension of this liquid is 718 erg/cm^2 (9). Its individual component surface tensions and interfacial tension with water are unknown at present but it appears to be a potentially useful probe fluid (along with mercury) for detecting solid polar surface free energy components that are larger than 50.8 erg/cm^2 in the aqueous environment. Thus, in principle, it seems feasible to detect γ_s^p values greater than 50.8 erg/ cm^2, by using some novel probe fluids such as mercury and gallium. It must be noted though that the probe fluids must be well characterized for their surface tension components and interfacial tension with water, prior to their use in underwater surface characterization. In the light of these considerations, the comment of Jho (6) that the limitation of underwater contact angle goniometry is due to the presence of water and has nothing to do with the choice of the probe fluid, seems applicable only to the conventional probe fluids (such as alkanes and air) which are in current use.

Regarding Jho's comments on the underwater surface characterization of glass, the following point is to be noted: the values obtained by both Jho (5) and Coleman *et al.* (10) for the polar and dispersion surface free energy components of glass under water were close to the maximum values of these components, i.e., 50.8 and 21.8 erg/cm^2, respectively, that could be detected by using air and octane as probe fluids in the two liquid underwater contact angle technique. Coleman et al. (10) found that the contact angles of both air and octane on the glass surface under water were less than 10° (see Table 1 of Ref. (10)), indicating that the contact angles of these two probe fluids were very close to their limiting values, i.e., 0°, on the glass surface. Based on the upper value of 10° for the contact angle of both these fluids, the estimated values of the dispersion and polar surface free energy components of glass were (10): $\gamma_s^d = 21.3$ erg/cm^2 and $\gamma_s^p = 49.7$ erg/cm^2. The results of Jho (5) showed slightly higher values of octane and air contact angles on the glass surface under water but the estimated surface free energy components were close to those of Coleman et al. It must be remembered, however, that the surface free energy component of glass was estimated from Eq, [1], by Jho. As discussed earlier, this equations does not permit the detection of γ_s^p values that are larger than 50.8 erg/cm^2. It can be seen from Eq. [1] that the value of the solid–octane–water contact angle (ϕ_0) is dependent only on the value of γ_s^p and independent of the value of γ_s^d. Therefore, even if the actual values of γ_s^p were higher than 50.8 erg/cm^2, then also the limiting value of $\phi_0 = 0°$ would have been observed. Consequently, the estimated value of γ_s^p will still be equal to 50.8 erg/cm^2 and thus

Surface Characterization of Solids in the Aqueous Environment 27

provide a misleading indication of the real value. Moreover, when air is used as a second probe fluid, the expression for the dispersion component of the surface free energy of glass under water, i.e., γ_s^d, can be written as

$$\gamma_S^d = \left(\gamma_o + \gamma_w \cos\phi_A - \gamma_{ow} \cos\phi_o\right)^2/4\gamma_w^d. \tag{2}$$

In this equation, γ_s^d is the dispersion surface free energy component of the solid under water, ϕ_A is the solid–air–water contact angle (measured through the water phase) and γ_w^d is the dispersion component of the surface tension of water. This equation shows that, for a solid on which $\phi_o = 0 \left(\text{i.e., } \gamma_s^p = 50.8 \text{ erg/cm}^2\right)$, the maximum value of γ_s^d that could be detected is 21.8 erg/cm², when $\phi_A = 0$. The values of γ_s^d for glass that were obtained by both Coleman *et al.* and Jho were very close to this upper limit. Here also it may be observed that higher values of γ_s^d could have given rise to the same limiting value of $\phi_A = 0°$ and thereby provided a misleading γ_s^d estimate of 21.8 erg/cm².

From the above discussion, it appears that the values reported by Coleman *et al.* and Jho for the surface free energy components of glass underwater, may be somewhat lower than the actual values. This is due to the limitation on the maximum values of both the polar and dispersion surface free energy components of a solid that can be detected by using octane and air as probe fluids in the underwater contact angle technique. In this context, it may be noted that the polar and dispersion surface free energy components of glass in air are (11): $\gamma_s^d = 80$ erg/cm and $\gamma_s^p = 90$ erg/cm². Realizing that the surface of glass is very rigid and therefore it is not expected to undergo any significant restructuring under water, it seems unlikely that its surface free energy components underwater will be drastically different from the above values. The underwater surface characterization of glass with higher surface tension probe fluids such as mercury and gallium may reveal if it does indeed possess larger values of polar and dispersion surface free energy components than those reported by Coleman *et al.* and Jho.

In conclusion, it can be stated that the main limitation of the techniques used to characterize the surfaces of solids in the aqueous environment, namely, the upper limit (of about 51 erg/cm²) on the detectable value of the solid's polar surface free energy component, can be overcome by resorting to the use of probe fluids like mercury and gallium, which are both very high surface tension liquids and therefore will bear high interfacial tensions with water, unlike the conventional probe fluids such as the alkanes and air, which are in current use. An important element in the use of this newer class of probe fluids is the fact that they must be well characterized for their individual surface tension components and their interfacial tension with water, prior to their use in underwater contact angle surface characterization.

REFERENCES

1. Ruckenstein, E., and Gourisankar, S. V., *J. Colloid Interface Sci.* **101**, 436 (1984) (Section 1.2 of this volume).
2. Hamilton, W. C., *J. Colloid Interface Sci.* **47**, 672 (1974).
3. Andrade, J. D., Ma, S. M., King, R. N., and Gregonis, D. E., *J. Colloid Interface Sci.* **72**, 488 (1979).
4. Ko, Y. C., Ratner, B. D., and Hoffman, A. S., *J. Colloid Interface* **82**, 25 (1981).
5. Jho, C., *J. Colloid Interface Sci.* **94**, 589 (1983).
6. Jho, C., *J. Colloid Interface Sci.* **109**, 000 (1986).
7. Fowkes, F. M., *Ind. Eng. Chem.* **56**, 40 (1964).
8. King, R. N., Andrade, J. D., Ma, S. M., Gregonis, D. E., and Brostrom, L. R., *J. Colloid Interface Sci.* **103**, 62 (1985).
9. Allen, B. C., *in* "Liquid Metals—Chemistry and Physics" (S. Z. Beer, Ed.), p. 186. Dekker, New York, 1972.
10. Coleman, D. L., Gregonis, D. E., and Andrade, J. D., *J. Biomed. Mater. Res.* **16**, 381 (1982).
11. Andrade, J. D., *Med. Instrum.* **7**, 110 (1973).

1.4 Preparation and Characterization of Thin Film Surface Coatings for Biological Environments[*]

Eli Ruckenstein and Sathyamurthy V. Gourisankar
Department of Chemical Engineering, State University
of New York at Buffalo, Buffalo, NY 14260, USA

(Received 20 December 1985; accepted 4 April 1986)

In the selection of synthetic materials for use in biological media, two important considerations are involved, namely, (i) the compatibility of the solid in the environment of the fluid and (ii) the suitability of its mechanical properties for a given application. For the first requirement it is necessary that when a foreign surface is placed in contact with a biological fluid, it must not provoke adverse responses such as excessive deposition of the fluid's components, or toxic and allergic reactions, or thrombosis in the case of blood contacting devices. For the suitability of their mechanical properties, the selected materials need to possess satisfactory levels of resistance to compression, tension or shear depending on the specific application. In some applications such as synthetic vascular prostheses or catheters for use in biomedical applications for instance, a high degree of flexibility is called for, while in some other applications such as artificial heart valves or heat exchange equipment used to process biological fluids (as in the milk pasteurization industry), one requires materials with some rigidity.

A closer examination of these two requirements reveals that they call for adequacy of two totally different aspects of material properties. The biocompatibility condition is really a reflection of the surface characteristics of solids. Contrary to this, the mechanical properties of solids are largely a manifestation of their bulk characteristics. Therefore, an 'ideal' synthetic material for use in biological fluid contact applications must be satisfactory in both its surface and bulk characteristics. This is a rather formidable demand on the currently available synthetic materials for it seems to be almost a paradox of nature that materials which possess suitable surface properties for biological fluid environments fail on account of their poor mechanical properties and *vice versa*. To illustrate this, it may be noted that some types of hydrogel biomaterials (which are crosslinked, hydrophilic, polymeric gels) are known to be highly compatible in the environment of blood though their poor mechanical properties precludes them from being successful as long-term artificial implant devices. On the other hand, materials such as metals and ceramics possess suitable mechanical characteristics for many applications but their surface properties are inadequate for use in a number of biological fluid environments. It is pertinent to note in this context that with the considerable advances in the areas of material processing and fabrication, it is now possible to prepare synthetic materials with suitable mechanical properties for almost any specific biological fluid contact application but what has really proven to be elusive is the identification of materials with suitable surface properties for use in biological media.

---

[*] *Biomaterials 1986*, Vol 7, p. 403 November. Republished with permission.

Biological fluids (e.g. blood, tear fluid, milk and sea water) generally contain a number of diverse components such as proteins, lipids, carbohydrates, salts and cellular elements. These multitudinous fluid components are present in a continuous medium, which is largely aqueous in nature in the case of most biological fluids. The main point to note here is that the various components of these fluids must interact with one another as well as with the continuous medium and yet maintain the stability of the fluid in terms of preventing particle aggregation, coalescence, chemical reaction, denaturation of enzymes, etc. From this highly simplified picture, it can be recognized that the stability of biological fluids is derived from a delicately maintained equilibrium between a plethora of interacting fluid components. Therefore, when a synthetic surface is placed in contact with such a fluid, its various components will tend to interact with the 'stranger' in their midst, thus upsetting the stability of the fluid and thereby promoting the onset of undesirable events on the surface of the solid (such as gross 'fouling') which can trigger such events in the bulk of the fluid as well. Nowhere is this scenario better illustrated than in the case of blood-biomaterials interactions. Over the past two decades, almost all types of synthetic materials, ranging from metals to plastics have been tried as blood contacting devices but none of them has emerged as an ideal blood contacting material on a long-term basis. The reason for this is only too obvious as blood plays host to a complex set of interactions between its myriad species and represents one of the most hostile milieu to foreign surfaces in terms of its biochemical activity and complex chemistry. Moreover, since blood is an archetypal biological fluid, the responses of its components to synthetic surfaces can be considered to typify the interaction of foreign surfaces with other biological fluids as well. Therefore, a closer examination of the physicochemical interactions involved in the encounter of synthetic surfaces with blood will be particularly relevant to the more general problem of identifying materials with suitable surface properties for use in any type of biological environment.

The goal of this investigation is to present an approach to the problem of selecting synthetic materials for use in any type of biological fluid environment (be it blood, tear fluid, sea water, milk or any other biological fluid). In the first part of this investigation, a surface energetic criterion of biocompatibility of foreign surfaces is suggested. This criterion, which is based on an analysis of the physico-chemical interactions between a typical biological fluid (i.e. blood) and a foreign surface, provides the range of values of the solid surface free energy components (both dispersion and polar) that can result in the optimal performance of a synthetic material in a biological environment. The second part of this paper addresses the issue of preparing synthetic substrates with suitable surfaces as well as bulk properties for use in biological environments. In this connection, a thin film approach, involving the radio frequency sputter deposition of appropriate polymeric coatings on substrates with any kind of bulk characteristics, is shown to be a promising tool to tailor the surface properties of many types of synthetic materials for use in biological fluid contact applications. The surface characterization of these polymeric coatings is then taken up, with the objective of estimating their relevant surface properties in biological media. In the final part of this investigation, the application of some surface modification techniques to effect drastic changes in the wetting properties of polymeric materials which are to be used in biological environments, is illustrated.

1 BLOOD–BIOMATERIAL ENCOUNTER: A CASE STUDY TYPIFYING THE INTERACTION OF FOREIGN SURFACES WITH BIOLOGICAL ENVIRONMENTS

Let us first understand how blood components respond to the presence of a foreign surface in their midst. Almost instantaneously upon encountering a foreign surface, specific fc blood proteins, consisting mainly of fibrinogen, adsorb onto the solid surface[1]. This pattern of initial protein adsorption is encountered on all types of synthetic surfaces. Then, after continued adsorption of proteinaceous components, the surface of the solid becomes covered by a 'conditioning' film of proteins. It is only after the build up of the conditioning protein film to generally about 200 Å in thickness, that one witnesses the participation of the other components of blood in the adsorption process. Platelets are known to be the first cellular components to adsorb onto the protein covered solid surface. These adherent platelets can then flatten, become activated and aggregate irreversibly with other platelets from the blood. Platelet adsorption and activation generally signal the onset of a number of highly unpredictable events involving several blood components and culminating in the formation of either a red thrombus (interwoven mesh of red blood cells in fibrin) or a white thrombus (interwoven mesh of white cells in fibrin) on the surface of the solid.

The interaction of proteins, cells and other blood components with synthetic surfaces is rather complex and numerous investigations have been directed toward their study. For this discussion, however, it is pertinent to note from the above simplified picture of the encounter between a thrombogenic foreign surface and blood that the cellular elements of blood, which are chiefly responsible for thrombus formation, do not interact directly with the solid surface. Rather, the influence of the substrate is relayed to the cellular elements only through the preadsorbed protein film. Thus, it can be seen that the fate of a biomaterial in the environment of blood is determined primarily by the initial events involving protein adsorption. This then brings us to two important questions with respect to the mechanism of adsorption of the initial blood proteins and their influence on the subsequent deposition events on solid surfaces: (1) How do proteins interact with solid surfaces? (2) How do surface induced conformational changes (i.e. denaturation) of the initially adsorbing plasma proteins affect the events of cellular adhesion and thrombosis? In an attempt to answer these questions, the surface interactions of a biomaterial with the initially adsorbing plasma proteins will be taken up next, with the objective of determining the surface property requirements of a synthetic material in order for it to remain in long-term compatibility with the environment of blood.

1A   SURFACE INTERACTIONS BETWEEN BLOOD AND BIOMATERIAL

The search for suitable synthetic biomaterials over the past two decades has been highlighted by several attempts to empirically correlate the performance of materials which have been tested as blood contacting devices (in either *in vivo, ex vivo* or *in vitro* tests), with certain characteristic surface properties of these materials. Prominent among such correlations include those proposed by Baier[2], Nyilas[3], Kaelble and Moacanin[4], Akers *et al.*[5] and Ratner *et al.*[6]. Though each of the above correlations has been successful in explaining the observed behavior of a limited set of biomaterials, none of them has been able to satisfactorily account for the performance of a wide variety of biomaterials tested to date. The reason for this may partly lie in the fact that all the above correlations (with the exception of Reference 4) consider only the surface properties of the biomaterials (such as their critical surface tension or their total surface free energy or the fractional contributions of their component surface free energies) as parameters to correlate with their blood compatibilities. However, since all blood contact applications of synthetic surfaces involve the interaction of biomaterials with the components of blood, it will be necessary to consider the surface properties of both the biomaterial and blood as important determinants of the performance of the former. Therefore, for the purpose of relating the performance of blood contacting synthetic materials with a characteristic surface property, it will be desirable to select a correlating parameter which can adequately portray the surface properties of both the interacting media, namely, blood and biomaterial. One of the most suitable of such parameters is the blood plasma–biomaterial interfacial free energy. This is a thermodynamic quantity, which incorporates the surface free energy contributions of both the solid and liquid phases and which provides a measure of the driving force for the adsorption of blood components on solid surfaces. Based on this parameter, Andrade proposed the minimum interfacial free energy hypothesis of biocompatibility[7]. Subsequently, we have suggested an optimum interfacial energetic criterion to relate the performance of synthetic materials with their surface energetic properties in the environment of blood[8]. The considerations of our treatment will be summarized below.

The introduction of a foreign surface in the environment of blood results in the creation of a new interface, namely, the blood plasma–biomaterial interface in the system. As a result, a thermodynamic driving force is generated for reducing the solid–liquid interfacial free energy at this new interface. Consequently, the most surface active components of blood plasma, namely, proteins (mainly fibrinogen) adsorb at the solid–liquid interface almost instantaneously following the contact of the foreign surface with blood. The configuration of the initially adsorbed proteins on the solid surface will be determined by the magnitude of the blood plasma–biomaterial interfacial free energy. If this interfacial free energy is high, the initially adsorbed proteins will tend to anchor at multiple sites on the solid surface in order to increase their interaction with the solid and thereby decrease the solid-liquid interfacial free energy. Therefore, the structure of the adsorbed proteins will be considerably altered (denatured) from their solution state configuration in this case. By the same token, if the blood plasma–biomaterial interfacial free energy is low, then the initially adsorbing plasma proteins will tend to undergo minimum distortion of their native configuration upon adsorption.

We will now consider the implications of the solid-liquid interfacial free energy on the structural alterations of the initially adsorbing plasma proteins and the subsequent events of cell adhesion and thrombus formation. Let us first consider the case in which the blood plasma-biomaterial interfacial free energy is high. In this case, the initially adsorbing plasma proteins will interact strongly with the solid surface in order to decrease the interfacial free energy (i.e. they will adsorb at multiple sites on the solid surface). In doing so, they will undergo structural alterations (denaturation) on the solid surface; this reduces their interaction with blood plasma, because some of their hydrophilic groups will no longer be exposed to the aqueous phase (blood plasma). Though the proteinated solid–liquid interfacial free energy will be decreased (in comparison to the blood plasma–biomaterial interfacial free energy), it will still have an appreciable value, because of the decrease in the interactions between the denatured (adsorbed) proteins and plasma. The relatively large interfacial free energy at this new interface will invite further adsorption (followed by denaturation) of a second layer of proteins, resulting in the creation of another interface with the medium and likewise, a succession of new interfaces will be presented to the components of blood within the first few minutes of their encounter with a foreign surface. Thus, a sequence of hierarchical adsorption of blood components will take place, with the most surface active components preferentially adsorbing at each stage, until finally, platelets begin to adsorb. Since the proteinated surface-blood plasma interfacial free energy is still relatively high, it is likely that the platelets will flatten upon adsorption and thereby become activated, thus releasing agents such as ADP. Due to this release, the adsorbed platelets will interact more strongly with the proteinated surface, resulting in their further activation and the stimulation of the deposition of other blood components, which ultimately results in thrombosis.

In contrast to the above scenario, let us consider the case in which the blood plasma-biomaterial interfacial free energy is relatively low. In this situation, the thermodynamic driving force for adsorption will be weaker, since the overall interactions in the adsorbed state (protein-plasma plus protein–solid) will not be too different from those in bulk plasma. For this reason, the adsorbed proteins will not undergo major structural alterations (i.e. they will not adsorb at multiple sites on the solid surface) because they prefer the greater entropie freedom which is afforded by a configuration which is not too different from that of their solution state structure. This means that the adsorbed proteins in this case will not appreciably decrease their interactions with blood plasma. Consequently, the interfacial free energy between the adsorbed proteins and blood plasma will be relatively small and, therefore, there will not be a significant driving force for the further adsorption of blood components on this protein covered solid surface. This provides a much higher blood compatibility to the foreign surface than the previous situation. However, if the blood plasma-biomaterial inter-facial free energy is very low, then the proteins may not even adsorb to any significant extent on the surface of the biomaterial. In such a situation, the components of blood are least affected by the presence of the solid surface and one will not witness the formation of a cascade of interfaces or ultimate thrombosis on such surfaces.

From the above simplified description of the interaction of foreign surfaces with blood, it is clear that a low blood plasma-biomaterial interfacial free energy can confer a high degree of blood compatibility to a foreign surface. However, while a very low (or better zero) interfacial free energy at the blood plasma-biomaterial interface appears to be ideal for blood compatibility, it suffers from the failing that such an interface will be mechanically unstable to perturbations. This is because, mechanically or thermally induced corrugations of such an interface will tend to grow rapidly with time when the solid-liquid interfacial free energy is very small[9]. This interfacial instability leads to adverse consequences such as the dissolution of the surface components of the solid or the absorption of blood components into the solid. The instability of the solid-liquid interface for very low interfacial free energies has been experimentally demonstrated for the case of a polymeric solid-water interface[10]. Therefore, the blood plasma-biomaterial interfacial free energy must be low enough to reduce the structural alterations of the initially adsorbing plasma proteins (so that the interactions between the adsorbed proteins and blood plasma remain quite strong) and at the same time, it should be high enough to maintain a mechanically stable blood plasma-biomaterial interface. In attempting to reconcile these two opposing requirements, let us note that the cellular elements of blood (whose compatibility with blood plasma is high and whose interface with blood plasma is mechanically stable as well), also bear an interfacial free energy with blood plasma in an optimal range of values which is generally of the order of 1–3 dyne/cm[11, 12]. It is pertinent to mention at this point that the high blood compatibility of the vascular surfaces of living systems is also known to arise from the unique surface properties of their outer coating of endothelial cellular elements which interface with the external environment of blood. Thus, the interfacial energetic characteristics of these supremely blood compatible surfaces, which have been designed by nature, seem to provide an endorsement of the premise that the twin requirements of high blood compatibility as well as mechanical stability of the blood plasma-biomaterial interface can be met only by maintaining an optimal value of the blood plasma-biomaterial interfacial free energy. In view of this analogy with the blood contacting surfaces of living systems, it seems reasonable therefore to suggest that a synthetic material-blood plasma interfacial free energy of the order of 1-3 dyne/cm could also provide a foreign surface with the two essential prerequisites for long-term blood contact applications, i.e. a high compatibility as well as a mechanically stable interface, with blood plasma. In this context, it is worth noting that the continuous medium of blood, namely blood plasma, is largely aqueous in nature like that of most other biological fluid environments. Moreover, the initial response of blood components to synthetic surfaces (involving the instantaneous adsorption and build up of a proteinaceous film of critical thickness prior to sub-sequent deposition events involving other components) is a pattern which is encountered in the interaction of many other biological fluids (e.g. milk, sea water and salivary fluid) with synthetic materials[13]. These considerations therefore permit us to view the above analysis of blood-biomaterial surface interactions as a rather typical illustration of the interaction of synthetic surfaces with any type of biological fluid. For this reason, the above criterion of blood compatibility of synthetic biomaterials can be considered as a generalized guideline to enable the selection of foreign surfaces for use in any type of biological environment.

1B BLOOD PLASMA-BIOMATERIAL INTERFACIAL FREE ENERGY

We will now derive an expression to relate the surface energetic properties of a biomaterial to its interfacial free energy with blood plasma. Assuming that the principle of additivity of dispersion, polar and specific interactions is valid[14], the equation for the work of adhesion between the solid and liquid phases, W_{SL}, is given by:

$$W_{SL} = \gamma_S + \gamma_L - \gamma_{SL} = W_{SL}^p + W_{SL}^d + W_{SL}^h + \cdots, \tag{1}$$

where $\gamma_s \left(= \gamma_S^p + \gamma_s^d + \gamma_s^h + \cdots \right)$ is the surface free energy of the solid, $\gamma_L \left(= \gamma_L^p + \gamma_L^d + \gamma_L^h \right)$ is the surface tension of the liquid, γ_{SL} is the interfacial free energy at the solid–liquid interface, the superscripts p, d and h denote the polar, dispersion, and hydrogen bonding contributions, respectively, and the subscript SL represents the solid–liquid interface. A geometric mean expression has been suggested to approximate the dispersion component of the work of adhesion $W_{SL}^d = 2\left[\left(\gamma_S^d \gamma_L^d\right)^{1/2}\right]^{14}$. Kloubek[15] has shown that the polar component of the work of adhesion can also be represented by a geometric mean approximation $\left[W_{SL}^p = 2\left(\gamma_S^p \gamma_L^p\right)^{1/2}\right]$. It has been shown, however, that a geometric mean approximation is not valid for the case of the hydrogen bonding component of the work of adhesion $\left(W_{SL}^h\right)^{16}$. Though hydrogen bonding and other types of specific interactions may also be contributing to the work of adhesion between a biomaterial and blood, it is currently not possible to evaluate their magnitudes when they are present alongside dispersion and polar interactions. It must be remembered that all these approximations represent only a simplified picture of the actual manner in which the two phases interact. Restricting this analysis, therefore, to only those types of biomaterials which interact with blood largely by means of non-specific interactions (i.e. dispersion and polar forces), one obtains the following expression for the blood plasma–biomaterial interfacial free energy[17]:

$$\gamma_{SL} = \left\{\left(\gamma_L^p\right)^{1/2} - \left(\gamma_S^p\right)^{1/2}\right\}^2 + \left\{\left(\gamma_L^d\right)^{1/2} - \left(\gamma_S^d\right)^{1/2}\right\}^2 \tag{2}$$

This expression is quite revealing from the point of view of relating the surface energetic properties of biomaterials with their blood compatibilities *via* the criterion suggested in Section 1 A. It shows that the solid–liquid interfacial free energy can attain a low value only when both the dispersion and polar surface free energy components of the solid are sufficiently close to those of their respective surface free energy counterparts of the liquid. For the case of the blood plasma-biomaterial interface, this means that the surface free energy characteristics of a synthetic material must be very close to those of water (since blood plasma is largely aqueous in nature) in order to reduce its interfacial free energy with blood plasma to a value of about 1–3 dyne/cm and enable it to remain in long-term compatibility with this fluid. This is a rather challenging demand on the surface properties of synthetic materials that are fit to be used in biological environments. A case in point is that of polymers, many of which possess some desirable properties for blood contact applications but their surface energetic properties are inadequate because they give rise to large interfacial free energies with the aqueous environment. This is because the polar surface free energy component of most polymeric solids is rather low in comparison to that of water $\left(\gamma_W^p = 50.8\,\text{dyne/cm}\right)$ and therefore, their interfacial free energy with water will perforce be high as can be seen from Equation 2. This consideration provides an explanation for the fact that many nonpolar materials (such as polymers) are incompatible with blood. Teflon is a classic illustration of such a nonpolar solid which is known to possess one of the highest levels of chemical inertness and yet whose blood compatibility is poor[18]. This discussion portrays the most formidable hurdle involved in the identification of synthetic surfaces for biological fluid environments, namely, to select a material which will be almost identical to the fluid in terms of its surface energetic properties and which will yet remain a mechanically viable and non-interacting contacting phase with the fluid.

2 PREPARATION OF THIN FILM COATINGS FOR BIOLOGICAL ENVIRONMENTS

As discussed earlier, synthetic materials which are to be used in biological environments must display an adequacy of both their surface and bulk characteristics in order to fulfil the dual requirements of biocompatibility and suitable mechanical properties for the given application. Depending on the biological environment involved, many types of materials, ranging from metals to plastics, have been tried as fluid contacting surfaces. In recent times, however, the use of polymers as candidate materials for a variety of biological fluid contact applications is becoming increasingly widespread. The most prominent of these is in the area of biomedical applications such as artificial implant devices in the environment of blood, contact lens materials in the ocular environment and dental prostheses in the oral cavity. Among the available synthetic materials, polymers seem best suited for such applications because of their comparatively high chemical inertness, nontoxicity in biological fluid environments and the fact that they lend themselves to relatively easy and well established processing techniques that enable one to prepare them with the desired mechanical properties for a given application. Additionally, their low surface free energies also enables them to remain less prone to 'fouling' by biological fluid components in comparison to the higher surface free energy materials such as metals and glasses. Polymeric materials do suffer from some limitations, however, such as their poor thermal and electrical conductivities, lack of extreme rigidity for some applications and their tendency to sometimes elute leachable impurities into the surrounding fluid phase as a function of time (this arises due to the diffusion of some of their surface active components from their bulk to the surface). The poor thermal conductivity of these materials is a handicap to their use in a number of important applications where severe 'fouling' is encountered on heat exchange surfaces such as in the milk pasteurization industry, electric power generation plants and the genetic engineering industry. It is therefore instructive to examine some means by which the relatively superior surface properties of polymeric materials can be complemented by more desirable bulk characteristics such as improved thermal conductivity and mechanical strength in order to enhance their versatility for a wide variety of biological fluid contact applications. This consideration provides the motivation for adopting the approach of tailoring only the surface properties of materials which possess otherwise desirable bulk characteristics for use in biological environments. One possible means of achieving this is by depositing very thin (of the order of 1000 Å) coatings of polymeric compounds onto substrates with the desired bulk properties. Moreover, with regard to the selection of a suitable thin film technique for the preparation of biocompatible surfaces, the following requirements must be borne in mind: (a) The selected technique must be versatile and capable of being applicable to a wide variety of polymeric compounds; (b) it must ensure the tight adhesion of the coated polymer film to substrates with any kind of mechanical characteristics (such as metals, ceramics, glasses or other polymers); (c) it must ensure uniform surface coverage of the substrates; (d) it must necessarily be a low temperature operation for substrates which cannot withstand high temperature conditions and (e) it must preferably offer a number of adjustable parameters so as to be able to prepare polymer films with the desired surface properties for the application in hand. Let us now see how the contemporary thin film deposition techniques for coating polymeric compounds fare with the above requirements.

The deposition of thin polymer films can be carried out by means of four different techniques, namely: (A) Solution deposition[19]; (B) vacuum evaporation[20-22]; (C) plasma polymerization[23-26] and (D) sputtering[27-33]. Solution deposition is ideally suited for coating thick polymer films (of the order of microns in thickness) but it suffers from the limitations that it does not give rise to tightly adhering films, and does not offer sufficient control over the rate and surface characteristics of the deposited films. Vacuum evaporation is a technique which employs elevated temperatures and is therefore unsuited for a number of polymers which will decompose under such conditions. Moreover, it also does not ensure the tight adhesion of the coatings to substrates and is not particularly suited to the deposition of multi-component compounds such as those of most polymers. Plasma polymerization is a technique which is widely employed for depositing thin films of polymeric compounds on substrates. This is a high vacuum technique which involves the deposition of the monomer on the substrate and the subsequent polymerization of the deposited film. Its use has become widespread in recent times for a number of biomedical applications (like the coating of polymer films on internally implanted electronic packages such as oxygen sensors, cardiac pacemakers, etc[34-37]). This technique relies on a vacuum of the order of 10^{-3} torr or less and requires the use of high purity monomers of the polymer. Its main limitation for applications involving biological environments, however, is that it does not give rise to very tightly adhering films on all kinds of substrates. This is because the substrate's surface properties are responsible for the strength of adhesion of the coated films in this technique. Sputtering seems to overcome many of the drawbacks of the above thin film deposition techniques and promises to be particularly well suited to the preparation of polymeric coatings for

Preparation and Characterization of Thin Film Surface Coatings for Biological Environments **35**

use in biological environments. This technique involves the creation of thin films by mechanically knocking out the atoms, molecules or fragments from the surface of a solid material by bombarding it with energetic, nonreactive ions. The ejection process occurs as a result of momentum transfer between the impinging ions and the atoms of the target being bombarded. The sputtered specimens are then made to travel towards the substrate where they condense to form a thin film. The sputtering process is best suited to the deposition of tightly adherent films of multicomponent compounds of either insulating or conducting materials. It can thus be used to coat substrates with thin films of either metallic, ceramic, glassy or polymeric compounds. Moreover, it can also be maintained as a low temperature operation for coating polymeric compounds and it offers a number of adjustable parameters to create polymer films with a variety of surface characteristics for different applications. Finally, it must be noted that this process of thin film formation requires considerable energy transfer and therefore the coated films are usually very tightly adherent to substrates with any type of surface properties. Thus, this potentially useful tool, belonging to the newly emerging technology of ion implantation methods, seems to be best suited to tailor the surface properties of many types of synthetic materials for use in biological media. There are several variants of the sputtering process to suit different purposes but one of the most important and useful of these for depositing thin polymer films is radio frequency sputtering.

2A RADIO FREQUENCY SPUTTER DEPOSITION OF THIN FILM COATINGS OF FLUOROCARBON COMPOUNDS

This technique, which was first developed in the early sixties[38], is applicable to the deposition of thin, solid films of either conducting, insulating or semiconducting materials. The principle of rf sputtering is illustrated in *Figure 1* and summarized below. The polymer material to be coated as a thin film, denoted as the target, is placed on a conducting plate at one end (the lower end in *Figure 1*) of a vacuum chamber and provided with water cooling to maintain lower temperatures on the target surface. The target is connected to a radio frequency power source. The substrate on which the coating is to be deposited is placed at the other end of the vacuum chamber and maintained at ground potential. In a typical experiment, the chamber is pumped down to a vacuum of about 10 μm or less and then a heavy, inert gas (such as argon) is let in at a carefully controlled rate by means of a needle valve and the pressure in the chamber is readjusted to the desired value (say 10 μm). Then the radio frequency power generator is switched on and a plasma of positively-charged argon ions is created in the chamber. These positively-charged heavy ions will be attracted towards the target surface during the negative cycle and upon impact, they will knock off the target material by a momentum transfer mechanism, resulting in the deposition of the ejected species on the substrate placed at the other end of the chamber. During the course of sputtering, the target surface acquires a net negative bias because the lighter electrons are initially attracted in greater abundance to the front surface of the target during the positive half cycle than the positive ions during the other half. This net negative bias on the target surface sustains the sputtering process and enables the uninterrupted ejection of the target material by the impacting ions. By the proper selection of the operating variables like the energy of impinging ions, the target to substrate spacing, the pressure in the chamber and the time of operation, it is possible to create polymer films with a wide variety of surface characteristics from the same target source by this technique. Moreover, since the process involves either direct or indirect momentum transfer from the incident ions to the surface atoms, the tenacity of adhesion of the coated films to substrates is also usually high and not solely determined by the surface properties of the latter as in most other thin film techniques.

Our studies were directed towards the preparation of thin film coatings of fluorocarbonaceous compounds and the subsequent modification of the surface properties of these coatings in order to enhance their compatibility with biological environments. For this, we selected Teflon FEP (obtained as 0.01 inch thick sheets from E.I. duPont de Nemours and Co.) as the target material and adopted the radio frequency sputtering technique to coat the surfaces of select substrates with thin solid films of fluorocarbon compounds. The same sputter-up configuration as shown in *Figure 1* was adopted in our experiments. The following conditions were found to be optimal for the preparation of high quality thin films with reproducible surface characteristics and tenacious adhesion to the substrates in our laboratory: (i) net radio frequency power input of 600 W; (ii) chamber pressure of 10-15 μm; (iii) target to substrate spacing of 4 cm; and (iv) deposition time of 10 h. The sputtering gas used was argon. The characterization of these deposited polymer films for their relevant surface properties in biological environments will be discussed next.

3. CHARACTERIZATION OF POLYMER SURFACES USED IN BIOLOGICAL ENVIRONMENTS

It is now common knowledge that the physical and chemical characteristics of the outermost surface layers of polymeric solids can be vastly different in comparison to their bulk. This unique aspect of polymer materials arises due to the ability of these solids to either alter their surface structures in response to their local environments or in some instances due to the migration of chain segments or pendant functional groups to or from their surface. In both these cases, the driving force responsible for the alteration of the surface properties of these solids is the thermodynamic requirement of minimizing the free energy of the system. This consideration therefore suggests that it is necessary to use highly surface sensitive techniques, which can sample only the outermost surface atoms of a specimen, in order to estimate the physicochemical properties of polymer surfaces in an environment of interest. The well known surface characterization techniques such as attenuated total reflectance infrared spectroscopy (ATR-IR), auger spectroscopy and electron and ion microprobe techniques are inapplicable to the study of polymer surfaces, though they can be used to probe the surface characteristics of other types of materials.

In the case of ATR-IR spectroscopy, its depth of penetration is of the order of a few microns[39], which therefore precludes its use for polymer surface characterization. On the other hand, auger spectroscopy and electron and ion microprobe techniques suffer from the added failing that they are likely to cause radiation damage to the surfaces of polymeric solids. Among the other surface characterization techniques, the ones which seem most relevant for polymer surfaces are: (a) electron spectroscopy for chemical analysis (ESCA) and (b) contact angle measurements. ESCA is a

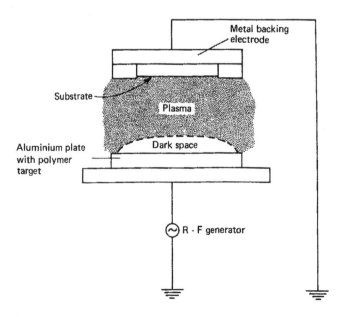

FIGURE 1 Principle of radio frequency sputtering.

Preparation and Characterization of Thin Film Surface Coatings for Biological Environments **37**

nondestructive technique which represents one of the most powerful tools to probe the chemical composition of the outermost surface layers of polymeric materials. It is generally known that the ESCA method can provide the elemental chemical composition of the outermost 10-100 Å of a polymer surface[40]. The contact angle technique, which is relatively simple to perform, is the only tool that is capable of sampling merely the surface atoms of a polymeric solid. It is generally used to estimate the wetting properties of solid surfaces. Together, ESCA and contact angle measurements are complementary and represent two of the most appropriate surface characterization techniques for polymer materials. At this point, it is necessary to mention that the conventional methods of estimating the wetting properties of solid surfaces by means of contact angle measurements are inapplicable to polymeric solids since the surfaces of these materials are not rigid and possess the unique ability to reorient their structures between different environments. Therefore, we will now address this aspect with the objective of developing a suitable contact angle procedure to estimate the wetting properties of polymeric solids in biological environments, i.e. in an aqueous medium.

3A ENVIRONMENTALLY INDUCED RESTRUCTURING OF POLYMER SURFACES AND ITS INFLUENCE ON THEIR WETTING PROPERTIES IN AN AQUEOUS MEDIUM

It is now evident that the surfaces of polymeric solids are relatively mobile, in comparison to those of their metallic, ceramic or glassy-solid counterparts[41,42]. This enables polymeric solids to adopt different surface configurations in different environments so as to increase their interaction with the latter and thereby minimize the free energy of the system. One direct consequence of this phenomenon of polymer surface relaxation from one equilibrium state to another is that the wetting characteristics of polymeric solids will also undergo a similar relaxation due to a change in environment. As a result of this, a polymer surface which appears to be hydrophobic in a non-polar environment, like air, may indeed prove to be hydrophilic in a polar environment like water. Such changes in the wetting characteristics of polymeric solids have been noted by means of contact angle hysteresis experiments by a number of researchers in the last few years[41-43]. It must be noted, however, that the dynamics of surface restructuring of polymeric solids between different environments will be largely determined by their surface structures. For example, the surfaces of some types of crosslinked, hydrophilic, methacrylate polymers (known as 'hydrogels') are highly mobile between the air and aqueous environments, with a relaxation time of the order of only 1 s or less[41]. On the other hand, the surfaces of some other polymers (like oxygen-plasma treated polypropylene[41] for instance) are more rigid and therefore, the time scales for their surfaces to orient themselves from the air to the aqueous environments can be of the order of several hours or even days.

In our studies, the two environments of interest are air and water. This is because polymeric solids are generally processed under ambient conditions but upon use in biological fluid environments, they encounter a largely aqueous medium. Therefore, in the contact angle characterization of polymeric surfaces which are to be used in biological fluid environments, it is necessary to estimate: (i) the instantaneous as well as equilibrium wetting properties of the polymeric solid in the aqeous environment and (ii) the time required for the polymer surface to attain its equilibrium surface configuration in the aqueous environment. While the former quantity is useful for correlating the surface energetic properties of polymeric solids with their performance in biological environments, the latter provides a measure of the time necessary for preconditioning a polymer surface in an aqueous environment prior to its final use, in order to ensure its optimal performance as a biological fluid contacting device.

Several contact angle techniques have been proposed in the literature for estimating the surface energetic properties of solids in an aqueous environment[44-51]. While some relied on the measurement of the contact angles of water immiscible fluids (e.g. octane[44] or octane and air[45]) on the solid surface equilibrated under water to obtain an estimate of the surface free energy components of the hydrated solid, other studies[46-51] chose to use water as a probe fluid on the solid surface immersed under two or more hydrocarbon fluids to estimate the surface energetic properties of the solid specimen. With respect to their applicability to polymeric solids, however, it must be pointed out that none of the above investigations provide a recognition of the fact that, in any type of contact angle experiment involving solids with some degree of surface mobility (such as polymers), the solid surface must be viewed as being heterogeneous in its wetting characteristics. As a result, the equilibrium surface free energy components of the solid portion which is in contact with the surrounding medium or environment will be different from the equilibrium surface free energy components of the solid portion which is in contact with the probe fluid, in the case of polymeric solids. This heterogeneity in the equilibrium wetting characteristics of a polymer surface in a contact angle experiment arises as a result of the environmentally induced restructuring of its surface structure in the two different fluid environments. This consideration of polymer surfaces seems particularly relevant in the light of recent evidence that the surfaces of almost all polymeric solids (including even those conventionally thought of as being 'rigid') are relatively mobile, although the degree of mobility can vary considerably among the different polymers.

Based on a recognition of the above mentioned aspect of environmentally induced restructuring of polymer surfaces, a four step contact angle procedure was recently suggested[52] for estimating the instantaneous and equilibrium wetting properties as well as the dynamics of surface restructuring of polymeric solids in an aqueous environment. The procedure will be summarized below.

3B ESTIMATION OF THE SURFACE FREE ENERGY COMPONENTS AND RELAXATION TIME OF POLYMERIC SOLIDS IN AN AQUEOUS ENVIRONMENT

In any contact angle experiment involving polymeric solids, it is important to note that, when the surface has attained its equilibrium wetting properties in the environments of both the immiscible fluids (i.e. surrounding medium and probe fluid), then its equilibrium interfacial free energies with these two fluids will be different from the instantaneous values. In this context, it is necessary to mention that the solid-fluid interfacial free energy is a clear indicator of the wetting properties of a solid in a particular fluid environment. Therefore, the objective of our contact angle procedure will be to estimate the instantaneous as well as equilibrium values of a polymeric solid's interfacial free energy with water and the time required for the polymeric solid to attain its equilibrium interfacial free energy with water. Let us first visualize the situation which will exist in a typical contact angle experiment involving polymeric solids. If the solid is allowed to attain its equilibrium surface configuration in the environment and then a probe fluid is placed on it and its instantaneous contact angle measured, the situation will correspond to that shown in *Figure 2 a*. In this *Figure,* θ represents the instantaneous contact angle, the environment and probe fluids are denoted as 1 and 2, respectively, and the interfacial free energies of the solid with these fluids are denoted as γ'_{s1} and γ_{s2}, where the prime symbol is used to depict the equilibrium value of the interfacial free energy (as opposed to the instantaneous value). Young's equation for this case is as follows:

$$\gamma'_{s1} = \gamma_{s2} + \gamma_{12} \cos\theta \qquad (3)$$

The value of γ_{s2} will, however, undergo a time dependent change if the solid surface can interact with the probe fluid. Consequently, the contact angle of the probe fluid on the solid surface will also undergo a time-dependent variation. If care is taken to ensure that the solid surface is smooth, non-porous, and chemically homogeneous and also to prevent the evaporation of fluids and adsorption of impurities at any of the three interfaces (i.e. γ'_{s1}, γ_{s2} and γ_{12}), then one can associate this change in θ solely with the surface restructuring of the solid part which is in contact with the probe fluid. After the solid surface attains its equilibrium configuration in the environment of the probe fluid, the situation will be as shown in *Figure 2b,* where the equilibrium contact angle is denoted by θ' and both the solid fluid interfacial free energies, representing their equilibrium values, are now denoted by primed symbols.

Preparation and Characterization of Thin Film Surface Coatings for Biological Environments   39

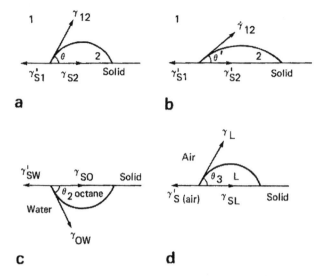

FIGURE 2 (a), Balance of interfacial free energies when the solid surface is in equilibrium with fluid 1 (environment) and then a drop of fluid 2 is placed on it and its instantaneous contact angle on the solid (θ) is measured; (b) same as (a) except that the solid surface has now attained its equilibrium surface configuration in the environments of both the fluids, namely 1 and 2; (c), this represents the interfacial free energy balance when the solid surface has been equilibrated in an aqueous environment and then a drop of octane is placed on it and the instantaneous value of the solid-octane-water contact angle (θ_2) is measured; (d), in this case, a drop of liquid L (which is nonpolar and nonhydrogen bonding) is placed on the solid surface which has been equilibrated in an air environment and its instantaneous contact angle (θ_3) is measured.

Young's equation for this case can be written as:

$$\gamma'_{s1} = \gamma'_{s2} + \gamma_{12}\cos\theta' \qquad (4)$$

It is pertinent to mention at this point that the solid-fluid interfacial free energy (unlike liquid-liquid interfacial free energies) is not a quantity that is amenable to direct experimental measurement. It can only be estimated indirectly, by making use of some semi-empirical expressions to relate the solid-fluid interfacial free energy to the measurable quantities from contact angle and surface tension experiments. In order to estimate the values of γ'_{s1}, γ_{s2} and γ'_{s2}, the following assumptions will once again be employed, (a) The solid and fluids interact by means of dispersion only or dispersion and polar forces[*], (b) The total surface free energies of the solid and fluid phases can each be considered as the sum of their dispersion and polar component surface free energies[14], (c) A geometric mean approximation can be used to describe the dispersion interactions between the solid and each of the fluid phases[14], (d) The geometric mean approximation also holds for describing the polar interactions between the solid and fluid phases[15, 54, 55]. Based on these assumptions, the following expressions can be written for γ'_{s1} γ_{s2} and γ'_{s2}:

$$\gamma'_{s1} = \gamma^{d'}_{s(1)} + \gamma^{p'}_{s(1)} + \gamma_1 - 2\left(\gamma^{d'}_{s(1)}\gamma^d_1\right)^{1/2}$$
$$-2\left(\gamma^{p'}_{s(1)}\gamma^p_1\right)^{1/2} \qquad (5)$$

---

[*] It is realized that many kinds of specific interactions (e.g. acid-base interactions[53]) may also be contributing to the overall interaction of the solid with the fluid phases. However, at the present time, it is not possible to estimate the magnitude of all such interactions (when they are present along with dispersion and polar interactions), for the case of polymeric solids interacting with different fluids. Therefore, this analysis, will be restricted to the simpler situation of nonspecific interactions (i.e. one involving only dispersion and polar forces) between the solid and fluid phases.

$$\gamma_{s2} = \gamma_{s(2)}^d + \gamma_{s(2)}^p + \gamma_2 - 2\left(\gamma_{s(2)}^d \gamma_2^d\right)^{1/2}$$

$$-2\left(\gamma_{s(2)}^p \gamma_2^p\right)^{1/2} \tag{6}$$

$$\gamma_{s2}' = \gamma_{s(2)}^{d'} + \gamma_{s(2)}^{p'} + \gamma_2 - 2\left(\gamma_{s(2)}^{d'} \gamma_2^d\right)^{1/2}$$

$$-2\left(\gamma_{s(2)}^{p'} \gamma_2^p\right)^{1/2} \tag{7}$$

In these equations, $\gamma_{s(1)}^{d'}$ and $\gamma_{s(2)}^{d'}$ represent the equilibrium values of the solid's dispersion surface free energy components in environments 1 and 2, respectively, and similarly $\gamma_{s(1)}^{p'}$ and $\gamma_{s(2)}^{p'}$ represent the equilibrium values of the solid's polar surface free energy components in environments 1 and 2, respectively. The unprimed quantities refer to the instantaneous values of the respective quantities.

Combining Equations (4), (5) and (7), the following expression results in the situation when the solid surface has attained its equilibrium wetting properties in the environments of both the fluids:

$$\gamma_{s(1)}^{d'} + \gamma_{s(1)}^{p'} + \gamma_1 - 2\left(\gamma_{s(1)}^{d'} \gamma_1^d\right)^{1/2}$$

$$-2\left(\gamma_{s(1)}^{p'} \gamma_1^p\right)^{1/2} = \gamma_{s(2)}^{d'} + \gamma_{s(2)}^{p'} + \gamma_1$$

$$-2\left(\gamma_{s(2)}^{d'} \gamma_2^d\right)^{1/2}$$

$$-2\left(\gamma_{s(2)}^{p'} \gamma_2^p\right)^{1/2} + \gamma_{12} \cos\theta' \tag{8}$$

In this equation, there are four unknown quantities, namely, $\gamma_{s(1)}^{d'}, \gamma_{s(1)}^{p'}, \gamma_{s(2)}^{d'}$ and $\gamma_{s(2)}^{p'}$. All the other quantities can either be measured or taken from the literature.

Let us now consider the more specific case of determining the instantaneous as well as equilibrium values of the solid–water interfacial free energy, i.e. γ_{sw} and γ_{sw}', respectively. Equations (6) and (7) are directly applicable for these two quantities, with water replacing fluid 2. From Equations (6) and (7), it can be seen that the estimation of γ_{sw} and γ_{sw}' involves the determination of four unknown quantities, namely, $\gamma_{s(w)}^d, \gamma_{s(w)}^p, \gamma_{s(w)}^{d'}$, and $\gamma_{s(w)}^{p'}$. This problem can be conveniently tackled by adopting the following procedure:

Step 1. Octane is selected as a nonpolar environmental fluid and the solid is allowed to equilibrate in it for a sufficiently long period of time. Then, a drop of water is placed on the solid surface under octane and the instantaneous value of the solid–water–octane contact angle (θ_1) is measured. Young's equation for this situation (*see Figure 2a*) becomes:

$$\gamma_{so}' = \gamma_{ow} \cos\theta_1 + \gamma_{sw} \tag{9}$$

In this expression, γ_{so}' is the equilibrium value of the solid–octane interfacial free energy, γ_{ow} is the octane–water interfacial tension and γ_{sw} is the instantaneous value of the solid–water interfacial free energy. Based on the assumptions stated earlier, the following expressions can be written for γ'_{so} and γ_{sw}:

$$\gamma'_{so} = \gamma^{d'}_{s(o)} + \gamma^{p'}_{s(o)} + \gamma_o - 2\left(\gamma^{d'}_{s(o)}\gamma^d_o\right)^{1/2} \tag{10}$$

$$\gamma_{sw} = \gamma^d_{s(w)} + \gamma^p_{s(w)} + \gamma_w - 2\left(\gamma^d_{s(w)}\gamma^d_w\right)^{1/2}$$

$$-2\left(\gamma^p_{s(w)}\gamma^p_w\right)^{1/2} \tag{11}$$

In these equations, $\gamma^{d'}_{s(o)}$ and $\gamma^{p'}_{s(o)}$ represent the equilibrium values of the dispersion and polar surface free energy components of the solid in an octane environment, respectively, and $\gamma^d_{s(w)}$ and $\gamma^p_{s(w)}$ represent the instantaneous values of the dispersion and polar surface free energy components of the solid in a water environment, respectively. Noting that the surface tension of octane is coincidentally equal to the dispersion component of water's surface tension and also also that $\gamma^{d'}_{s(o)} = \gamma^d_{s(w)}$ and $\gamma^{p'}_{s(o)} = \gamma^p_{s(w)}$, Equations (9), (10) and (11) can be simplified to give the following expression for $\gamma^p_{s(w)}$:

$$\gamma^{p'}_{s(o)} = \gamma^p_{s(w)} = \left(\gamma_{ow}\cos\theta_1 + \gamma_w - \gamma_o\right)^2/4\gamma^p_w \tag{12}$$

It is important to note in this step that Equation (12) is valid only if the solid surface does not reorient itself to any significant extent in the aqueous environment, in the time it takes to make the contact angle measurement of θ_1.

Step 2. After the instantaneous value of the solid-water-octane contact angle is measured in Step 1, the change in this angle is noted as a function of time until the angle reaches its equilibrium value $\left(\theta'_1\right)$. Let the time required for the angle to change from θ_1 to θ'_1 be τ. This change in the water contact angle on the solid specimen under octane can be attributed solely to the surface restructuring of the solid which is in contact with the aqueous environment, provided care is taken to eliminate other possible causes like surface roughness and surface chemical heterogeneity, surface porosity, impurity adsorption at any of the solid–liquid or liquid–liquid interfaces and evaporation of the probe fluid. Assuming that these occurrences can be prevented in our experiments, the equilibrium contact angle θ'_1 is measured along with the time required for the angle to reach its equilibrium value. In this situation (see *Figure 2b*), Young's equation becomes:

$$\gamma'_{so} = \gamma_{ow}\cos\theta'_1 + \gamma'_{sw} \tag{13}$$

where γ'_{sw} represents the equilibrium solid–water interfacial free energy. It must be noted that, in this instance, the solid surface has attained its equilibrium wetting properties in both the octane and water environments. Equation (8) is directly applicable for this case, with octane representing fluid 1 and water representing fluid 2. Equation (8) can thus be written for this case as:

$$\gamma^{d'}_{s(o)} + \gamma^{p'}_{s(o)} + \gamma_o - 2\left(\gamma^{d'}_{s(o)}\gamma_o\right)^{1/2}$$

$$= \gamma^{d'}_{s(w)} + \gamma^{p'}_{s(w)} + \gamma_w - 2\left(\gamma^{d'}_{s(w)}\gamma^d_w\right)^{1/2}$$

$$-2\left(\gamma^{p'}_{s(w)}\gamma^p_w\right)^{1/2} + \gamma_{ow}\cos\theta'_1 \tag{14}$$

Of the four unknown quantities in this equation, namely, $\gamma^{d'}_{s(o)}, \gamma^{p'}_{s(w)}, \gamma^{d'}_{s(w)}$ and $\gamma^{p'}_{s(o)}$, only $\gamma^{p'}_{s(o)}$, has been determined in Step 1. Two more equations [in addition to Equation (14)] are therefore needed to determine the remaining three unknown quantities. This can be done as outlined in Steps 3 and 4.

Step 3. Selecting octane now as the probe fluid and water as the environment, let the specimen be equilibrated under water for a period of time equal to τ determined in Step 2. Then, a drop of octane is placed on the solid surface under water and the instantaneous value of the solid–octane–water contact angle, i.e. θ_2, is measured. Now the equilibrium value of the solid's polar surface free energy component in the aqueous environment $\left(\gamma^{p'}_{s(w)}\right)$ can be calculated by noting that, in this case, the instantaneous values of the surface free energy components of the solid in the octane environment will be equal to the equilibrium values of their respective surface free energy counterparts in the water environment, i.e. $\gamma^{d}_{s(o)} = \gamma^{d'}_{s(w)}$ and $\gamma^{p}_{s(o)} = \gamma^{p'}_{s(w)}$. Referring to *Figure 2c,* Young's equation for this case can be written as:

$$\gamma'_{sw} = \gamma_{so} + \gamma_{ow} \cos\theta_2 \tag{15}$$

Writing the expressions for γ'_{sw} and γ_{so} on the basis of the assumptions used to derive Equations (5–7), the following equation can be derived to provide an estimate of $\gamma^{p'}_{s(w)}$

$$\gamma^{p'}_{s(w)} = \left(\gamma_w - \gamma_o - \gamma_{ow} \cos\theta_2\right)^2 / 4\gamma^p_w \tag{16}$$

As in Step. 1, it is important to note here also that $\gamma^{p'}_{s(w)}$ can be estimated from Equation (16) only if the solid surface in contact with the probe fluid (i.e. octane) does not significantly reorient its structure, in the time it takes to measure θ_2.

Step 4. Two of the four unknown quantities of Equation (14) have already been determined. The remaining quantities to be determined are the equilibrium values of the solid's dispersion surface free energy components in the octane and water environments, i.e. $\gamma^{d'}_{s(o)}$ and $\gamma^{d'}_{s(w)}$, respectively. This can be carried out along the following lines. For more details on the rationale of this procedure, one can consult Reference 52. Since octane is a nonpolar and nonhydrogen bonding liquid with a relatively low surface tension ($\gamma_o = 21.8$ dyne/cm), it can be assumed that it is almost akin to air in its surface tension properties and therefore, the solid surface will not undergo a significant change in its wetting characteristics between the air and octane environments. If this assumption is made, the equilibrium dispersion surface free energy component of the solid in an octane environment $\gamma^{d'}_{s(o)}$ can be equated to its equilibrium dispersion surface free energy component in air $\gamma^{d'}_{s(air)}$). Then it is possible to estimate $\gamma^{d'}_{s(o)}$ directly, by measuring the instantaneous contact angle of one or more suitably nonpolar and nonhydrogen bonding liquids, on the solid surface which is equilibrated in air. Referring to *Figure 2d,* Young's equation for this case can be written as:

TABLE 1
Surface Characterization of Sputtered End Control Teflon FEP Films

Specimen	Thickness	Surface morphology	ESCA analysis*		
			Carbon	Fluorine	Oxygen
Control Teflon FEP film	0.01″ (according to manufacturer)	Textured (see Figure 3 a)	1.0	2.18	0
Sputtered polymeric film	700-800 Å (measured by ellipsometry)	Very smooth and free of pinholes (see Figure 3 b)	1.0	0.54	0.38

*Surface elemental composition expressed as number of atoms relative to total carbon defined as 1.0

$$\gamma'_{s(air)} = \gamma'_{s(o)} = \gamma_{s(w)} = \gamma_{sL} + \gamma_L \cos\theta_3 \tag{17}$$

In this equation, fluid L represents an appropriate nonpolar and nonhydrogen bonding liquid and θ_3 is its instantaneous contact angle on the solid specimen which is equilibrated in air. Based on the assumptions used to arrive at Equations (5-7), the simplified expression for the equilibrium dispersion component of the solid's surface free energy in air can be written as:

$$\gamma^{d'}_{s(air)} = \gamma^{d'}_{s(o)} = \gamma^{d}_{s(w)} = \gamma_L \left(1 + \cos\theta_3\right)^2/4 \tag{18}$$

It may be noted that this value of $\gamma^{d'}_{s(o)}$ will also be equal to the instantaneous value of the solid's dispersion surface free energy component in the aqueous environment, i.e. $\gamma^{d}_{s(w)}$.

Now, three of the four unknowns of Equation (14) are already determined. Therefore, by using the already measured value of θ'_1 from Step 2, the only remaining quantity of interest, namely, $\gamma^{d'}_{s(w)}$, can be determined from Equation (14). It is pertinent to mention at this point that this entire procedure is based on the implicit assumption that Young's equation holds good for all the situations in Steps 1 to 4.

3C   RESULTS AND DISCUSSION

In this study, the deposited polymer films were characterized for their surface morphology, thickness, elemental surface chemical composition and wetting properties in an aqueous environment, by the techniques of SEM, ellipsometry, ESCA and contact angle measurements, respectively. The details of these characteristics are listed in *Tables 1–3* and *Figures 3–6*. The SEM of the sputtered films in *Figure 3b* show their surfaces to be very smooth and free of pinholes, in contrast to the surfaces of the commercially available control Teflon FEP films (*Figure 3 a*). ESCA spectra for the control Teflon FEP specimen revealed the characteristic C_{1s} and F_{1s} peaks (*Figure 4*). The carbon to fluorine atomic ratio of the control Teflon FEP surface was then calculated by using atomic sensitivity values of 0.205 and 1.0 for carbon and fluorine, respectively, by the method outlined by Wagner *et al.*[56]. The experimentally determined fluorine to carbon (F/C) ratio of the control surface was 2.18, which is in fairly close agreement with the theoretical (F/C) value of 2 that is to be expected for this specimen based on its chemical formula, i.e. $[CF(CF_3)–CF_2(CF_2–CF_2)_n]_m$. As for the sputtered specimen, the ESCA spectra revealed a distinct oxygen peak in addition to the carbon and fluorine peaks (see *Figure 5*). Using an atomic sensitivity value of 0.63 for oxygen and the same values as before for carbon and fluorine, the atomic ratio of fluorine to carbon was 0.54 and that of oxygen to carbon was 0.38, for the sputtered polymeric specimen. This characterization confirms that the outer surface layers of the sputtered polymeric specimen are partially oxidized and therefore more polar than the control Teflon FEP films. The ellipsometric data showed the sputtered polymer films to be sufficiently thin (of the order of 700–800 Å in thickness), so that their surfaces can be assumed to be chemically homogeneous.

TABLE 2
Contact Angle Measurements on the Sputtered Polymeric Films

Reading	θ_1	θ'_1	θ_2	θ_3 (bromobenzene)	τ,(h)
1	165	124	117	24	24
2	170	125	124	25	24
3	170	116	123	27	24
4	166	113	122	23	24
Average value	168	120	122	25	24

TABLE 3
Estimated Wetting Characteristics of the Sputtered Polymeric Specimen in the Aqueous Environment

No.	Quantity	Magnitude, dyne/cm	Remarks
1	Instantaneous polar surface free energy component of the solid in the aqueous environment $\left(\gamma^p_{s(w)} = \gamma^{p'}_{s(o)}\right)$	~0	Estimated from Equation 12
2	Equilibrium polar surface free energy component of the solid in the aqueous environment $\left(\gamma^{p'}_{s(w)}\right)$	30	Estimated from Equation 16
3	Instantaneous dispersion surface free energy component of the solid in the aqueous environment $\left(\gamma^d_{s(w)} = \gamma^{d'}_{s(o)}\right)$	33	Estimated from Equation 18
4	Equilibrium dispersion surface free energy component of the solid in the aqueous environment $\left(\gamma^{d'}_{s(w)}\right)$	91	Estimated from Equation 14
5	Equilibrium solid-octane interfacial free energy $\left(\gamma'_{so}\right)$	1	Estimated from Equation 10 assuming $\gamma^{p'}_{s(o)} = 0$
6	Instantaneous solid–water interfacial free energy $\left(\gamma_{sw}\right)$	51	Estimated from Equation 9 using γ'_{so} value from row 5
7	Equilibrium solid-water interfacial free energy $\left(\gamma'_{sw}\right)$	27	Estimated from Equation 13, using the γ'_{sw} value from row 5
8	Equilibrium wetting characteristics of the sputtered polymeric specimen in the aqueous environment	$\gamma^{d'}_{s(w)} = 91$, $\gamma^{p'}_{s(w)} = 30$ and $\gamma'_{sw} = 27$	

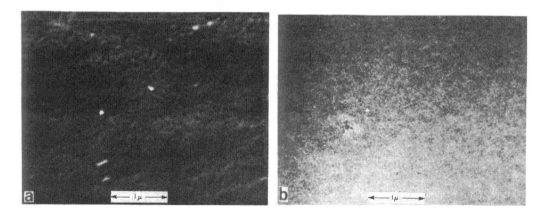

FIGURE 3 Electron photomicrographs of (a) control Teflon FEP surface and (b) sputtered polymer film.

Contact angles of liquids were measured on polymeric surfaces by using a Rame-Hart contact angle goniometer (Model A-100). The water used in these experiments was doubly-distilled and deionized. In all measurements, the contact angles of both sides of a drop were measured. In the initial experiments involving Steps 1 and 2 of the procedure outlined in Section 3B, the sputtered polymeric specimen was placed in a glass cell containing water saturated n-octane and allowed to equilibrate in the octane environment for at least 24 h. Then a small drop (about 2 µl in vol.) of doubly-distilled and deionized water was formed on the solid specimen under octane and the initial contact angle, θ_I, was measured. The changes in the solid–water–octane contact angles were then

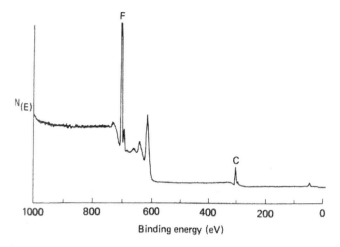

FIGURE 4 ESCA spectrum of control Teflon FEP film.

noted as a function of time, until the angle reached its equilibrium value, θ_i'. *Figure 6* shows the time-dependent variation of the contact angles of different drops of water placed on different parts of one of the sputtered polymeric surfaces. Generally, the angles continually decreased with time, before reaching their equilibrium values in a period of about 24 h. It was confirmed that these contact angles did not change much with considerably increased exposure times. The decrease in these contact angles was also accompanied by a continual increase in the width of the octane drop's base in contact with the solid, thus confirming that the drop was actually spreading on the solid surface with the passage of time. After the attainment of the equilibrium angle, the width of the octane drop's base also remained constant with time. The trend of variation of the solid–water–octane contact angles shown in *Figure 6* was reproducible on all the other sputtered specimens investigated as well, although the magnitudes of θ_i and θ_i' differed to some extent from one specimen to another, presumably due to differences in their extent of surface oxygen content. This decrease in the solid–water–octane contact angles on the sputtered specimen can arise due to a decrease in the solid–water interfacial free energy, which means that the buried polar groups in the solid begin to reorient themselves to the solid–water interface and interact with water. In order to confirm this possibility, the above experiment was repeated on the control Teflon FEP specimen. The variation

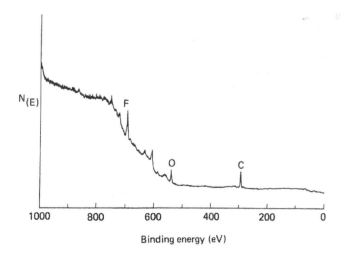

FIGURE 5 ESCA spectrum of sputtered polymer film.

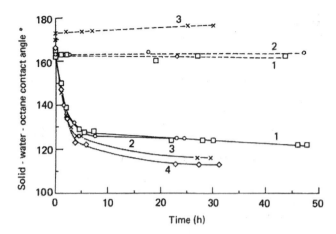

FIGURE 6 Variation of the contact angle of water on the solid specimen, under octane, as a function of time. Solid lines represent the curves for the sputtered polymeric specimen while the broken lines represent the curves for the control Teflon FEP surface. The numbers of the curves represent different parts of the surface on which the readings were measured.

of the solid-water-octane contact angles on one of the control specimen is also shown in *Figure 6*. While the angles on the sputtered films changed by about 40–50° during a 24 h period, the angles on the control Teflon FEP surface remained practically constant over periods as long as 96 h (not shown in *Figure 6*). The same result was observed on a number of specimens of the control Teflon FEP surface. The reason for this difference between the sputtered polymer films and the control Teflon FEP films can be understood by analyzing the contact angle results in conjunction with the ESCA data (*Table 1*). As noted earlier, the ESCA technique provides an average chemical composition of the surface layer of the order of 100 Å in depth, while the contact angle technique provides the characteristics of the outermost surface atoms of a specimen. The ESCA spectra reveal a distinct oxygen peak for the sputtered films, while they show no trace of oxygen for the control Teflon FEP surface. This suggests that, in the case of the sputtered polymeric films, the polar groups containing oxygen atoms remain buried in the bulk of the solid when they are equilibrated in a nonpolar environment such as air or octane but on exposure to a strongly polar environment like water, they begin to undergo a time-dependent reorientation to become exposed to the solid-water interface. In the case of the control Teflon FEP surface, however, since there are no polar groups in the outer surface layers, there will be no significant reorientation of its surface structure in the aqueous environment and therefore, neither the solid-water interfacial free energy nor the solid–water–octane contact angle can be expected to change as a function of time, on this specimen. In this context, it is necessary to mention that liquid penetration into the pores of solids may also be responsible for the time-dependent variation of their contact angles on solid surfaces[57]. In the case of the sputtered specimen of this study, however, this possibility seems highly unlikely since the sputtering technique generally provides uniform surface coverage of the substrates and does not give rise to porous surface structures in the deposited films.

In order to complete the characterization of the sputtered polymeric films for their surface energetic properties in the aqueous environment, the procedure outlined in Steps 3 and 4 were then carried out. All the measured contact angles and the calculated values of the surface free energy components are shown in *Tables 2 and 3*, respectively. The values used for the surface and interfacial tensions of the liquids in the above calculations are listed in *Table 4*.

The wetting properties of the sputtered specimen in the aqueous environment, are as follows: the equilibrium solid-water interfacial free energy $\left(\gamma'_{sw}\right)$ is 27 dyne/cm (as opposed to the initial solid-water interfacial free energy of 51 dyne/cm) and the time (τ) required for the solid surface to attain

TABLE 4
Surface Tension Properties of Liquids Used in Contact Angle Experiments

Liquid	γ_{LV} or $\gamma_{LL'}$ dyne/cm	γ_{LV}^{d} dyne/cm	γ_{LV}^{p} dyne/cm	Reference
Water	72.6	21.8	50.8	55
Water saturated octane	21.8	21.8	0	55
Bromobenzene	36.5	35.44	1.06	64
Octane-water	50.8			55

its equilibrium wetting characteristics in the aqueous environment is of the order of 24 h. From the results in *Table 3,* it can be seen that the sputtered polymeric surfaces are almost 'Teflon'-like in their initial wetting properties in an aqueous environment but on continued contact with water, they undergo a considerable decrease in their interfacial free energy with water, to attain an equilibrium solid-water interfacial free energy of about 27 dyne/cm. On the other hand. Teflon undergoes practically no change in its interfacial free energy with water on continued contact with the aqueous environment. This example illustrates the importance of taking into account the dynamics of restructuring of polymeric surfaces between different environments and its implications for the use of these types of materials in biological environments. In the case of the sputtered polymeric surfaces of our study, the solid–water interfacial free energy can be reduced to 27 dyne/cm by pre-equilibration under water for 24 h but even this interfacial free energy is too high for their successful use in biological environments. As outlined in Section 1 A, our objective must be to minimize the solid-water interfacial free energy of these sputtered polymeric surfaces to a value of about 1–3 dyne/cm, in order to ensure their compatibility with biological fluid environments. In order to effect this reduction in their interfacial free energy with the aqueous medium, it will be necessary to modify the wetting properties of these surfaces in biological environments by the application of some surface modification techniques. This will be taken up in Section 4.

It may be noted here that there can be several variants to the procedure suggested in Section 3B for estimating the wetting characteristics of polymeric solids in an aqueous environment. However, the point under emphasis is that, in any type of contact angle characterization of mobile surfaces such as those of polymeric solids, the dynamics of surface restructuring of the solid, caused by the change in environment, must be taken into account.

4 SURFACE MODIFICATION OF POLYMERIC MATERIALS FOR ENHANCING THEIR BIOCOMPATIBILITY

In order to satisfy the criterion of biocompatibility suggested in Section 1 B, it is necessary to either find materials whose individual surface free energy components belong to the biocompatible range ($\gamma_{sw} \simeq 1 - 3$ dyne/cm) or else, it is necessary to modify the surface properties of existing materials for improving their compatibility with biological fluids. The former alternative places a high premium on a trial and error search for suitable synthetic materials. On the other hand, it is a more attractive proposition to attempt an improvement in the surface properties of existing materials which possess otherwise desirable qualities for biological fluid contact applications. In pursuit of this, we will now consider the techniques that can be used to effect improvements in the surface properties of polymeric solids which are to be used in biological environments.

The goal of this surface modification procedure is to be able to reduce the polymeric solid-water interfacial free energy to a low value of about 1–3 dyne/cm. To maintain such a low interfacial free energy with water, a solid must possess surface free energy components which are very close to those of water [see Equation (2)]. This means that the solid's polar surface free energy component must be fairly high ($\simeq$50.8 dyne/cm) and its dispersion surface free energy component must be fairly low ($\simeq$21.8 dyne/cm). The requirement on the polar surface free energy component is virtually impossible to be realized on the existing polymeric materials, though many of them come close to satisfying the dispersion surface free energy component requirement. Therefore, the main purpose of attempting a surface modification of polymeric materials for use in biological environments is to considerably increase their polar surface free energy component (without markedly altering their dispersion surface free energy component) and thus, minimize their interfacial free energy with water. To effect such a considerable change in the polar surface free energy component of polymeric solids, two approaches seem promising, namely, (i) ultraviolet (u.v.)-irradiation of polymer surfaces and (ii) chemical etching of the surfaces of polymeric solids. In the former category, irradiation with u.v.-light of suitable wavelength range is known to cause a marked increase in the polar surface free energy component of many types of polymers[58]. This is believed to take place as a result of atmospheric photo-oxidation of the polymer surface in the presence of u.v.-radiation.

Esumi *et al.*[58] have investigated the effect of u.v.-irradiation on the surface free energy components of polymeric solids. They selected 6 hydrophobic polymers, three of medium surface free energy and three of low surface free energy. The former group was comprised of polystyrene, poly(methyl vinyl ketone) and poly(diphenyl siloxane). The latter group consisted of three different copolymers of dimethyl siloxane (a) PS 255,1–3% comethyl vinyl siloxane, (b) PS054, 2–4% coaminoalkyl methyl siloxane, and (c) PS-264, 3–7% codiphenyl siloxane plus 0.5–1% comethyl vinyl siloxane. *Figures 7–10* and *Table 5* show some of the results of Esumi *et al.*[58] regarding the effect of u.v.-irradiation on polymer surfaces. The surface free energy components of the polymeric solids, evaluated by two different techniques', namely, Kaelble's method[17, 59] and the Fowkes-Hamilton method[60], are shown in these *Figures*. While the former technique was used to calculate the solid's polar and dispersion surface free energy components in air, i.e. the unhydrated state (γ_s^p and γ_s^d) the latter technique provided the value of the polar surface free energy component in water, i.e. the hydrated state $\left(\gamma_s^{p'}\right)$ and the dispersion surface free energy component in the un hydrated state $\left(\gamma_s^{d'}\right)$. For each combination of solid surface free energy components, the value of the solid–water interfacial free energy [calculated from Equation (2)] is also shown in *Figures 7–10* and *Table 5*. It is important to note that, in these calculations, the interfacial free energies of the solid with both the environmental and probe fluids used in contact angle experiments were equated to the solid–environmental fluid interfacial free energy.

Figure 7 and *Table 5* show that, exposure to u.v.-radiation in air for 1 h increases the polar surface free energy component of poly(diphenyl siloxane) in the hydrated state, from an initial value of 0.37 dyne/cm to a value of 28 dyne/cm. The dispersion component changed far less dramatically (from an initial value of 45.6 dyne/cm to a final value of 40.6 dyne/cm after 1 h of exposure), though this component was estimated only for the unhydrated solid. As a result of these changes in the individual component surface free energies, the solid–water interfacial free energy changed considerably, from an initial value of 47 dyne/cm to a value of 6 dyne/cm after only 1 h of exposure to u.v.-radiation. Though these values of γ_{sw} may not exactly represent the values attained by such surfaces when they are placed in contact with a biological fluid (as a result of using the unhydrated solid's dispersion surface free energy component), they are expected to reasonably approximate the actual values of γ_{sw} (that are based on the hydrated solid's surface free energy components, $\gamma_s^{p'}$ and $\gamma_s^{d'}$).

Figures 8–10 show the effect of u.v.-irradiation on other types of polymers. Even though these materials were relatively nonpolar initially, exposure to u.v.-radiation caused an increase in their component surface free energies (both $\gamma_s^{p'}$ and γ_s^d), so as to cause a marked decrease in their interfacial free energy with water. It must be noted that, Esumi *et al.*[58] have characterized the surfaces of the u.v.-irradiated polymers, after 2 d of dark aging (to enable the escape of volatile irradiation products). However, the long-term stability of these irradiated surfaces in an aqueous environment (such as that of a biological fluid) is not yet known.

Preparation and Characterization of Thin Film Surface Coatings for Biological Environments 49

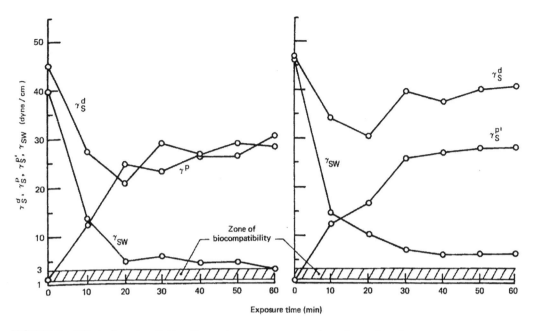

FIGURE 7 Effect of u.v.-irradiation (for various exposure times) on the surface free energy components of poly(diphenyl siloxane)[58]. The solid-water interfacial free energy is also plotted for different exposure times. The left hand side shows the values $\gamma_s^{p'}$ and γ_s^d [evaluated by Kaelble's method], while the right hand side shows the values of $\gamma_s^{p'}$ and γ_s^d [evaluated by the Fowkes-Hamilton method].

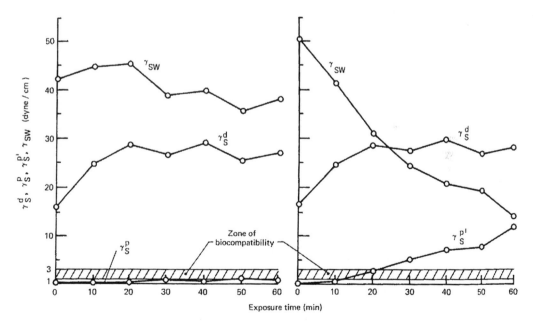

FIGURE 8 Effect of u.v.-irradiation (for various exposure times) on the surface free energy components of copolymer dimethyl siloxane PS255[58]. The solid-water interfacial free energy is also plotted for different exposure times. The left hand side shows the values γ_s^p and γ_s^d (evaluated by Kaelble's method) while the right hand side shows the values of $\gamma_s^{p'}$ and γ_s^d (evaluated by the Fowkes-Hamilton method).

It can be seen from *Figures 7–10* that the components surface free energies of various polymers are affected to different extents, by this surface modification technique. These examples are discussed only to illustrate the possibility that such a technique may be useful in improving the biocompatibilities of those polymers that can undergo significant surface modification, as a result of this treatment.

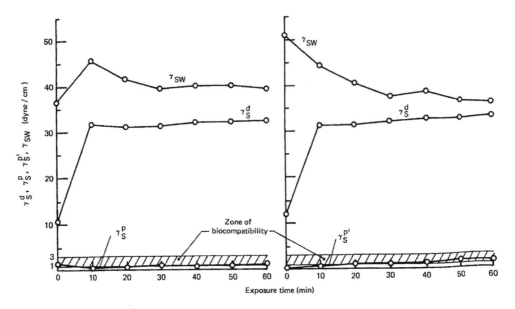

FIGURE 9 Effect of u.v.-irradiation (for various exposure times) on the surface free energy components of copolymer dimethyl siolxane PS054[58]. The solid–water interfacial free energy is also plotted for the different exposure times. The left hand side shows the values of γ_s^p and γ_s^d (evaluated by Kaelble's method), while the right side shows the values of $\gamma_s^{p'}$ and γ_s^d (evaluated by the Fowkes-Hamilton method).

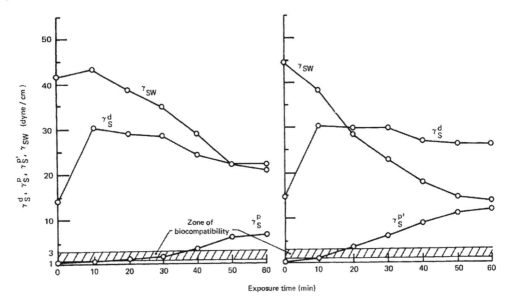

FIGURE 10 Effect of u.v.-irradiation (for various exposure times) on the surface free energy components of copolymer dimethyl siloxane PS264[58]. The solid–water interfacial free energy is also plotted for different exposure times. The left hand side shows the values of γ_s^p and γ_s^d (evaluated by Kaelble's method), while the right side shows the values of $\gamma_s^{p'}$ and γ_s^d (evaluated by the Fowkes-Hamilton method).

TABLE 5

Changes in Surface Free Energy Components and Solid-Water Interfacial Free Energies of Polymers as a Result of Exposure to U.v.-radiation (Taken from Reference 58)

Polymer	Exposure time (min)	Kaelble surface energies (dyne/cm)			Fowkes-Hamilton surface energies (dyne/cm)		
		γ_s^d	γ_s^p	γ_{sw}	$\gamma_s^{d'}$	$\gamma_s^{p'}$	γ_{sw}
Poly(diphenyl siloxane)	0	45.32	1.27	40.26	46.50	0.37	47
	10	27.36	12.19	13.54	34.27	12.26	15
	20	20.74	24.58	4.72	30.30	16.48	10
	30	29.03	23.40	5.76	39.61	25.70	7
	40	26.58	26.25	4.25	37.54	26.85	6
	50	28.81	26.16	4.54	40.11	28.00	6
	60	28.32	30.34	3.05	40.60	28.00	6
Copolymer dimethyl siloxane (PS255)	0	16.09	0.41	45.52	16.48	0.0012	51
	10	24.84	0.18	45.03	24.56	0.51	41
	20	28.92	0.18	45.43	28.58	2.63	31
	30	26.62	0.84	38.82	27.42	4.97	24
	40	29.13	0.72	39.95	29.73	6.98	21
	50	25.47	1.36	35.68	26.85	7.63	19
	60	27.09	0.95	38.14	28.00	11.83	14
Copolymer dimethyl siloxane (PS054)	0	10.58	1.49	36.90	11.83	0.046	49
	10	31.58	0.18	45.88	31.45	0.29	44
	20	31.15	0.53	41.78	31.45	0.70	40
	30	31.32	0.80	39.71	32.02	1.18	37
	40	32.01	0.73	40.33	32.59	1.00	39
	50	32.01	0.73	40.33	32.59	1.39	36
	60	32.37	0.85	39.55	33.15	1.50	36
Copolymer dimethyl siloxane (PS264)	0	14.00	0.54	41.72	14.53	0.26	45
	10	30.28	0.37	43.19	30.30	1.00	38
	20	28.85	0.91	38.60	29.73	3.37	29
	30	28.17	1.59	34.82	29.73	5.78	23
	40	24.13	3.19	28.59	26.85	8.67	18
	50	22.21	6.00	21.88	26.28	10.99	15
	60	21.99	6.52	20.92	26.28	11.83	14

Another technique to increase the polar surface free energy component of polymeric solids is by chemical etching of some of their surface functional groups. In this technique, the surface functional groups of polymeric materials react with the etching solution and the subsequently etched surfaces become more reactive to oxygen and thereby get oxidized on exposure to air. This technique has been particularly successful in the treatment of fluoropolymer surfaces (such as Teflon) to improve their bondability to other materials. Benderley[61] has reviewed several methods of treating fluoropolymer surfaces to promote bondability, including treatment with solutions of sodium in either naphthalene/tetrahydrofuran or liquid ammonia. Immersion of fluoropolymers in these solutions darkens their surface and it has been postulated that the surfaces are graphitized and defluorinated by this treatment[61]. Kaelble and Circlin[62] have measured the advancing contact angles of several liquids on Na-etched Teflon (PTFE) in air and concluded that the surface is highly polar. Dwight and Riggs[63] have characterized the surfaces of Teflon FEP, before and after etching with sodium in liquid ammonia. They found that the ESCA results for the etched specimen revealed an intense oxygen peak and showed an absence of fluorine, whereas for the unetched specimen, only carbon and fluorine peaks were noticed. Moreover, the advancing contact angle of water on the etched fluoropolymer specimen was only 52° in comparison to the value of 109° for the unetched surface. This is a confirmation of the fact that such a chemical treatment can indeed lower the solid-water interfacial free energy of fluoropolymers by a significant extent.

In this investigation, the chemical etching technique was employed to alter the wetting properties of sputtered fluorocarbon coatings in biological media. The wetting properties of the unmodified coatings in an aqueous medium are listed in *Table 3*. The chemical etching of the sputtered specimen was performed by dipping the samples in a proprietary fluorocarbon etchant (Poly-Etch, Matheson, USA) for 30 s and then rinsing the specimen in acetone and water (to remove the residual etchant), prior to air drying. The chemically treated surfaces of the sputtered polymer were than characterized for their wetting properties in an aqueous medium, by the four step procedure discussed in Section 3B.

The results of contact angle measurements on the chemically-etched sputtered specimen of this study are shown in *Figure 11* and *Table 6*. For comparison, the results on the sputtered and control Teflon FEP specimen are also provided. *Figure 11* shows the time-dependent variation of the solid-water-octane contact angles on the three different specimens. For the control Teflon FEP surface, it can be seen that the solid-water-octane contact angles remain almost constant with time. For the sputtered polymeric surfaces, the contact angles continually decrease with time (between points A to B in *Figure 11*) before attaining their equilibrium values in a period of about 24 h (corresponding to point B in *Figure 11*) and then remain constant thereafter. In the case of the chemically-etched sputtered polymeric specimen, the solid-water-octane contact angles decrease rapidly initially (between points A' and B' in *Figure 11*), followed by a period of quasi-equilibrium (between B' to C') and then again decrease more gradually (from C' to D'), until the drop totally disappears from sight in a period of about 72 h (corresponding to point D' in *Figure 11*). These contrasting trends in the time-dependent variation of the solid-water-octane contact angles can be explained by considering the physicochemical characteristics of the three different types of polymeric surfaces used in this study. Let us first recall the comparison between the control Teflon FEP surface and the sputtered specimen. In the case of the former, the ESCA spectra showed that their outer surface layers contained only carbon and fluorine atoms, with a fluorine to carbon atomic ratio of 2.18.

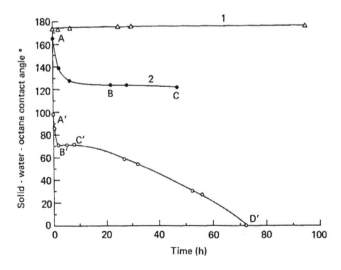

FIGURE 11 Variation of the solid-water-octane angle on different polymeric surfaces, as a function of time: 1, the control Teflon FEP surface; 2, the sputtered polymeric surface; and 3, the chemically-etched sputtered polymeric surface.

TABLE 6
Contact Angle Measurements on the Polymeric Surfaces

Specimen	θ_1	θ_1'	θ_2	θ_3
Sputtered polymer	168	120	122	25 (bromobenzene)
Chemically-etched sputtered polymer	99	72	138	20 (methylene iodide)

For the sputtered specimen, the ESCA spectra showed that their outer surface layers contain a fair amount of oxygen atoms in addition tc carbon and fluorine atoms. Based on this information, it is now possible to interpret the contact angle results of *Figure 11* as follows: The continual decrease in the solid-water-octane contact angles on the sputtered specimen arises due to a decrease in the solid-water interfacial free energy. In the case of these specimens (which represent model surfaces in that they are very smooth in surface texture, chemically homogenous and nonporous), a decrease in the solid-water interfacial free energy can be brought about by a reorientation of the oxygen containing polar groups from the subsurface layers of the solid to its surface when it is placed in contact with a strongly polar liquid like water. Moreover, the decrease in the solid-water-octane contact angles on these specimen was also accompanied by a continual increase in the width of the water drop's base in contact with the solid, thus confirming that the drop was indeed spreading on the solid surface in a time-dependent manner. After the equilibrium angle $\left(\theta_1'\right)$ was reached in a period of about 24 h, neither the solid-water-octane contact angle nor the width of the water drop's base in contact with the solid changed with time on this specimen. Contrary to this behaviour, both the solid-water-octane contact angles and the width of the water drop's base in contact with the solid remained practically constant with time for the control Teflon FEP surface. This is because, as shown by the ESCA characterization, the control Teflon FEP specimen contains no polar moieties in its outer surface layers that are capable of orienting themselves to the surface and interacting with the aqueous environment. Let us now consider the contact angle results of *Figure 11* for the case of the chemically-etched sputtered specimen. It must be noted that the surface layers of these specimens are likely to be more oxidized and therefore more polar than those of the sputtered specimen.

This is because the chemical etching process adopted in this study is known to considerably increase the oxygen content as well as the water wettability of fluorocarbon surfaces such as Teflon FEP[63]. It must also be noted that the etching process imparts a certain amount of roughness to the surface texture of the sputtered specimen. With this picture of their physicochemical surface characteristics, the contact angle results of *Figure 11* on the chemically-etched sputtered specimen can be explained as follows: During the initial period of rapidly decreasing solid-water-octane contact angles (from A' to B'), the polar functionalities of the specimen are reorienting towards the surface to interact with water. In this period, it was also observed that the width of the water drop's base continually increased with time. After the reorientation of the polar groups towards the surface is essentially completed (at B' in *Figure 11*), a period of quasi-equilibrium exists between B' and C', when neither the solid-water-octane contact angle nor the width of water drop's base in contact with the solid changed. Following this, the angles once again begin to decrease, albeit more gradually this time (from C' to D'), until the drop completely penetrates into the solid at point D'. The important point to note about the variation in the contact angles during this period is that the width of the water drop's base in contact with the solid also decreases continuously (from C' to D'), along with the decreasing angles. This indicates that penetration of water into the solid was probably taking place between C' and D'. Moreover, during this period (C' to D'), it was also observed that the drop's curved periphery became more planar and turbid near the point where the drop totally penetrated into the solid. This suggests that some dissolution of the solid as well as penetration of water into the solid's pores might be taking place simultaneously. From this time-dependent variation of the solid-water-octane contact angles on this specimen, one can conclude that, as a result of the initial decrease in the solid-water interfacial free energy to very small values (from A′ to B′), the solid-water interface becomes quite unstable at about the point C′ and this interfacial instability is responsible for effects such as the penetration of water into the solid and the possible dissolution of the solid in water. It may also be noted that, as a result of the reorientation of the polar groups of the solid from its bulk to its surface (from A′ to B′), the porosity of the surface is likely to be increased (over and above that of the initial surface). This increased porosity of the surface, coupled with the reduced solid-water interfacial free energy, may have accelerated the penetration of water into the solid and the possible dissolution of the solid in water (between C′ and D′). In addition, it may be noted that the chemical etching process might have caused changes in the crosslink density and molecular weight of the sputtered surface, which in turn could also have affected the penetration of water into the solid and/or the dissolution of the solid in water.

In connection with the estimation of the wetting characteristics of the sputtered and chemically-etched sputtered surfaces, it will be useful to evaluate the solid–water interfacial free energies at various times relevant to *Figure 11* and relate these values to the physical explanations proposed here concerning the surface mobility and the stability of the solid–water interface of the different surfaces. For the sputtered specimen, the instantaneous solid-water interfacial free energy, γ_{sw}, (at point A in *Figure 11*) and the equilibrium solid–water interfacial free energy, γ'_{sw}, (at point B in *Figure 11*) were evaluated by the four step contact angle procedure suggested earlier. In the case of the chemically-etched sputtered specimen, the quantities of interest are: (i) the magnitude of the instantaneous solid–water interfacial free energy, γ_{sw} (at point A′ in *Figure 11*) and (ii) the magnitude of the solid–water interfacial free energy at the onset of instability of the solid-water interface, i.e. γ'_{sw} (corresponding to point C′ in *Figure 11*).

The measured values of various contact angles are shown in *Table 6* and the estimated values of the various interfacial free energies are shown in *Table 7*. The contact angle measurements were repeated on at least three specimens of each polymer surface to check for repro-ducibility. Advancing contact angles of the fluids were measured on at least four different regions of each specimen. The contact angle values reported in *Table 6* represent the average readings for the different surfaces.

The results shown in *Table 7* and *Figure 11* indicate that the solid–water interfacial free energy of the sputtered specimen decreases from an instantaneous value of 51 dyne/cm to an equilibrium

---

TABLE 7

Estimated Wetting Characteristics of the Sputtered and Chemically-Etched Sputtered Polymeric Surfaces in the Aqueous Environment

	Quantity	Magnitude (dyne/cm)	Remarks
Sputtered polymer	(a) Equilibrium solid–octane interfacial free energy $\left(\gamma'_{so}\right)$	1	Estimated from Equation 10
	(b) Instantaneous solid–water interfacial free energy $\left(\gamma_{sw}\right)$	51	Estimated from Equation 9 for θ_1 value corresponding to point A in *Fig. 11*
	(c) Equilibrium solid–water interfacial free energy $\left(\gamma'_{sw}\right)$	27	Estimated from Equation 13 for θ_1 value corresponding to point B in *Fig. 11*
Chemically-etched sputtered polymer	(a) Equilibrium solid–octane interfacial free energy $\left(\gamma'_{so}\right)$	14	Estimated from Equation 10
	(b) Instantaneous solid–water interfacial free energy $\left(\gamma_{sw}\right)$	22	Estimated from Equation 9 for θ_1 value corresponding to point A′ in *Fig. 11*
	(c) Quasi-equilibrium solid-water interfacial free energy $\left(\gamma'_{sw}\right)$	−2	Estimated from Equation 13 for θ'_1 value corresponding to point C′ in *Fig. 11*

value of 27 dyne/cm, in a duration of about 24 h. In the case of the chemically-etched sputtered surfaces, the solid–water interfacial free energy decreases from an initial value of 22 dyne/cm to a low quasi-equilibrium value of − 2 dyne/cm in a period of about 2 h (A′ to B′), then it remains constant for about 6 h (B′ to C′), before the solid–water interface becomes visibly unstable, giving rise to the gradual penetration of water into the solid's pores and the possible dissolution of the solid in water. From *Table 7* it can be seen that the solid-water interfacial free energy of this specimen decreases to a low value of − 2 dyne/cm at the onset of instability of the polymeric solid-water interface. Of course, the above magnitude and even sign of this threshold interfacial free energy must not be taken too seriously as the precision of contact angle goniometry is not sufficiently high to report interfacial free energies to an accuracy of ±1 dyne/cm. Apart from this, the estimation of the solid–liquid interfacial free energies was based on a number of simplifying assumptions stated in Section 3B. Moreover, the calculated value of γ'_{sw} [from Equation (13)] is highly sensitive to the measured value of θ'_1. For example if θ'_1 is 74° rather than 72°, the estimated value of γ'_{sw} will be a small positive value instead of − 2 dyne/cm. Since the accuracies of contact angle measurements are usually about ±2°, it can be seen that the estimated value of γ'_{sw} must only be considered as an approximate measure of the solid-water interfacial free energy of the chemically-etched sputtered specimen. It must also be noted that the surfaces of the chemically-etched sputtered specimens were more rough than those of the sputtered specimen and therefore, the measured contact angles on the former could differ to some extent from the angles that must be used in Young's equation to estimate γ'_{sw}. From these considerations, the message that is conveyed *via* this estimate of γ'_{sw} is that the solid–water interfacial free energy attains a very small value at the onset of instability of the solid–water interface (corresponding to point C′ in *Figure 11*) and continues to decrease to much smaller values (including possibly negative values) until the total penetration of the water drop into the solid's pores takes place. This result provides an illustration of the premise that very low values of polymeric solid-water interfacial free energies can indeed lead to an instability of the solid–water interface and thus promote the occurrence of undesirable effects such as the penetration of water into the solid and/or the possible dissolution of the solid in water. It may be noted that, under the conditions of thermodynamic equilibrium, the solid–liquid interfacial free energy cannot attain negative values. However, under the nonequilibrium conditions of dynamic processes, if the traditional thermodynamic concept of a solid–fluid interfacial free energy is assumed to hold, then it is possible to arrive at negative values for this interfacial free energy in situations wherein the solid–fluid interface is unstable.

The above example illustrates that it is possible to drastically minimize the solid-water interfacial free energy of rf sputtered fluorocarbon coatings by means of a chemical surface modification technique. Of course, a trial and error approach is necessary to manipulate the solid-water interfacial free energy of these polymeric coatings to remain in the narrow range of 1–3 dyne/cm that is considered suitable for their use in biological media. This involves the use of an appropriate etching solution and control over the strength of the solution and the rate of etching. In this connection, it must be borne in mind that the chemical nature of the surface modified specimen, i.e. whether the surface functional groups are carbonyl or hydroxyl or other types of groups, can also play an important role in the performance of such materials in biological media. Thus, this approach of depositing tightly adherent coatings of fluorocarbon compounds by rf sputtering and subsequently modifying their wetting properties by chemical means to improve their compatibility with biological fluid environments, seems potentially well suited to the task of imparting biocompatible surface properties to many types of synthetic materials which possess suitable bulk characteristics for biological fluid contact applications.

CONCLUSIONS

The present work was aimed at the identification of a unified approach to the problem of selecting synthetic materials for use in any type of biological fluid environment (be it blood, tear fluid, milk or any other biological fluid). In the first part of this investigation, the surface interactions between a typical biological fluid, namely, blood and a synthetic surface were analysed with a view to deriving a criterion to relate the performance of synthetic materials in biological environments with some of their characteristic surface properties. The results of this treatment showed that a low blood plasma-biomaterial interfacial free energy and a mechanically stable blood plasma-biomaterial interface, two essential pre-requisites for permitting a foreign surface to remain in long-term compatibility with the environment of blood, can both be realized only when the dispersion and polar surface free energy components of the biomaterial are sufficiently near to those of their respective surface free energy components of blood plasma so that the blood plasma-biomaterial interfacial free energy can be reduced to an optimally low value of about 1–3 dyne/cm. This range of interfacial free energy values ($\gamma_{sL} \simeq$ 1-3 dyne/cm) was selected on the basis of the fact that the cellular elements, which form the outer surface coating of the highly blood compatible vascular surfaces of living systems, also bear an optimum interfacial free energy with blood plasma, generally of the order of 1–3 dyne/cm. It is also noted that the encounter of blood components with synthetic surfaces is fairly representative of the interaction of many other biological fluids with foreign materials. For this reason, the above interfacial energetic criterion of blood compatibility of biomaterials is seen as a generalized guideline to enable the selection of synthetic materials for use in any type of biological fluid environment.

In the second part of this investigation, an experimental approach to prepare synthetic substrates with suitable surface as well as bulk properties for use in biological environments is identified. In this connection, the radio frequency sputter deposition of tighly adherent coatings of appropriate polymeric compounds on substrates with the desired bulk characteristics, is shown to be a promising method of tailoring the surface properties of many types of synthetic materials for use in biological environments. This approach is illustrated by the deposition of thin, solid films of oxidized fluorocarbon compounds (from a Teflon FEP target) onto the surfaces of select substrates. The deposited polymer films were found to be smooth in surface texture, nonporous and chemically homogeneous. With regard to the estimation of their wetting properties in an aqueous medium (which is the environment encountered in most biological fluids), it was seen that the sputtered polymeric specimen exhibited a considerably different wetting character between air and aqueous environments. This arises as a result of the environmentally induced restructuring of polymer surfaces between diverse fluid environments. Based on a recognition of this aspect of polymer surfaces, a suitable contact angle procedure was developed to estimate the following quantities of interest: (a) the instantaneous as well as equilibrium values of the solid-water interfacial free energy and (b) the time required for the polymeric solid to attain its equilibrium wetting properties in an aqueous environment. In the case of the sputtered polymeric specimen, it was seen that the solid-water interfacial free energy decreased from an instantaneous value of 51 dyne/cm to an equilibrium value of 27 dyne/cm. The time required for this change to take place was approximately 24 h. This change in the solid–water interfacial free energy of such surfaces can arise if the solid specimen contains some polar groups which remain buried in their bulk in a nonpolar environment, such as air, but reorient towards the surface in a time-dependent manner when they are placed in contact with a strongly polar liquid like water. This possibility was confirmed by means of the ESCA characterization of the surfaces of these sputtered polymeric specimens, which showed that their outer surface layers contained a fair amount of the polar oxygen atoms which are capable of orienting themselves from either the bulk of the solid to its surface or *vice versa,* depending on the nature of the environment in which they are placed.

Finally, the possibility of markedly improving the wetting properties of polymeric surfaces in biological environments is discussed. In this regard, a chemical surface modification of the sputtered polymeric specimen, involving the etching of some of the surface functional groups of the solid by a proprietary fluoropolymer etchant, was seen to be suitable for effecting a drastic reduction in the solid–water interfacial free energy of such surfaces.

ACKNOWLEDGEMENT

This work was supported by the National Science Foundation.

REFERENCES

1 Baier, R.E. and Dutton, R.C., Initial events in interactions of blood with a foreign surface, *J. Biomed. Mater. Res.* 1969, **3**, 191–206.

2 Baier, R.E., in *Adhesion in Biological Systems,* (Ed. R.S. Manly), Academic Press, New York, 1970, p 15.

3 Nyilas, E., Morton, W.A., Cumming, R.D., Lederman, D.M., Chiu, T.H. and Baier, R.E., Effects of polymer surface molecular structure and force-field characteristics on blood interfacial phenomena. I., *J. Biomed. Mater. Res. Symp.* 1977, **8**, 51–68.

4 Kaelble, D.H. and Moacanin, J., A surface energy analysis of bioadhesion. *Polymer* 1977, **18**, 475–482.

5 Akers, C.K., Dardik, I., Dardik, H. and Wodka, M,, Computational methods comparing the surface properties of the inner walls of isolated human veins and synthetic biomaterials, *J. Coll. Interface Sci.* 1977. **59**. 461–467.

6 Ratner, B.D., Hoffman, A.S., Hanson, S.R., Harkar, L.A. and Whiffen, J.D., Blood-compatibility-water-content relationships for radiationgrafted hydrogels, *J. Polym. Sci., Polym. Symp.* 1979, **66**, 363–375.

7 Andrade, J.D., Interfacial phenomena and biomaterials, *Med. Instru.* 1973, **7**, 110–120.

8 Ruckenstein, E. and Gourisankar, S.V., A surface energetic criterion of blood compatibility of foreign surfaces, *J. Coll. Interface Sei.* 1984, **101**, 436–451 (Section 1.2 of this volume).

9 Ruckenstein, E. and Dunn, C.S., Stability of thin solid films, *Thin Solid Films* 1978, **51**, 43–75 (Section 5.2 of Volume I).

10 Ruckenstein, E. and Gourisankar, S.V., Surface restucturing of polymeric solids and its effect on the stability of the polymer-water interface, *J. Coll. Interface Sci.* 1986, **109**, 557–566 (Section 2.2 of this volume).

11 Troshin, AS., *Problems of Cell Permeability,* Pergamon Press, London, 1966.

12 Weiss, L., *The Cell Periphery, Metastasis and Other Contact Phenomena, North Holland, Amsterdam, 1967.*

13 Baier, R.E., Conditioning surfaces to suit the biomedical environment: Recent progress, *J. Biomech. Eng.* 1982, 104, 257–271.

14 Fowkes, F.M., Attractive forces at interfaces, *Ind. Eng. Chem.* 1964, **56**, 40–52.

15 Kloubek, J., Interaction of polar forces and their contribution to the work of adhesion, *J. Adhesion* 1974, **6**, 293–301.

16 Fowkes, F.M., in *Physicochemical Aspects of Polymer Surfaces,* (Ed. K.L. Mittal), Vol. 2, Plenum Press, New York, 1983, p 583.

17 Kaelble, D.H. and Uy, K.C., A reinterpretation of organic liquid-polytetrafluoroethylene surface interactions, *J. Adhesion* 1970, **2**, 50–60.

18 Bruck, S.D., Physicochemical aspects of the blood compatibility of polymeric surfaces, *J. Polym. Sci., Polym. Symp.* 1979, **66**, 283–312.

19 Chopra, K.L., Rao, T.V. and Rastogi, A.C., Solution growth of metallopolymer films, *Appl. Phys. Lett.* 1976, **29**, 340.

20 Murakami, Y. and Shintani, T., Vacuum deposition of Teflon-FEP, *Thin Solid Films,* 1972, **9**, 301–304

21 Luff, P.P. and White, M., The structure and properties of evaporated polyethylene thin films. *Thin Solid Films* 1970, **6**, 1 75–195.

22 Suzuki, M., Tanaka, Y. and Ito, S., Formation of semiconductive thin film of polyacrylonitrile by vacuum deposition. *Jpn. J. Appl. Phys.* 1976,**10**,817–818.

23 Morosoff, N., Newton, W. and Yasuda, H., Plasma polymerization of ethylene by magnetron discharge, *J. Vac. Sei. Technol.* 1978, **15**, 1815.

24 Nakajima, K., Bell, A.T., Shen, M. and Millard, M.M., Plasma polymerization of tetrafluoroethylene, *J. Appl. Polym. Sci.* 1979, **23**, 2627–2737.

25 Turban, G. and Catherine, Y., A kinetic model for radio frequency plasma-activated chemical vapour deposition. *Thin Solid Films* 1978, **48**, 57–65.

26 Phadke, S., Dielectric properties of plasma-polymerized ferrocene films. *Thin Solid Films* 1978, **48**, 319–324.

27 Harrop, R. and Harrop, P.J., Friction of sputtered PTFE films. *Thin Solid Films* 1969, **3**, 109–117.

28 Pratt, I.H. and Lausman, T.C., Some characteristics of sputtered polytetrafluoroetylene films. *Thin Solid Films* 1972, **10**, 151–154.

29 Morrison, D.T. and Robertson, T., R.F., sputtering of plastics. *Thin Solid Films* 1973, **15**, 87–101.

30 Tibbit, J.M., Shen, M. and Bell, AT., A comparison of r.f. sputtered and plasma polymerized thin films of tetrafluoroethylene, *Thin Solid Films* 1975, **29**, L43–L45.

31 Biederman, H., Ojha, S.M. and Holland, L., The properties of fluorocarbon films prepared by r.f sputtering and plasma polymerization in inert and active gas. *Thin Solid Films* 1977, **41**, 329–339.

32 Holland, L., Biederman, H. and Ojha, S.M., Sputtered and plasma polymerized fluorocarbon films. *Thin Solid Films* 1976, **35**, L19–L21.

33 Holland, L., Some characteristics and uses of low pressure plasmas in materials science, *J. Vac. Sci. Technol.* 1977, **14**, 5.

34 Havens, M.R., Biolsi, M.E. and Mayhan, K.G., Survey of low temperature r.f. plasma polymerization and processing, *J. Vac. Sci. Technol.* 1976, **13**, 575.

35 Dynes, P.J. and Kaelble, D.H., in *Plasma Chemistry of Polymers,* (Ed. M. Shen), Plenum Press, New York, 1976.

36 Yasuda, H.K., in *Contemporary Topics in Polymer Science,* (Ed. M. Shen), Vol. **3**, Plenum Press, New York, 1979.

37 Hahn, A.W., Nicholos, M.F., Sharma, A.K. and Hellmuth, E.W., in *Biomedical and Dental Applications of Polymers,* (Eds C.G. Gebelein and F.K. Koblitz), Plenum Press, New York, 1981.

38 Anderson, G.S., Maya, W.N. and Wehner, G.K., Sputtering of dielectrics by high-frequency fields, *J. Appl. Phys.* 1962, **33**, 2991.

39 Harrick, N.J., *Internal Reflection Spectroscopy,* Interscience, New York, 1967.

40 Ratner, B.D., in *Biomaterials: Interfacial Phenomena and Applications,* (Eds. S.L. Cooper and N.A. Peppas), Vol. **199**, Advances in Chemistry Series, American Chemical Society, Washington, DC, **1982**, p **9**.

41 Yasuda, H., Sharma, A.K. and Yasuda, T., Effect of orientation and mobility of polymer molecules at surfaces on contact angle and its hysteresis, *J. Polym, Sci. Poly. Phys. Ed.* 1981, **19**, 1285–1291.

42 Andrade, J.D., Gregonis, D.E. and Smith, L.M., in *Physicochemical Aspects of Polymer Surfaces,* (Ed. K.L. Mittal), Vol. **2**, Plenum Press, New York, 1983.

43 Holly, F.J. and Refujo, M.F., Wettability of hydrogels. I. Poly(2-hydroxyethyl methacrylate), *J. Biomed. Mater. Res.* 1975, **9**, 315–326.

44 Hamilton, W.C., A technique for the characterization of hydrophilic solid surfaces, *J. Coll. Interface Sci.* 1972, **40**, 219–222.

45 Andrade, J.D., Ma, S.M., King, R.N. and Gregonis, D.E., Contact angles at the solid-water interface, *J. Coll. Interface Sci.* 1979, **72**, 488–494.

46 Tamai, Y., Makuuchi, K. and Suzuki, M., Experimental analysis of interfacial forces at the plane surface of solids, *J. Phys. Chem.* 1967, **71**, 4176–4179.

47 Bargeman, D., Contact angles on nonpolar solids, J. *Coll. Interface Sci.* 1972, **40**, 344–348.

48 Matsunaga, T., Surface free energy analysis of polymers and its relation to surface composition, *J. Appl. Polym. Sci.* 1977, **21**, 2847–2854.

49 Tamai, Y., Matsunaga, T. and Horiuchi, K., Surface energy analysis of several organic polymers: comparison of thetwo-liquid-contact-angle method with the one-liquid-contact-angle method, *J. Coll. Interface Sci.* 1977, **60**, 112–116.

50 Schultz, J., Tsutsumi, K. and Donnet, J.B., Surface properties of high-energy solids: I. Determination of the dispersive component of the surface free energy of mica and its energy of adhesion to water and n-alkanes, *J. Coll. Interface Sci.* 1977, **59**, 272–282.

51 Lavielle, L. and Schultz, J., Surface properties of graft polyethylene in contact with water: I. Orientation phenomena, *J. Coll. Interface Sci.* 1985, **106**, 438–445.

52 Ruckenstein, E. and Gourisankar, S.V., Environmentally induced restructuring of polymer surfaces and its influence on their wetting characteristics in an aqueous environment, *J. Coll. Interface Sci.* 1985, **107**, 488–502 (Section 2.1 of this volume).

53 Fowkes, F.M. and Mostafa, A.M., Acid-base interactions in polymer adsorption, *Ind. Eng. Chem. Prod. Res. Dev.* 1978, **17**, 3–7.

54 Schultz, J., Tsutsumi, K. and Donnet, J.R., *J. Coll. Interface Sci.* 1977, **59**, 277–282.

55 Bagnal, R.D. and Green, G.F., Some observations on octane contact angles in aqueous media, *J. Coll. Interface Sci.* 1979, **68**, 387–388.

56 Wagner, C.D., Riggs, W.M., Davis, L.E., Moulder, J.F. and Muilenberg, J.E., *Handbook of XPS,* Perkin-Elmer, Physical Electronics, Palo-Alto, Ca, 1979.

57 Timmons, C.O. and Zisman, W.A., The effect of liquid structure on contact angle hysteresis, *J. Coll. Interface Sci.* 1966, **22**, 165–171.

58 Esumi, K., Schwartz, A.M. and Zettlemoyer, A.C., Effects of ultra-violet radiation on polymer surfaces, *J. Coll. Interface Sci.* 1983, **95**, 102–107.

59 Kaelble, D.H., Dispersion-polar surface tension properties of organic solids, *J. Adhesion* 1970, **2**, 66–81.

60 Hamilton, W.C., Measurement of the polar force contribution to adhesive bonding, *J. Coll. Interface Sci.* 1974, **47**, 672–675.

61 Benderley, A.A., Treatment of Teflon to promote bondability, *J. Appl. Polymer Sci.* 1962, **6**, 221–225.

62 Kaelble, D.H. and Circlin, E.H., Dispersion and polar contributions to surface tension of poly(methylene oxide) and Na-treated polytetra-fluoroethylene, *J. Polymer Sci.* 1971, **A-29**, 363–368.

63 Dwight, D.W. and Riggs, W.M., Fluoropolymer surface studies, *J. Coll. Interface Sci.* 1974, **47**, 650–660.

64 *CRC Handbook of Chemistry and Physics*, (62nd Edn), CRC Press, Florida, 1981–1982.

65 Gardon, A., *in* Encyclopedia of Polymer Science and Technology, (*Eds* H.F. Mark, N.G. Gaylord and N.M. Bikales), Vol. **3**, John Wiley & Sons Inc., New York, 1965.

APPENDIX

Notations

$\gamma_{s(o)}^{d'}$ Equilibrium value of the solid's dispersion surface free energy component in an octane environment, dyne/cm.

$\gamma_{so}^{p'}$ Equilibrium value of the solid's polar surface free energy component in an octane environment, dyne/cm.

γ_o Surface tension of octane, dyne/cm.

γ_{so}' Equilibrium value of solid-octane interfacial free energy, dyne/cm.

$\gamma_{s(w)}^{d}$ Instantaneous value of the solid's dispersion surface free energy component in an aqueous environment, dyne/cm.

$\gamma_{s(w)}^{d'}$ Equilibrium value of the solid's dispersion surface free energy in an aqueous environment dyne/cm.

$\gamma_{s(w)}^{p}$ Instantaneous value of the solid's polar surface free energy component in an aqueous environment, dyne/cm.

$\gamma_{s(w)}^{p'}$ Equilibrium value of the solid's polar surface free energy component in an aqueous environment, dyne/cm.

γ_w Surface tension of water, dyne/cm.

γ_{sw} Instantaneous value of solid–water interfacial free energy, dyne/cm.

γ_{ow}' Equilibrium value of solid–water interfacial free energy, dyne/cm.

γ_{ow} Octane-water interfacial tension, dyne/cm.

$\gamma_{s(air)}'$ Equilibrium value of the solid's total surface free energy in an air environment, dyne/cm.

$\gamma_{s(air)}^{d'}$ Equilibrium value of the solid's dispersion surface free energy component in an air environment, dyne/cm.

θ_1 Instantaneous contact angle of water on the solid surface equilibrated in an octane environment.

θ_1' Equilibrium contact angle of water on the solid surface equilibrated in an octane environment.

θ_2 Instantaneous contact angle of octane on the solid surface equilibrated in an aqueous environment.

θ_3 Instantaneous contact angle of a nonpolar and nonhydrogen bonding liquid (e.g. bromobenzene) on the solid surface equilibrated in an air environment.

τ Time (h) required for the solid surface to attain its equilibrium surface configuration in the aqueous environment.

2 Experiments on Wetting of Polymers

Eli Ruckenstein and Gersh Berim

INTRODUCTION TO CHAPTER 2

Chapter 2 contains experimental studies of wetting of polymer surfaces. The main goal was to estimate experimentally various surface characteristics. In Sec. 2.1, the instantaneous as well as equilibrium surface energetic properties of a polymeric solid and the time required for the polymeric surface to attain its equilibrium wetting characteristics were investigated in an aqueous environment. The ability of polymeric solids to alter their surface structures between different environments in order to minimize their interfacial free energy with a surrounding medium and thereby minimize the free energy of the system was studied in Sec. 2.2. In Sec. 2.3, the stability of a polymeric surface when subjected to ultraviolet irradiation for various times was examined. The equilibrium values of the polar and dispersion components of the surface free energies of polymeric surfaces equilibrated in both polar and nonpolar environments were estimated in Sec. 2.4. In Sec. 2.5, the dynamics of surface modifications of various polymeric surfaces of different hydrophilicities is investigated. Finally, in Sec. 2.6, the radiolabeling technique was employed to examine the wetting of several polymeric surfaces and of glass by bovine serum albumin.

2.1 Environmentally Induced Restructuring of Polymer Surfaces and Its Influence on Their Wetting Characteristics in an Aqueous Environment[*]

Eli Ruckenstein and Sathyamurthy V. Gourisankar
Department of Chemical Engineering, State University
of New York at Buffalo, Buffalo. New York 14260

Received January 8, 1985; accepted March 12, 1985

1. INTRODUCTION

It is now becoming increasingly evident that the surfaces of polymeric solids are relatively mobile, in comparison to those of their metallic, ceramic, or glassy solid counterparts (1, 2). This means that polymeric solids can adopt different surface configurations in different environments, so as to increase their interaction with the latter and thereby minimize the total free energy of the system. The relaxation time for a polymer surface to orient itself and equilibrate to a new environment will, however, be highly dependent on its surface structure. For example, the surfaces of cross-linked, hydrophylic, methacrylate polymers (known as "Hydrogels") are highly mobile between the air and aqueous environments, with a relaxation time of the order of only a second or less (1). In contrast, the surfaces of some other polymers (like oxygen plasma treated polypropylene (1) for instance) are more rigid and therefore, the time scales for their surfaces to orient themselves from the air to the aqueous environments can be of the order of several hours or even days.

A direct consequence of the phenomenon of polymer surface relaxation from one equilibrium state to another is that the wetting characteristics of polymeric solids will also undergo a similar relaxation due to a change in environments. As a result of this, a polymer surface which appears to be hydrophobic in a nonpolar environment like air may indeed prove to be hydrophylic in a polar environment like water. Such changes in the wetting characteristics of polymers have been experimentally observed by a number of researchers in the last few years (3, 1, 2). This dependence of their wetting characteristics on the environment in which they are placed, bears important implications for the successful use of polymers in several applications such as in the biomedical, food and pharmaceutical industries.

In our attempts to understand the wetting behavior of polymeric solids in an environment of interest therefore, two important issues must be addressed: (a) estimation of the instantaneous as well as equilibrium wetting characteristics of the polymer in the environment of interest and (b) estimation of the time required for the polymer surface to attain its equilibrium surface configuration in

---

[*] *Journal of Colloid and Interface Science.* Vol. 107, No. 2, p. 488, October 1985. Republished with permission.

the environment of interest. While the former quantity is useful for correlating the wetting behavior of polymers with their antifouling properties, the latter quantity can provide a measure of the time required for preconditioning a polymer surface in the environment of interest, prior to its final use.

In several applications involving polymers, one encounters an aqueous medium as the environment of interest. The widespread use of polymers as artificial implant devices in the biomedical industry is a striking example of such an application. It is therefore particularly important to be able to estimate the equilibration time as well as the instantaneous and equilibrium wetting characteristics of a polymeric surface, when it is placed in contact with an aqueous environment. Some prior investigations have sought to characterize the surface energetic properties of solids which have been equilibrated in an aqueous environment (4, 5). These studies relied on the measurement of the contact angles of water immiscible fluids (such as octane (4) or octane and air (5) for example) on the solid surface which was immersed under water, to obtain an estimate of the individual surface free energy components of the hydrated solid. Some other studies chose to use water as a probe fluid on the solid surface immersed under two or more hydrocarbon fluids, to estimate the surface energetic properties of the solid specimen (6-10). In the estimation of the individual surface free energy components (i.e., dispersion only or both the dispersion and polar components) of the solids from these contact angle experiments, however, none of the above investigations provide a recognition of the fact that, in any type of contact angle experiment involving solids with some degree of surface mobility (such as polymers), the solid surface must be viewed as being heterogeneous in its wetting characteristics. In other words, the equilibrium surface free energy components of the solid portion which is in contact with the surrounding medium or environment will be different from the equilibrium surface free energy components of the solid portion which is in contact with the probe fluid, in the case of polymeric solids. This heterogeneity in the equilibrium wetting characteristics of a polymer surface in a contact angle experiment arises as a result of the environmentally induced reorientation of its surface structure in the two different fluid environments. The above consideration of polymer surfaces seems particularly relevant in the light of increasing evidence in recent times that the surfaces of almost all polymeric solids (including even those conventionally thought of as being "rigid") are relatively mobile, although the degree of mobility can vary considerably among the different polymers. It must also be noted that the effect of this mobility can be particularly pronounced in the case of polar polymeric solids interacting with the aqueous environment. This is because water has a sufficiently high polarity to provoke a large scale reorientation of the polar functionalities of the solid, which will tend to remain buried in its bulk when it is maintained in a nonpolar environment such as air.

It is the objective of this paper to recognize the above-mentioned aspect of polymer surface mobility and suggest a suitable procedure based on contact angle measurements, to estimate: (i) the instantaneous as well as equilibrium surface energetic properties of a polymeric solid in an aqueous environment and (ii) the time required for the polymeric solid to attain its equilibrium surface configuration in the aqueous environment. The suggested procedure will then be illustrated by preparing model polymeric surfaces (which are smooth in surface texture, nonporous, and chemically homogeneous) and estimating their wetting characteristics in the aqueous environment.

2. ESTIMATION OF THE EQUILIBRIUM SURFACE FREE ENERGY COMPONENTS AND RELAXATION TIME OF POLYMERIC SOLIDS IN AN AQUEOUS ENVIRONMENT

Let us first visualize the situation which will exist in a typical contact angle experiment involving polymeric solids. If the solid is allowed to attain its equilibrium surface configuration in the environment and then a probe fluid is placed on it and its instantaneous contact angle measured, the situation will correspond to that shown in Fig. 1a. In this figure, θ represents the instantaneous contact angle, the environment and probe fluids are denoted as 1 and 2, respectively, and the interfacial free energies of the solid with these fluids are denoted as γ'_{s1} and γ_{s2}, where the prime symbol is used to depict the equilibrium value of the interfacial free energy (as opposed to the instantaneous value). Young's equation for this case is

$$\gamma'_{s1} = \gamma_{s2} + \gamma_{12} \cos\theta. \tag{1}$$

The value of γ_{s2} will, however, undergo a time dependent change if the solid surface can interact with the probe fluid. Consequently, the contact angle of the probe fluid on the solid surface will also undergo a time dependent variation. If care is taken to ensure that the solid surface is smooth, non-porous, and chemically homogeneous and also to prevent the evaporation of fluids and adsorption of impurities at any of the three interfaces, (i.e., and γ'_{s1}, γ_{s2}, and γ_{12}), then one can associate this change in θ solely with the surface restructuring of the solid part which is in contact with the probe fluid. After the solid surface attains its equilibrium configuration in the environment of the probe fluid, the situation will be as shown in Fig. 1b, where the equilibrium contact angle is denoted by θ' and both the solid fluid interfacial free energies, representing their equilibrium values, are now denoted by primed symbols. Young's equation for this case can be written as

$$\gamma'_{s1} = \gamma'_{s2} + \gamma_{12} \cos\theta'. \tag{2}$$

In order to estimate the individual surface free energy components of the solid specimen, the following assumptions are necessary: (a) The solid and fluids interact by means of dispersion only or dispersion and polar forces.[1] (b) The total surface free energies of the solid and fluid phases can each be considered as the sum of their dispersion and polar component surface free energies (12). (c) A geometric mean approximation can be used to describe the dispersion interactions between the solid and each of the fluid phases (12). (d) The geometric mean approximation also holds for describing the polar interactions between the solid and fluid phases (13-15). Based on these assumptions, the following expressions can be written γ'_{s1} and γ'_{s2}:

$$\gamma'_{s1} = \gamma^{d'}_{s(1)} + \gamma^{p'}_{s(1)} + \gamma_1 - 2\left(\gamma^{d'}_{s(1)}\gamma^d_1\right)^{1/2} - 2\left(\gamma^{p'}_{s(1)}\gamma^p_1\right)^{1/2} \tag{3}$$

$$\gamma'_{s2} = \gamma^{d'}_{s(2)} + \gamma^{p'}_{s(2)} + \gamma_2 - 2\left(\gamma^{d'}_{s(2)}\gamma^d_2\right)^{1/2} - 2\left(\gamma^{p'}_{s(2)}\gamma^p_2\right)^{1/2}. \tag{4}$$

In these equations, $\gamma^{d'}_{s(1)}$ and $\gamma^{d'}_{s(2)}$ represent the equilibrium values of the solid's dispersion surface free energy components in environments 1 and 2, respectively, and similarly $\gamma^{p'}_{s(1)}$ and $\gamma^{p'}_{s(2)}$ represent the equilibrium values of the solid's polar surface free energy components in environments 1 and 2, respectively.

Combining Eqs. [2]–[4],

$$\gamma^{d'}_{s(1)} + \gamma^{p'}_{s(1)} + \gamma_1 - 2\left(\gamma^{d'}_{s(1)}\gamma^d_1\right)^{1/2} - 2\left(\gamma^{p'}_{s(1)}\gamma^p_1\right)^{1/2}$$

$$= \gamma^{d'}_{s(2)} + \gamma^{p'}_{s(2)} + \gamma_2 - 2\left(\gamma^{d'}_{s(2)}\gamma^d_2\right)^{1/2} - 2\left(\gamma^{p'}_{s(2)}\gamma^p_2\right)^{1/2} + \gamma_{12} \cos\theta'. \tag{5}$$

---

[1] It is realized that many kinds of specific interactions (such as acid-base interactions (11) for example) may also be contributing to the overall interaction of the solid with the fluid phases. However, at the present time, it is not possible to estimate the magnitude of all such interactions (when they are present along with dispersion and polar interactions), for the case of polymeric solids interacting with different fluids. Therefore, this analysis will be restricted to the simpler situation of nonspecific interactions (i.e., one involving only dispersion and polar forces) between the solid and fluid phases.

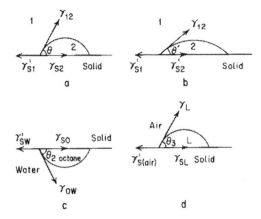

FIG. 1. (a) Balance of interfacial free energies when the solid surface is in equilibrium with fluid 1 (environment) and then a drop of fluid 2 is placed on it and its instantaneous contact angle on the solid (θ) is measured, (b) Same as (a) except that the solid surface has now attained its equilibrium surface configuration in the environments of both the fluids, namely, 1 and 2. (c) This represents the interfacial free energy balance when the solid surface has been equilibrated in an aqueous environment and then a drop of octane is placed on it and the instantaneous value of the solid-octane-water contact angle (θ_2) is measured, (d) In this case, a drop of liquid L (which is nonpolar and nonhydrogen-bonding) is placed on the solid surface which has been equilibrated in an air environment and its instantaneous contact angle (θ_3) is measured.

In this equation, there are four unknown quantities, namely, $\gamma^{d'}_{s(1)}$, $\gamma^{p'}_{s(1)}$, $\gamma^{d'}_{s(2)}$ and $\gamma^{p'}_{s(2)}$. All the other quantities can either be measured or taken from the literature.

We will now consider the more specific case of determining the instantaneous as well as equilibrium surface energetic properties of a polymeric solid in an aqueous environment. This problem can be conveniently tackled by adopting the following procedure:

Step 1

Let us select octane as the environmental fluid and equilibrate the solid in it for a sufficiently long period of time. Then, if a drop of water is placed on the solid surface under octane and the instantaneous value of the solid-water-octane contact angle (θ_1) is measured, Young's equation for this situation (Fig. 1a) becomes

$$\gamma'_{so} = \gamma_{ow} \cos\theta_1 + \gamma_{sw}. \qquad [6]$$

In this expression, γ'_{so} is the equilibrium value of the solid-octane interfacial free energy, γ_{ow} is the octane-water interfacial tension, and γ_{sw} is the instantaneous value of the solid-water interfacial free energy. Based on the assumptions stated earlier, the following expressions can be written for γ'_{so} and γ_{sw}:

$$\gamma'_{so} = \gamma^{d'}_{s(o)} + \gamma^{p'}_{s(o)} + \gamma_o - 2\left(\gamma^{d'}_{s(o)}\gamma^d_o\right)^{1/2} \qquad [7]$$

$$\gamma_{sw} = \gamma^d_{s(w)} + \gamma^p_{s(w)} + \gamma_w - 2\left(\gamma^d_{s(w)}\gamma^d_w\right)^{1/2} - 2\left(\gamma^p_{s(w)}\gamma^p_w\right)^{1/2}. \qquad [8]$$

In these equations, $\gamma_{s(o)}^{d'}$ and $\gamma_{s(o)}^{p'}$ represent the equilibrium values of the dispersion and polar surface free energy components of the solid in an octane environment, respectively, and $\gamma_{s(w)}^{d}$ and $\gamma_{s(w)}^{p}$ represent the instantaneous values of the dispersion and polar surface free energy components of the solid in a water environment, respectively. Noting that octane is a nonpolar liquid and its surface tension is equal to the dispersion component of water's surface tension and also noting that $\gamma_{s(o)}^{d'} = \gamma_{s(w)}^{d}$ and $\gamma_{s(o)}^{p'} = \gamma_{s(w)}^{p}$, Eqs. [6]–[8] can be simplified to give the following expression for $\gamma_{s(w)}^{p}$:

$$\gamma_{s(o)}^{p'} = \gamma_{s(w)}^{p}$$
$$= \left(\gamma_{ow}\cos\theta_1 + \gamma_w - \gamma_o\right)^2 / 4\,\gamma_w^p. \qquad [9]$$

It is important to note in this step that Eq. [9] is valid only if the solid surface does not reorient itself to any significant extent in the aqueous environment, in the time it takes to measure the contact angle θ_1.

Step 2

After measuring the instantaneous value of the solid-water-octane contact angle in step 1 let us then note the change in this angle as a function of time until the angle reaches its equilibrium value (θ'_1). Let the time required for the angle to change from θ_1 to θ'_1 be τ. This change in the water contact angle on the solid specimen under octane can be associated solely with the restructuring of the solid surface which is in contact with the aqueous environment, provided care is taken to eliminate other possible causes like surface roughness and surface chemical heterogeneity, surface porosity, impurity adsorption at any of the solid-liquid or liquid-liquid interfaces, and evaporation of the probe fluid. Assuming that these occurrences can be prevented in our experiments, let us then measure the equilibrium contact angle θ'_1 and also note the time required for the angle to reach its equilibrium value. In this situation (ref. Fig. lb), Young's equation becomes

$$\gamma'_{so} = \gamma_{ow}\cos\theta'_1 + \gamma'_{sw}, \qquad [10]$$

where γ'_{sw} represents the equilibrium solid- water interfacial free energy. It must be noted that, in this instance, the solid surface has attained its equilibrium wetting characteristics in both the octane and water environments. Equation [5] is directly applicable for this case, with octane representing fluid 1 and water representing fluid 2. Equation [5] can thus be written for this case as

$$\gamma_{s(o)}^{d'} + \gamma_{s(o)}^{p'} + \gamma_o - 2\left(\gamma_{s(o)}^{d'}\gamma_o\right)^{1/2} = \gamma_{s(w)}^{d'} + \gamma_{s(w)}^{p'} + \gamma_w - 2\left(\gamma_{s(w)}^{d'}\gamma_w^{d}\right)^{1/2}$$
$$-2\left(\gamma_{s(w)}^{p'}\gamma_w^{p}\right)^{1/2} + \gamma_{ow}\cos\theta'_1. \qquad [11]$$

In this equation, it is important to note that the equilibrium values of both the dispersion and polar surface free energy components of the solid in the water environment, i.e., $\gamma_{s(w)}^{d'}$ and $\gamma_{s(w)}^{p'}$, respectively, are different from their instantaneous values in the aqueous environment. Of the four unknown quantities in this equation, namely, $\gamma_{s(o)}^{d'}$, $\gamma_{s(o)}^{p'}$, $\gamma_{s(w)}^{d'}$, and $\gamma_{s(w)}^{p'}$, only $\gamma_{s(o)}^{p'}$ has been determined in step 1. Two more equations (in addition to Eq. [11]) are therefore needed to determine the remaining three unknown quantities. This can be done as outlined in steps 3 and 4.

Step 3

Selecting octane now as the probe fluid and water as the environment, let the specimen be equilibrated under water for a sufficiently long period. Since the time τ required for the solid surface to equilibrate to the water environment is known from step 2, it can be maintained under water for that period in this step. Then, if a drop of octane is placed on the solid surface under water and the instantaneous value of the solid-octane-water contact angle, i.e., θ_2, is measured, the equilibrium value of the solid's polar surface free energy component in the aqueous environment $\left(\gamma_{s(w)}^{p'}\right)$ can be calculated by noting that, in this case, the instantaneous values of the surface free energy components of the solid in the octane environment will be equal to the equilibrium values of their respective surface free energy counterparts in the water environment, i.e., $\gamma_{s(o)}^{d} = \gamma_{s(w)}^{d'}$ and $\gamma_{s(o)}^{p} = \gamma_{s(w)}^{p'}$. Referring to Fig. lc, Young's equation for this case can be written as

$$\gamma_{sw}' = \gamma_{so} + \gamma_{ow} \cos\theta_2. \qquad [12]$$

Writing the expressions for γ_{sw}' and γ_{so} on the basis of the assumptions used to derive Eqs. [3] and [4], the following equation can be derived to provide an estimate of $\gamma_{s(w)}^{p'}$:

$$\gamma_{s(w)}^{p'} = \left(\gamma_w - \gamma_o - \gamma_{ow} \cos\theta_2\right)^2 / 4\gamma_w^p. \qquad [13]$$

As in step 1, it is important to note here also that $\gamma_{s(w)}^{p'}$ can be estimated from Eq. [13] only if the solid surface in contact with the probe fluid (i.e., octane) does not significantly reorient its structure, in the time it takes to measure θ_2.

Step 4

Now, two of the four unknown quantities of Eq. (11) have been determined. The remaining quantities to be determined are the equilibrium values of the solid's dispersion surface free energy components in the octane and water environments, i.e., $\gamma_{s(o)}^{d'}$ and $\gamma_{s(w)}^{d'}$, respectively. If the equilibrium value of the solid-octane-water contact angle, i.e., θ_2', can be measured, then it will be possible to obtain one more equation (in addition to Eq. [11]) involving $\gamma_{s(o)}^{d'}$ and $\gamma_{s(w)}^{d'}$ and thus the two unknown quantities can be estimated from the two equations. However, it must be noted that the change in the contact angle value from θ_2 to θ_2' will arise due to a decrease of γ_{so}, which involves an overturning of the polar groups from the surface of the solid to its bulk. For most polymeric solids (with the possible exception of some types of hydrogels), this process is generally orders of magnitude slower than the opposite process, namely, the overturning of the buried polar groups of the specimen from within its bulk to its surface in the presence of a strongly polar liquid like water. This fact is well illustrated by the experiments of Yasuda, *et al.* (1). They noted that while the contact angle hysterisis of water drops in air was pronounced on oxygen plasma treated polypropylene surfaces, the decay in the water wettability of the same specimen was a very gradual process and took place in approximately 20 days. Since the latter result arises due to the burying of the solid's polar groups into its bulk while the former arises due to the overturning of the buried polar groups from the interior of the polymer to its surface when it is placed in contact with water, it can clearly be seen that the time scales of the two processes differ by a considerable extent. Therefore, in the interests of practical considerations, the measurement of θ_2' will not be carried out and instead, the following procedure will be adopted to estimate the values of $\gamma_{s(o)}^{d'}$ and $\gamma_{s(w)}^{d'}$. Since octane is a nonpolar and nonhydrogen-bonding liquid with a relatively low surface tension $\left(\gamma_o = 21.8 \,\text{dyn}/\text{cm}\right)$, it can be assumed that the solid surface will not undergo a significant change in its wetting characteristics between the air and octane environments. If this assumption is made, the equilibrium dispersion surface free energy component of the solid in an octane environment $\left(\gamma_{s(o)}^{d'}\right)$ can be equated to its equilibrium dispersion surface free energy component in air $\left(\gamma_{s(air)}^{d'}\right)$. Then it is possible to estimate $\gamma_{s(o)}^{d'}$ directly, by measuring the instantaneous contact angle of one or more suitably nonpolar and nonhydrogen-bonding liquids, on the solid surface which is equilibrated in air. Referring to Fig. 1d, Young's equation for this case can be written as

$$\gamma_{s(air)}' = \gamma_{s(o)}' = \gamma_{s(w)} = \gamma_{sL} + \gamma_L \cos\theta_3. \tag{14}$$

In this equation, fluid L represents an appropriate nonpolar and nonhydrogen bonding liquid and θ_3 is its instantaneous contact angle on the solid specimen which is equilibrated in air. Based on the assumptions stated earlier, the simplified expression for the equilibrium dispersion component of the solid's surface free energy in air can be written as

$$\gamma_{s(air)}^{d'} = \gamma_{s(o)}^{d'} = \gamma_{s(w)}^{d} = \gamma_L \left(1 + \cos\theta_3\right)^2/4. \tag{15}$$

It may be noted that this value of $\gamma_{s(o)}^{d'}$ will also be equal to the instantaneous value of the solid's dispersion surface free energy component in the aqueous environment, i.e., $\gamma_{s(w)}^{d}$.

Now, three of the four unknowns of Eq. [11] are already determined. Therefore, by using the value of θ_1' already measured in step 2, the only remaining quantity of interest, namely, $\gamma_{s(w)}^{d'}$ can be determined from Eq. [11]. It is pertinent to mention at this point that this entire procedure is based on the implicit assumption that Young's equation is valid for all the situations in steps 1 through 4.

3. RESULTS AND DISCUSSION

3A. Preparation and Characterization of Model Polymeric Surfaces

To illustrate the applicability of the procedure suggested in steps 1-4 of Section 2, it is desirable to obtain polymeric surfaces which are very smooth in texture, nonporous, chemically homogeneous, and sufficiently polar so as to be able to undergo a discernable interaction with the aqueous environment. Since the surfaces of most commercially available polymers are neither very smooth nor chemically homogeneous, we decided to prepare model polymeric surfaces by depositing thin solid films of polymers onto the surfaces of smooth substrates. The polymer selected for this study was Teflon FEP (obtained as 0.01- inch-thick sheets from E. I. duPont de Nemours and Company). The substrates on which the polymer films are to be deposited must also be very smooth and ultraclean, so as to ensure both the smoothness as well as the tight adherence of the deposited films. For these reasons, we decided to use highly polished, single crystal silicon wafers, which are normally used in semiconductor applications, as the substrates for this study. The surfaces of these substrates presented a very smooth morphology (free from any cracks or irregularities) when they were examined by scanning electron microscopy. Radio frequency sputtering was the technique used to deposit the thin solid films of the polymer onto the surfaces of the silicon substrates. This deposition technique was selected because it offers a number of advantages for the purpose of this study: (i) It enables one to obtain a more or less constant rate of deposition on any substrate. Thus, the deposited films will also be smooth provided the underlying substrate does not present major surface cracks or irregularities. (ii) By adjusting the time and power of the deposition process, one can coat sufficiently thin polymer films (of the order of some hundreds of angstroms) on the substrates. In the case of such thin films, there is no reason to suspect that the surfaces will be chemically heterogeneous. Moreover, the films coated by this technique are usually nonporous and more tightly adherent to the substrates in comparison to the other contemporary techniques employed for the deposition of thin solid films of polymers. (iii) By varying a number of adjustable parameters of this technique (like the radio frequency power input, the pressure in the chamber, the composition of the plasma environment, and the target to substrate spacing), it is possible to prepare polymer films with a great diversity of surface stoichiometries and wetting characteristics. Particularly useful to this study is the fact that this technique enables the preparation of oxidized fluorocarbon films, which can be expected to possess a reasonable degree of surface polarity in an aqueous environment and yet reveal a nonpolar character (characteristic of a typical fluorocarbon) in an air environment. It may also be noted that this method of preparing oxidized polymeric surfaces offers a distinct advantage over the conventional methods of oxidizing the surfaces of commercial polymers (such as by glow discharge or corona discharge treatments), since the latter techniques are usually known to cause an increase in the surface roughness of the specimen (16, 17).

The conditions employed for the radio frequency sputter deposition of oxidized fluorocarbon films in this study were as follows:(i) net radio frequency power input of 600 W: (ii) pressure of 10 μm in the chamber, (iii) target to substrate spacing of 4 cm, and (iv) deposition time of 10 h. The sputtering gas used was argon.

The deposited polymer films were then characterized for their surface morphology, thickness, and their elemental surface chemical composition, by the techniques of scanning electron microscopy, ellipsometry, and electron spectroscopy for chemical applications (ESCA), respectively. The details of these characterizations are listed in Table I and Fig. 2. The scanning electron micrographs of the sputtered films in Fig. 2b show their surfaces to be very smooth and free of pinholes, in contrast to the surfaces of the commercially available control Teflon FEP films (Fig. 2a). Prior to analyzing the ESCA data of Table I, it must be noted that this technique represents one of the most powerful tools to probe the chemical composition of the outermost surface layers of polymeric materials. It is generally recognized that the ESCA technique can provide quantitative elemental composition of the outermost 5-100 Å of a polymer surface. ESCA spectra for the control Teflon FEP specimen revealed the characteristic C_{1s} and F_{1s} peaks. The carbon to fluorine atomic ratio of the control Teflon FEP surface was then calculated by using atomic sensitivity values of 0.205 and 1.0 for carbon and fluorine, respectively. The method of calculating the atom fraction of any constituent in the sample is outlined by Wagner *et al.* (18). The experimentally determined fluorine to carbon (F/C) ratio of the control surface was 2.18, which is in fairly close agreement with the theoretical (F/C) value of 2 that is to be expected for this specimen based on its chemical formula, i.e.. $\left[CF\left(CF_3\right) - CF_2\left(CF_2 - CF_2\right)_n \right]_m$. As for the sputtered specimen, the ESCA spectra revealed a distinct oxygen peak in addition to the carbon and luorine peaks. Using an atomic sensitivity value of 0.63 for oxygen and the same values as before for carbon and fluorine, the atomic ratio of fluorine to carbon was 0.54 and that of oxygen to carbon was 0.38, for the sputtered polymeric specimen. This characterization confirms that the outer surface layers of the sputtered polymeric specimen are partially oxidized and therefore more polar than the control Teflon FEP films. The ellipsometric

TABLE I
Surface Characterization of Sputtered and Control Teflon FEP Polymer Films

No.	Specimen	Thickness	Surface morphology	ESCA analysis		
				Surface elemental composition expressed as number of atoms relative to total carbon defined as 1.0.		
				Carbon	Fluorine	Oxygen
1.	Control teflon FEP film	00.01″ (according to manufacturer's specification)	Textured (Fig.2a)	1.0	2.18	0
2.	Sputtered polymeric film	700-800 Å (measured by ellipsometry)	Very smooth and free of pinholes (Fig. 2b)	1.0	0.54	0.38

3B. Contact Angle Measurements

Contact angles of liquids were measured on polymeric surfaces by using a Rame-Hart contact angle goniometer (Model A-100). The water used in these experiments was doubly distilled and deionized. All the chemicals used in these experiments were of the highest available purity specifications. In all measurements, the contact angles on both sides of each drop were measured. The liquid drop volume was controlled between 2-5μl to eliminate gravitational effects. In the first set of experiments, the sputtered polymeric specimen was placed in a glass cell containing water saturated n-octane and allowed to equilibrate in the octane environment for at least 24 h. Then a small drop (about 2 μl in volume) of doubly distilled and deionized water was formed on the solid specimen under octane and the initial contact angle, θ_1 was measured. The changes in the solid-water-octane contact angles were then noted as a function of time, until the angle reached its equilibrium value, θ'_1. Figure 3 shows the time-dependent variation of the contact angles of different drops of water placed on different parts of one of the sputtered polymeric surfaces. In general, the angles continually decreased with time, before reaching their equilibrium values in a period of about 24 h. It was confirmed that these contact angles did not change much with considerably increased exposure times. The decrease in these contact angles was also accompanied by a continual increase in the width of the water drop's base in contact with the solid, thus confirming that the drop was actually spreading on the solid surface with the passage of time. After the attainment of the equilibrium angle, the width of the water drop's base also remained constant with time. The trend of variation of the solid-water-octane contact angles shown in Fig. 3 was reproducible on all the other sputtered specimen investigated as well, although the magnitudes of θ_1 and θ'_1 differed to some extent from one specimen to another, presumably due to differences in their extent of surface oxidation. This decrease in the solid-water-octane contact angles on the sputtered specimen can arise due to a decrease in the solid-water interfacial free energy, which means that the buried polar groups in the solid begin to reorient themselves to the solid-water interface and interact with water. In order to confirm this possibility, the above experiment was repeated on the control Teflon FEP specimen. The variation of the solid- water-octane contact angles on one of the control specimens is also shown in Fig. 3. While the angles on the sputtered films changed by about 40-50° during a 24 h. period, the angles on the control Teflon FEP surface remained practically constant over periods as long as 96 h (not shown in Fig. 3). The same result was observed on a number of specimens of the control Teflon FEP surface. The reason for this difference between the sputtered polymer films and the control Teflon FEP films can be understood by analyzing the contact angle results in conjunction with the ESC A data (Table I). As noted earlier, the ESCA technique provides an average chemical composition of a surface layer of the order of 100 Å in depth, while the contact angle technique provides the characteristics of the outermost surface atoms of a specimen. The ESCA spectra reveal a distinct oxygen peak for the sputtered films, while they show no trace of oxygen for the control Teflon FEP surface. This suggests that, in the case of the sputtered polymeric films, the polar groups containing oxygen atoms remain buried in the bulk of the solid when it is equilibrated in a nonpolar environment such as air or octane but on exposure to a strongly polar environment like water, they begin to undergo a time-dependent reorientation to become exposed to the solid-water interface. In the case of the control Teflon FEP surface, however, since there are no polar groups in the outer surface layers, there will be no significant reorientation of its surface structure in the aqueous environment and therefore, neither the solid-water interfacial free energy nor the solid-water-octane contact angle can be expected to change as a function of time, on this specimen. In this context, it is necessary to mention that liquid penetration into the pores of solids may also be responsible for the time-dependent variation of their contact angles on solid surfaces [19]. In the case of the sputtered

Environmentally Induced Restructuring of Polymer Surfaces and Its Influence 71

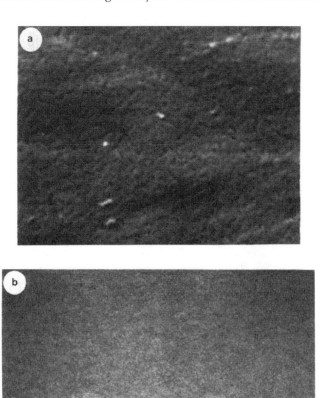

FIG. 2. Electron photomicrographs: (a) control Teflon FEP surface and (b) Sputtered polymer film.

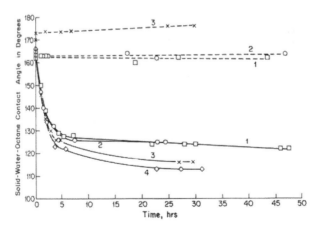

FIG. 3. Variation of the contact angle of water on the solid specimen under octane, as a function of time. Solid lines represent the curves for the sputtered polymeric specimen while the broken lines represent the curves for the control Teflon FEP surface. The numbers of the curves represent different parts of the surface on which the readings were measured.

specimen of this study, however, this possibility seems highly unlikely since the sputtering technique generally provides uniform surface coverage of the substrates and does not give rise to porous surface structures in the deposited films.

In order to complete the characterization of the sputtered polymeric films for their surface energetic properties in the aqueous environment, the procedure outlined in steps 3 to 4 of Section 2 were then carried out. All the measured contact angles and the calculated values of the surface free energy components are shown in Tables II and III, respectively. The values used for the surface and interfacial tensions of the liquids in the above calculations are listed in Table IV.

From the results of Table III, it can be seen that the instantaneous value of the polar surface free energy component of the solid in the water environment, i.e., $\gamma^p_{s(w)} \left(= \gamma^{p'}_{s(o)} \right)$ is nearly zero, indicating that the sputtered specimen exhibits an almost nonpolar surface character in an octane or air environment. This fact can be used to provide independent estimates of both the instantaneous as well as equilibrium values of the solid-water interfacial free energies, without requiring a knowledge of either the instantaneous or equilibrium values of the solid's polar surface free energy component in the aqueous environment, as follows:

Based on the estimate of the solid's equilibrium dispersion surface free energy component in an octane or air environment, i.e., $\gamma^{d'}_{s(o)}$, and assuming the solid to be nonpolar in an octane environment, it is possible to calculate the equilibrium value of the solid-octane interfacial free energy $\left(\gamma'_{so} \right)$ from Eq. [7]. This value of γ'_{so} can then be used in Eqs. [6] and [10] to provide the values of γ_{sw} and γ'_{sw}, respectively. Since the values of θ_1, and θ'_1 are already known from experiment, the values of γ_{sw} and γ'_{sw} can thus be calculated without requiring a knowledge of either $\gamma^p_{s(w)}$ or $\gamma^{p'}_{s(w)}$. These estimated values of γ_{sw} and γ'_{sw} are listed in rows 6 and 7 of Table III and it can be seen that the solid-water interfacial free energy decreases from an initially high value of 50.88 dyn/cm to an equilibrium value of 26.59 dyn/cm. This decrease in the solid-water interfacial free energy is also well manifested by the experimentally observed changes (of about 40-50°) in the contact angles of water drops on the solid specimen maintained in an octane environment, i.e., from θ_1 to θ'_1.

Row 4 of Table III shows that the equilibrium dispersion surface free energy component of the sputtered specimen in the aqueous environment $\left(\gamma^{d'}_{s(w)} \right)$ can take on either of the two values indicated, i.e., 91.13 or 0.04 dyn/cm. This happens because $\gamma^{d'}_{s(w)}$ is the only unknown quantity to be calculated from Eq. [11], which is a quadratic equation in this variable since all the other unknowns of this equation have already been determined. In order to determine which of the values of $\gamma^{d'}_{s(w)}$ is correct, the following procedure can be adopted: Two values of γ'_{sw} can be computed by using $\gamma^{p'}_{s(w)}$ and each of the two values of $\gamma^{d'}_{s(w)}$ in

$$\gamma'_{sw} = \gamma^{d'}_{s(w)} + \gamma^{p'}_{s(w)} + \gamma_w - 2\left(\gamma^{d'}_{s(w)} \gamma^d_w \right)^{1/2} - 2\left(\gamma^{p'}_{s(w)} \gamma^p_w \right)^{1/2}. \qquad [16]$$

These computed values of γ'_{sw} can then be compared with the value of γ'_{sw} which was estimated from γ'_{so} and Eq. [10] (shown in row 7 of Table III). The value of $\gamma^{d'}_{s(w)}$ which gives rise to the closest match

---

TABLE II

Contact Angle Measurements on the Sputtered Polymeric Films

Reading	θ_1	θ'_1	θ_2	θ_3(Bromobenzene)	τ(h)
1	165	124	117	24	24
2	170	125	124	25	24
3	170	116	123	27	24
4	166	113	122	23	24
Average value	168	120	122	25	24

TABLE III

Estimated Wetting Characteristics of the Sputtered Polymeric Specimen in the Aqueous Environment

No.	Quantity	Magnitude(dyn/cm)	Remarks
1.	Instantaneous polar surface free energy component of the solid in an aqueous environment $\left(\gamma_{s(w)}^{p} = \gamma_{s(o)}^{p'}\right)$	~0	Estimated from Eq. [9]
2.	Equilibrium polar surface free energy component of the solid in the aqueous environment $\gamma_{s(w)}^{p'}$	29.73	Estimated from Eq. [13]
3.	Instantaneous dispersion surface free energy component of the solid in the aqueous environment $\left(\gamma_{s(w)}^{d} = \gamma_{s(o)}^{d'}\right)$	33.16	Estimated from Eq. [15]
4.	Equilibrium dispersion surface free energy component of the solid in the aqueous environment $\left(\gamma_{s(w)}^{d'}\right)$	91.13 or 0.04	Estimated from Eq. [11]
5.	Equilibrium solid-octane interfacial free energy $\left(\gamma_{so}'\right)$	1.19	Estimated from Eq. [7], assuming $\gamma_{s(o)}^{p'} = 0$
6.	Instantaneous solid-water interfacial free energy $\left(\gamma_{sw}\right)$	50.88	Estimated from Eq. [6], using the γ_{so}' value from row 5.
7.	Equilibrium solid-water interfacial free energy $\left(\gamma_{sw}'\right)$	26.59	Estimated from Eq. [10], using the γ_{so}' value from row 5.
8.	Equilibrium wetting characteristics of the sputtered polymeric specimen in the aqueous environment	$\gamma_{s(w)}^{d'} = 91.13$, $\gamma_{s(w)}^{p'} = 29.37$ and $\gamma_{sw}' = 26.59$	γ_{sw}' value estimated in row 7 compared with the values estimated by using each of the $\gamma_{s(w)}^{d'}$ values of row 4 and the $\gamma_{s(w)}^{p'}$ value of row 2 in Eq. [16]

between these independent methods of estimating γ_{sw}' must then be selected as the appropriate value. For the sputtered specimen, it is found that, if $\gamma_{s(w)}^{p'} = 91.13$ dyn/cm and $\gamma_{s(w)}^{p'} = 29.37$ dyn/cm, the γ_{sw}' value is 26.59 dyn/cm, while if $\gamma_{s(w)}^{d'} = 0.04$ dyn/cm and $\gamma_{s(w)}^{p'} = 29.37$ dyn/cm, the γ_{sw}' value is 22.78 dyn/cm. However, since the former value is closer to the γ_{sw}' value of 26.59 dyn/cm (row 7 of Table III) that is obtained from an estimate of γ_{so}' and Eq. [10], it can be concluded that the equilibrium value of the dispersion surface free energy component of the sputtered polymeric specimen in the aqueous environment, $\gamma_{s(w)}^{d'}$, is 91.13 dyn/cm. The same conclusion can also be reached along the following lines: It is well known that the sputtering of polytetrafluoroethylene (PTFE) gives rise to polymer films which are highly cross-linked and harder than their bulk PTFE counterparts (20). In view of this and the fact that Teflon FEP bears a close resemblance to PTFE in its chemical and surface characteristics, it can be expected that the sputtered specimen prepared in this study (from a Teflon FEP target) will also be highly cross-linked and hard, and on this basis one can infer that their surface functional groups will be fairly dense and tightly packed. Therefore a $\gamma_{s(w)}^{d'}$ value of 91.13 dyn/cm seems more plausible than the alternate value of 0.04 dyn/cm, for these sputtered specimens. To summarize then, the wetting characteristics of the sputtered specimen in the aqueous environment, are as follows: the equilibrium solid-water interfacial free energy $\left(\gamma_{sw}'\right)$ is 26.59 dyn/cm (as opposed to the initial solid-water interfacial free energy of 50.88 dyn/cm) and the time (τ) required for the solid surface to attain its equilibrium wetting characteristics in the aqueous environment is of the order of 24 h.

It may be noted here that there can be several variants to the procedure suggested in this paper for estimating the wetting characteristics of polymeric solids in an aqueous environment. However, the

TABLE IV
Surface Tension Properties of Liquids Used in Contact Angle Experiments

Liquid	γ_{LV} or γ_{LL} (dyn/cm)	γ_{LV}^d (dyn/cm)	γ_{LV}^p (dyn/cm)	Reference
Water	72.6	21.8	50.8	(15)
Water-saturated octane	21.8	21.8	0	(15)
Bromobenzene	36.5	35.44	1.06	(21, 22)
Octane-water	50.8			(15)

point under emphasis is that, in any type of contact angle characterization of mobile surfaces such as those of polymeric solids, the differences in the equilibrium wetting characteristics of the solid surface in contact with the environment and probe fluids must be taken into account.

It must also be borne in mind that the surface of a polymeric solid may never be in true thermodynamic equilibrium with its surroundings. Therefore, the perceived equilibrium condition of a polymer surface may only correspond to a quasi-equilibrium state. As a result of this, the relaxation of polymeric surfaces from one environment to another is often only a partially reversible process. Consequently, the estimated "equilibrium" wetting characteristics of a polymer surface in an environment of interest may be influenced to some extent by the choice of the environmental and probe fluids in the contact angle experiments involving such solids.

CONCLUSION

In estimating the wetting characteristics of solids which possess some degree of surface mobility, such as polymers, it is important to take into account the fact that in any type of contact angle experiments involving such surfaces, the equilibrium surface energetic properties of the solid part which is in contact with the environmental fluid will be different from the equilibrium surface energetic properties of the solid part which is in contact with the probe fluid. This arises due to the environmentally induced rearrangement of the solid surface in the two different fluid environments. Realizing this dependence of a polymeric solid's wetting properties on its surrounding environment, the suggested four-step contact angle procedure permits the estimation of the following quantities of interest: (i) the instantaneous as well as equilibrium values of the polymeric solid's dispersion and polar surface free energy components in an aqueous environment and (ii) the time required for the polymeric solid surface to attain its equilibrium wetting characteristics in the aqueous environment.

An application of this procedure to estimate the wetting characteristics of thin solid films of sputtered polymeric specimens (which represented model polymeric surfaces in that their surfaces presented a very smooth texture and were also nonporous and chemically homogeneous) confirmed the belief that the equilibrium values of their individual surface free energy components (i.e., dispersion and polar) in the aqueous environment will differ considerably from those of their respective surface free energy counterparts in a nonpolar environment such as octane or air. In the case of these sputtered polymeric specimens, which were initially equilibrated in an octane environment, it was seen that the solid-water interfacial free energy decreased from an instantaneous value of 50.88 dyn/cm to an equilibrium value of 26.59 dyn/cm. The time required for this change to take place was approximately 24 h. This change in the solid- water interfacial free energy of such model surfaces can arise if the solid specimen contains some polar groups which remain buried in its bulk in a nonpolar environment like air or octane, but reorient toward the surface in a time-dependent manner when it is placed in contact with a strongly polar liquid like water. This possibility was confirmed by means of the ESCA characterization of the surfaces of these sputtered polymeric specimens, which showed that their outer surface layers contained a fair amount of the polar oxygen atoms that are capable of reorienting themselves from either the solid's surface to its bulk or vice versa, depending on the nature of the environment in which they are placed.

APPENDIX: NOMENCLATURE

$\gamma_{s(o)}^{d'}$	Equilibrium value of the solid's dispersion surface free energy component in an octance environment, dyn/cm
$\gamma_{s(o)}^{p'}$	Equilibrium value of the solid's polar surface free energy component in an octane environment, dyn/cm
γ_o	Surface tension of octane, dyn/cm
γ_{so}'	Equilibrium value of solid-octane in terfacial free energy, dyn/cm
$\gamma_{s(w)}^{d}$	Instantaneous value of the solid's dispersion surface free energy component in an aqueous environment, dyn/cm
$\gamma_{s(w)}^{d'}$	Equilibrium value of the solid's dispersion surface free energy component in an aqueous environment, dyn/cm
$\gamma_{s(w)}^{p}$	Instantaneous value of the solid's polar surface free energy component in an aqueous environment, dyn/cm
$\gamma_{s(w)}^{p'}$	Equilibrium value of the solid's polar surface free energy component in an aqueous environment, dyn/cm
γ_w	Surface tension of water, dyn/cm
γ_{sw}	Instantaneous value of solid-water in terfacial free energy, dyn/cm
γ_{sw}'	Equilibrium value of solid-water interfacial free energy, dyn/cm
γ_{ow}	Octane-water interfacial tension, dyn/ cm
$\gamma_{s(air)}'$	Equilibrium value of the solid's total surface free energy in an air environment, dyn/cm
$\gamma_{s(air)}^{d'}$	Equilibrium value of the solid's dispersion surface free energy component in an air environment, dyn/cm
θ_1	Instantaneous contact angle of water on the solid surface equilibrated in an octane environment
θ_1'	Equilibrium contact angle of water on the solid surface equilibrated in an octane environment
θ_2	Instantaneous contact angle of octane on the solid surface equilibrated in an aqueous environment
θ_3	Instantaneous contact angle of a nonpolar and nonhydrogen-bonding liquid (e.g., bromobenzene) on the solid surface equilibrated in an air environment
τ	Time required for the solid surface to attain its equilibrium surface configuration in the aqueous environment, h

ACKNOWLEDGMENT

The authors thank Prof. R. J. Good for the use of his contact angle apparatus at the State University of New York at Buffalo. The authors are also indebted to Dr. M. K. Chaudhury for his help with the ESCA analysis, conducted at the Sinclair Laboratory, Lehigh University, Bethlehem, Pennsylvania.

REFERENCES

1. Yasuda, H., Sharma, A. K., and Yasuda, T., *J. Polym. Sci. Polym. Phys. Ed.* **19**(9), 1285 (1981).
2. Andrade. J. D., Gregonis, D. E., and Smith, L. M., *in* "Physicochemical Aspects of Polymer Surfaces" (K. L. Mittal, Ed.), Vol. 2. Plenum New York, 1983.
3. Holly, J., and Refujo, M. F., *J. Biomed. Mater. Res.* **9**, 315 (1975).
4. Hamilton, W. C., *J. Colloid Interface Sci.* **40**, 219 (1972).
5. Andrade, J. D., Ma, S. M., King, R. N., and Gregonis, D. E., *J. Colloid Interface Sci.* **72**, 488 (1979).
6. Tamai. Y., Makuuchi, K., and Suzuki, M., *J. Pirn.Chem.* **71**, 4176 (1967).
7. Bargeman, D., *J Colloid Interface Sci.* **40**, 344 (1972).
8. Matsunaga, T., *J. Appl. Polym. Sci.* **21**, 2847 (1977).
9. Tamai, Y., Matsunaga, T., and Horiuchi, K., *J. Colloid Interface Sci.* **60**, 112 (1977).
10. Schultz, J., Tsutsumi, K., and Donnet, J. B,, *J Colloid Interface Sci.* **59**, 272 (1977).
11. Fowkes, F. M., and Mostafa, A. M., *Ind. Eng. Chem. Prod. Res. Dev.* **17**, 3 (1978).
12. Fowkes, F. M., *Ind. Eng. Chem.* **56**, 40 (1964).
13. Kloubek, J., *J. Adhesion* **6**, 293 (1974).
14. Schultz, J., Tsutsumi, K., and Donnet, J. B., *J. Colloid Interface Sci.* **59**, 277 (1977).
15. Bagnal, R. D., and Green. G. F., *J. Colloid InterfaceSci.* **68**, 387 (1979).
16. Wrobel, A. M., Kryszewski, M., Rakowski, W., Okoniewski, M., and Kubacki, Z, *Polymer* **19**, 908 (1978).
17. Hansen, R. H., Pascale, J. V., DeBenedictis, T., and Rentzekis, P. M., *J. Polym. Sci. A* **3**, 2205 (1965).
18. Wagner, C. D., Riggs, W. M., Davis, L. E., Moulder, J. F., and Muilenberg, J. E., "Handbook of XPS." Perkin-Elmer, Physical Electronics, Palo Alto, Calif., 1979.
19. Timmons, C. O., and Zisman, W. A., *J. Colloid Interface Sci.* **22**, 165 (1966).
20. White, M., *Thin Solid Films* **18**, 157 (1973).
21. CRC Handbook of Chemistry and Physics, 62nd edition, CRC Press, Boca Raton, Fla., 1981-1982.
22. Gardon, J. L., *in* "Encyclopaedia of Polymer Science and Technology" (H. F. Mark, N. G. Gaylord, and N. M. Bikales, Eds.), Vol. 3. Wiley, New York, 1965.

2.2 Surface Restructuring of Polymeric Solids and Its Effect on the Stability of the Polymer-Water Interface*

Eli Ruckenstein and Sathyamurthy V. Gourisankar
Department of Chemical Engineering, State University
of New York at Buffalo, Buffalo, New York 14260

Received April 29, 1985; accepted June 28, 1985

1. INTRODUCTION

The antifouling characteristics of a solid surface in a biological environment are dictated to a large extent by the magnitude of its interfacial free energy with the continuous phase of the biological fluid. In connection with the selection of blood compatible surfaces, it was pointed out recently (1) that the interfacial free energy between the solid surface and the continuous phase of blood must be sufficiently low but not too low in order to ensure the long term compatibility of a foreign surface in the environment of blood. This criterion, which is based on thermodynamic considerations, is of general applicability to the problem of selecting antifouling surfaces for use in other biological environments (such as contact lenses in the tear fluid environment, and heat exchange surfaces in contact with milk and sea water, for example) as well. The basic features of this criterion can be summarized as follows: If the interfacial free energy between a solid and the continuous phase of a biological fluid is high, then the thermodynamic driving force for adsorption will be high and therefore the solid will be "fouled" to a considerable extent by the surface active components of the biological fluid. On the other hand, at very low values of solid-fluid interfacial free energies, the interface between the solid and the biological fluid is likely to become unstable and highly susceptible to mechanical and/or thermal perturbations. This interfacial instability can promote the occurrence of undesirable consequences such as the dissolution of the solid in the fluid and/or the absorption of the fluid and its components into the solid.

Though the above considerations are applicable to all types of solid surfaces, they are especially relevant for the case of polymeric solids, since these are some of the most appropriate candidates that could be used as antifouling surfaces in biological environments. This is because, many polymers are chemically inert and nontoxic in biological environments and moreover, some of them can maintain lower interfacial free energies with biological fluids than most other types of solids and thus decrease the extent of deposition of the fluids' components on their surfaces. In connection with the latter aspect, it is necessary to discuss one important characteristic of polymeric solids, namely, their surface mobility (2,3), which sets these solids apart from most other types of solid surfaces encountered in nature. Due to their surface mobility, polymeric solids possess the capacity to alter their surface structures between different environments and minimize their interfacial free energy with a surrounding medium, in order to decrease the free energy of the system. As a result of this environmentally induced surface restructuring, many polymeric solids which present a

---

* *Journal of Colloid and Interface Science*, Vol. 109, No. 2, p. 557, February 1986. Republished with permission.

hydrophobic surface character in a nonpolar environment such as air are known to actually display a reasonable degree of hydrophilicity in a more polar environment such as water (2–4). It must be noted, however, that the dynamics of surface restructuring of polymers in a particular environment will be largely dependent on their surface structures. In the case of highly crosslinked and rigid polymeric surfaces for example, the surface restructuring between environments may be too slow to be perceptible on a practical time scale, while in the case of highly mobile polymeric surfaces, the process may be almost instantaneous. Examples abound for both the above types of polymer surface structures. One of the important consequences of the surface restructuring of polymeric solids between different environments is the fact that these solids can give rise to very low interfacial free energies with an environment in which they have been equilibrated. Therefore, in the case of polymeric solids, there is a distinct possibility that their interfacial free energy with a surrounding medium can become so low as to adversely affect the stability of their interface with the medium and thereby lead to undesirable effects as mentioned earlier.

It can now be realized that the dynamics of polymer surface restructuring plays an important role in their use as antifouling surfaces in biological environments. In this context, it is necessary to mention that most of the biological fluids contain an aqueous medium as their continuous phase and therefore, a study of the polymer-water interface will be fairly representative of the interaction of polymeric solids with biological fluids. One of the main questions that remains unanswered with regard to the polymer-water interface is: How does polymer surface restructuring affect the stability of the polymer-water interface? More specifically, it is important to study the changes in the stability of the polymer-water interface as the polymer-water interfacial free energy reduces to very low values (as a result of surface restructuring of these solids in water). An answer to these questions can be particularly useful to the successful use of polymers as antifouling surfaces in a multitude of biological environments.

It is the objective of this paper to suggest a simple experimental procedure, based primarily on contact angle measurements of select fluids on solid surfaces in specified environments, to study the stability of the polymeric solid-water interface vis-à-vis the dynamics of their environmentally induced surface restructuring between the air and aqueous environments.

2. PREPARATION AND CHARACTERIZATION OF POLYMERIC SURFACES WITH WIDELY VARYING WETTING CHARACTERISTICS IN THE AQUEOUS ENVIRONMENT

We will now take up the task of identifying three types of polymeric surfaces, which will give rise to high, medium, and low interfacial free energies with water as a result of their structural alterations in the aqueous environment. Commercially available Teflon FEP (E.I. du Pont de Nemours and Co.) is a good example of a surface which is nonpolar and undergoes practically no structural alteration in the aqueous environment (5, 6). Therefore, its interfacial free energy with water is of the order of 50 dyn/cm. In order to prepare a more polar polymeric surface, which will undergo some structural alteration in the aqueous environment to maintain a medium interfacial free energy with water (i.e., of the order of 30 dyn/cm), we resorted to the deposition of thin solid films (of about 700–800 Å in thickness) of partially oxidized fluorocarbon compounds (from a Teflon FEP target) onto the smooth surfaces of highly polished, single crystal silicon substrates. A radio frequency sputtering technique was selected for the deposition of these polymer films. Details of the preparation and characterization of these films are contained in Ref. (6). Finally, in order to prepare polymeric surfaces which will maintain still lower interfacial free energies with water as a result of their equilibration in the aqueous environment, a chemical surface modification of the above sputtered specimen was carried out. This was performed by chemically etching the sputtered surfaces in order to increase their polarity and thereby decrease their interfacial free energy with water. The etching was carried out by dipping the sputtered specimen in the etching solution (Poly-Etch, Matheson, Mass.) for 30 s and then rinsing it in acetone and water (to remove the etchant), followed by air drying. This etchant has been known to cause a marked increase in the wettabilities and

Surface Restructuring of Polymeric Solids

polarities of fluorocarbon compounds (7, 8). The estimation of the wetting characteristics of these surfaces in the aqueous environment will be the topic of the next section.

3. CONTACT ANGLE CHARACTERIZATION OF THE POLYMERIC SURFACES

Prior to estimating the wetting characteristics of the sputtered and chemically etched- sputtered specimen in the aqueous environment, it must be noted that, in any type of contact angle experiment involving solids with some degree of surface mobility such as polymers, the equilibrium surface energetic properties of the solid portion which is in contact with the environmental fluid will be different from the equilibrium surface energetic properties of the solid portion which is in contact with the probe fluid. This arises due to the environmentally induced restructuring of the polymeric surface in the two different fluid environments. Based on this consideration, a four-step contact angle procedure was suggested in Ref. (6) for the estimation of the wetting characteristics of polymeric solids in aqueous environments. A summary of the procedure is provided here.

If it is assumed that the total surface free energy of the solid can be expressed as the sum of its dispersion and polar component surface free energies (9), then there are four quantities to be estimated, namely, the instantaneous values of the solids dispersion and polar surface free energy components in the aqueous environment, i.e., $\gamma^{d}_{s(w)}$ and $\gamma^{p}_{s(w)}$, respectively, and the equilibrium values of the solid's dispersion and polar surface free energy components in the aqueous environment, i.e., $\gamma^{d'}_{s(w)}$ and $\gamma^{p'}_{s(w)}$, respectively. This procedure also enables the estimation of the instantaneous and equilibrium values of the solid-water interfacial free energies, i.e., γ_{sw} and γ'_{sw}, respectively. The estimation of the solid's instantaneous polar surface free energy component in the aqueous environment ($\gamma^{p}_{s(w)}$) can be carried out by measuring the instantaneous contact angle of a water drop on the solid surface equilibrated in an octane environment. For this case, the value of $\gamma^{p}_{s(w)}$ (which is also equal to the equilibrium polar surface free energy component of the solid in an octane environment, i.e., $\gamma^{p'}_{s(o)}$ is given by

$$\gamma^{p}_{s(w)} = \gamma^{p'}_{s(o)}$$

$$= \left(\gamma_{ow}\cos\theta_{1} + \gamma_{w} - \gamma_{o}\right)^{2}/4\gamma^{p}_{w}. \tag{1}$$

In this equation, γ_{ow} is the octane-water interfacial tension, θ_{1} is the instantaneous value of the solid-water-octane contact angle, γ_{w} is the surface tension of water, γ_{o} is the surface tension of octane and γ^{p}_{w} is the polar component of water's surface tension. Except θ_{1}, which needs to be measured, all the other quantities of Eq. [1] are known. It must be noted that this simplified expression for $\gamma^{p}_{s(w)}$ results from the coincidence that the dispersion component of the surface tension of water is equal to the surface tension of octane. The estimation of the solid's equilibrium polar surface free energy component in the aqueous environment, i.e., $\gamma^{p'}_{s(w)}$, can be carried out by measuring the instantaneous contact angle of octane on the solid surface which is equilibrated in water. In this case, the expression for $\gamma^{p'}_{s(w)}$ is

$$\gamma^{p'}_{s(w)} = \left(\gamma_{w} - \gamma_{o} - \gamma_{ow}\cos\theta_{2}\right)^{2}/4\gamma^{p}_{w} \tag{2}$$

In the above equation, θ_{2} is the instantaneous contact angle of octane on the solid surface under water. In order to determine the instantaneous value of the solid's dispersion surface free energy component in the aqueous environment, i.e., $\gamma^{d}_{s(w)}(=\gamma^{d'}_{s(o)})$, it is necessary to measure the instantaneous contact angle of a nonpolar and nonhydrogen bonding liquid on the solid surface equilibrated in air. The expression to calculate $\gamma^{d}_{s(w)}$ is

$$\gamma^{d}_{s(w)} = \gamma^{d'}_{s(o)} = \gamma^{d'}_{s(air)} = \gamma_{L}\left(1 + \cos\theta_{3}\right)^{2}/4. \tag{3}$$

In this expression, $\gamma_{s(o)}^{d'}$ is the equilibrium value of the solid's dispersion surface free energy component in octane, $\gamma_{s(air)}^{d'}$ represents the equilibrium value of the solid's dispersion surface free energy component in air, γ_L is the surface tension of a fluid L which is nonpolar and nonhydrogen bonding, and θ_3 is the instantaneous contact angle of fluid L on the specimen in air. The final quantity of interest, i.e., the equilibrium value of the solid's dispersion surface free energy component in water ($\gamma_{s(w)}^{d'}$) can now be calculated by placing a drop of water on the solid surface which is equilibrated under octane and noting the change in the solid-water-octane contact angle until it attains its equilibrium value (θ_1'). Then, $\gamma_{s(w)}^{d'}$ can be calculated from

$$\gamma_{s(o)}^{d'} + \gamma_{s(o)}^{p'} + \gamma_o - 2\left(\gamma_{s(o)}^{d'} \, \gamma_o\right)^{1/2}$$

$$\gamma_{s(w)}^{d'} + \gamma_{s(w)}^{p'} + \gamma_w - 2\left(\gamma_{s(w)}^{d'} \, \gamma_w^d\right)^{1/2}$$

$$-2\left(\gamma_{s(w)}^{p'} \, \gamma_w^p\right)^{1/2} + \gamma_{ow}\cos\theta_1'. \qquad [4]$$

In the above equation, the only unknown quantity is $\gamma_{s(w)}^{d'}$ and therefore, by measuring θ_1', it can be evaluated. Having obtained the magnitudes of the various solid surface free energy components, it is now possible to estimate the values of the solid-water interfacial free energies γ_{sw} and γ_{sw}', from the measurements of θ_1 and θ_1', respectively, as follows: For the situations involving the measurements of θ_1, and θ_1', Young's equation can be written as

$$\gamma_{so}' = \gamma_{sw} + \gamma_{ow}\cos\theta_1 \qquad [5]$$

and

$$\gamma_{so}' = \gamma_{sw} + \gamma_{ow}\cos\theta_1'. \qquad [6]$$

In these equations, γ_{so}' is the equilibrium value of the solid-octane interfacial free energy. It can be estimated from the known values of $\gamma_{s(o)}^{d'}$ and $\gamma_{s(o)}^{p'}$ by means of (6)

$$\gamma_{so}' = \gamma_{s(o)}^{d'} + \gamma_{s(o)}^{p'} + \gamma_o - 2\left(\gamma_{s(o)}^{d'} \, \gamma_o\right)^{1/2}. \qquad [7]$$

Once the value of γ_{so}' is known, it can be used in Eqs. [5] and [6] to provide the values of γ_{sw} and γ_{sw}' from the measured values of θ_1 and θ_1'. The derivation of Eqs. [1] to [7] is contained in Ref. 6.

4.   RESULTS AND DISCUSSION

The results of contact angle measurements on the three different specimens of this study are shown in Fig. 1 and Table I. Figure 1 shows the time dependent variation of the solid- water-octane contact angles on the three different specimen. For the control Teflon FEP surface, it can be seen that the solid-water- octane contact angles remain almost constant with time. For the sputtered polymeric surfaces, the contact angles continually decrease with time (between points A and B in Fig. 1) before attaining their equilibrium values in a period of about 24 h (corresponding to point B in Fig. 1) and then remain constant thereafter. In the case of the chemically etched- sputtered polymeric surfaces, the contact angles decrease rapidly initially (between points A' and B' in Fig. 1), followed by a period of quasiequilibrium (between B' and C') and then again decrease more gradually (from C to D'), until the drop totally disappears from sight in a period of about 72 h (corresponding to point D' in Fig. 1). These contrasting trends in the time dependent variation of the solid-water-octane contact angles can be explained by considering the physicochemical characteristics of the three different types of polymeric surfaces used in this study. Let us first compare the control Teflon FEP surface and the

Surface Restructuring of Polymeric Solids

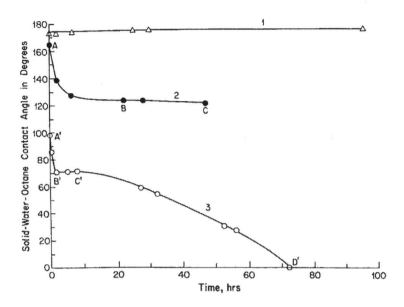

FIG. 1. Variation of the solid-water-octane angle on different polymeric surfaces, as a function of time. Curve 1 is for the control Teflon FEP surface, Curve 2 is for the sputtered polymeric surface, and Curve 3 is for the chemically etched-sputtered polymeric surface.

sputtered specimen. In the case of the former, the ESCA spectra show that their outer surface layers contain carbon and fluorine atoms, with a fluorine to carbon atomic ratio of 2.18. For the sputtered specimen, the ESCA spectra show that their outer surface layers contain a fair amount of oxygen atoms in addition to carbon and fluorine atoms. The atomic ratios for the sputtered specimen were as follows: (F/C) = 0.54 and (O/C) = 0.38. Since the ESCA technique probes the chemical composition of a surface layer of the order of 100 Å in thickness, this characterization confirms that the outer surface layers of the sputtered specimen contain sufficient polarity (due to their oxygen content), unlike their control Teflon FEP counterparts. Based on this information, it is now possible to interpret the contact angle results of Fig. 1 as follows: The continual decrease in the solid- water-octane contact angles on the sputtered specimen arises due to a decrease in the solid- water interfacial free energy. In the case of these specimen (which represent model surfaces in that they are very smooth in surface texture, chemically homogeneous, and non- porous (6)), a decrease in the solid-water interfacial free energy can be brought about by a reorientation of the oxygen containing polar groups from the sub-surface layers of the solid to its surface when it is placed in contact with a strongly polar liquid like water. Moreover, the decrease in the solid-water-octane contact angles on these specimen was also accompanied by a continual increase in the width of the water drop's base in contact with the solid, thus confirming that the drop was indeed spreading on the solid surface in a time-dependent manner. After the equilibrium angle θ_1' was reached in a period of about 24 h, neither the solid-water-octane contact angle nor the width of the water drop's base in contact with the solid changed with time on this specimen. Contrary to this behavior, both the solid-water-octane contact angles and the width of the water drop's base in contact with the solid remained practically constant with time for the control Teflon FEP surface. This is because, as shown by the ESCA characterization, the control Teflon FEP specimen contains no polar moieties in its outer surface layers that are capable of orientating themselves to the surface and interacting with the aqueous environment. Let us now consider the contact angle results of Fig. 1 for the case of the chemically etched-sputtered specimen. It must be noted that the surface layers of these specimen are likely to be more oxidized and therefore more polar than those of the sputtered specimen. This is because the chemical etching process adopted in this study is known to considerably increase the oxygen content as well as the water wettability of fluorocarbon surfaces such as Teflon FEP (8). It must also be noted that the etching process imparts a certain amount of roughness to the surface texture of the sputtered specimen (compare Figs. 2a and b). With this picture of their physicochemical surface characteristics, the contact angle results of Fig. 1 on the chemically etched-sputtered specimen can be explained as follows: During the initial period of rapidly decreasing solid-water-octane contact angles (from A′ to B′), the polar functionalities of the specimen are reorienting toward the surface to interact with water. In this period, it was also observed that the width of the water drop's base continually increased in time. After the reorientation of the polar groups toward the surface is essentially completed (at B′ in Fig. 1), a period of quasiequilibrium exists between B′ and C′, when neither the solid-water-octane contact angle nor the width of the water drop's base in contact with the solid change. Following this, the angles once again begin to decrease, albeit more gradually this time (from C′ to D′), until the drop completely penetrates into the solid at the point D′. The important point to note about the variation in the contact angles during this period is that the width of the water drop's base in contact with the solid also decreases continuously (from C′ to D′), along with the decreasing angles. This indicates that penetration of water into the solid was taking place between C′ and D′. Moreover, during this period (C′ to D′), it was also observed that the drop's curved periphery became more planar and turbid near the point when the drop totally penetrated into the solid. This suggests that some dissolution of the solid as well as penetration of water into the solid's pores might be taking place simultaneously. From this time-dependent variation of the solid-water-octane contact angles on this specimen, one can conclude that, as a result of the initial decrease in the solid-water interfacial free energy to very small values (from A′ to B′), the solid-water interface becomes quite unstable at about the point C′ and this interfacial instability is responsible for effects such as the penetration of water into the solid and the possible dissolution of the solid in water. It may also be noted that, as a result of the reorientation of the polar groups of the solid from its bulk to its surface (from A′ to B′), the porosity of the surface is likely to be increased (over and above that of the initial surface). This increased porosity of the surface, coupled with the reduced solid-water interfacial free energy, may have accelerated the penetration of water into the solid and the possible dissolution of the solid in water (between C′ and D′). In addition, it may be noted that the chemical etching process might have caused changes in the crosslink density and molecular weight of the sputtered surface, which in turn could also have affected the penetration of water into the solid and/or the dissolution of the solid in water.

In connection with the estimation of the wetting characteristics of the sputtered and chemically etched-sputtered surfaces, it will be useful to evaluate the solid-water interfacial free energies at various times relevant to Fig. 1 and relate these values to the physical explanations proposed here concerning the surface mobility and the stability of the solid-water interface of the different surfaces. For the sputtered specimen, the instantaneous solid-water interfacial free energy, θ_{sw},

 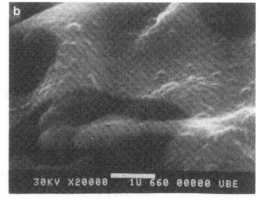

FIG. 2. Electron photomicrographs: (a) sputtered polymeric surface and (b) chemically etched-sputtered polymeric surface.

(at point A in Fig. 1) and the equilibrium solid-water interfacial free energy, θ'_{sw} (at point B in Fig. 1) were evaluated by the four step contact angle procedure suggested earlier (6). In the case of the chemically etched-sputtered specimen, the quantities of interest are: (i) the magnitude of the instantaneous solid-water interfacial free energy, θ_{sw} (at point A' in Fig. 1) and (ii) the magnitude of the solid-water interfacial free energy at the onset of instability of the solid-water interface, i.e., θ'_{sw} (corresponding to point C' in Fig. 1).

The measured values of various contact angles are shown in Table I and the estimated values of the various interfacial free energies are shown in Table II. The contact angle measurements were repeated on at least three specimen of each polymer surface to check for reproducibility. Advancing contact angles of the fluids were measured on at least four different regions of each specimen. The contact angle values reported in Table I represent the average readings for the different surfaces. The surface tensions of the liquids used in the contact angle experiments are shown in Table III.

The results of Table II and Fig. 1 indicate that the solid-water interfacial free energy of the sputtered specimen decreases from an instantaneous value of 51 dyn/cm to an equilibrium value of 27 dyn/cm, in a duration of about 24 h. In the case of the chemically etched-sputtered surfaces, the solid-water interfacial free energy decreases from an initial value of 22 dyn/cm to a low quasiequilibrium value of -1.6 dyn/cm in a period of about 2 h (A' to B'), then it remains constant for about 6 h (B' to C'), before the solid-water interface becomes visibly unstable, giving rise to the gradual penetration of water into the solid's pores and the possible dissolution of the solid in water. From Table II it can be seen that the solid-water interfacial free energy of this specimen decreases to a low value of -1.6 dyn/ cm at the onset of instability of the polymeric solid-water interface. Of course, the above magnitude and even sign of this threshold interfacial free energy must not be taken too seriously as the precision of contact angle

TABLE I
Contact Angle Measurements on the Polymeric Surfaces

No.	Specimen	θ_1	θ'_1	θ_2	θ_3
1	Sputtered polymer	168	120	122	25 (bromobenzene)
2	Chemically etched-sputtered polymer	99	72	138	20 (methylene iodide)

TABLE II
Estimated Wetting Characteristics of the Sputtered and Chemically Etched-Sputtered Polymeric Surfaces in the Aqueous Environment

No.	Specimen	Quantity	Magnitude (dyn/cm)	Remarks
1	Sputtered polymer	a. Equilibrium solid-octane interfacial free energy (γ'_{so})	1.2	Estimated from Eq. [7]
		b. Instantaneous solid-water interfacial free energy (γ_{sw})	51	Estimated from Eq. [5] for θ_1, value corresponding to point A in Fig. 1
		c. Equilibrium solid-water interfacial free energy (γ'_{sw})	27	Estimated from Eq. [6] for θ'_1 value corresponding to point B in Fig. 1
2	Chemically etched-sputtered polymer	a. Equilibrium solid-octane interfacial free energy (γ'_{so})	14	Estimated from Eq. [7]
		b. Instantaneous solid-water interfacial free energy (γ_{sw})	22	Estimated from Eq. [5] for θ_1, value corresponding to point A' in Fig. 1
		c. Quasiequilibrium solid-water interfacial free energy	-1.6	Estimated from Eq. [6] for θ'_1 value corresponding to point C in Fig. 1

TABLE III
Surface Tension Properties of Liquids Used in Contact Angle Experiments

Liquid	γ_{LV} or γ_{LL} (dyn/cm)	γ_{LV}^d (dyn/cm)	γ_{LV}^P (dyn/cm)	Reference
Water-air	72.6	21.8	50.8	(12)
Water saturated octane-air	21.8	21.8	0	(12)
Bromobenzene-air	36.5	35.44	1.06	(13,14)
Methylene iodide- air	50.8	48.5	2.3	(7)
Octane-water	50.8			(12)

goniometry is not sufficiently high to report interfacial free energies to an accuracy of ± 1 dyn/cm. Apart from this, the estimation of interfacial free energies was based on a number of simplifying assumptions such as: (a) the total surface free energies of the solid and fluids are the sum of their dispersion and polar surface free energy components (9), (b) the solid and fluids interact only by means of dispersion and polar forces, and (c) a geometric mean rule holds for describing the dispersion (9) as well as polar interactions (10–12) between the dissimilar phases. Moreover, the calculated value of γ_{sw} (from Eq. [6]) is highly sensitive to the measured value of γ'_{sw}. For example, if γ_{sw} is 74° rather than 72°, the estimated value of γ'_{sw} will be +0.1 dyn/cm instead of -1.6 dyn/cm. Since the accuracies of contact angle measurements are usually about $\pm 2°$, it can be seen that the estimated value of γ'_{sw} must only be considered as an approximate measure of the solid-water interfacial free energy of the chemically etched-sputtered specimen. It must also be noted that the surfaces of the chemically etched-sputtered specimen were more rough than those of the sputtered specimen and therefore, the measured contact angles on the former could differ to some extent from the angles that must be used in Young's equation to estimate θ'_1. From these considerations, the message that is conveyed via this estimate of θ'_1 is that the solid-water interfacial free energy attains a very small value at the onset of instability of the solid-water interface (corresponding to point C′ in Fig. 1) and continues to decrease to much smaller values (including possibly negative values) until the total penetration of the water drop into the solid's pores takes place. This result provides an illustration of the premise that very low values of polymeric solid-water interfacial free energies can indeed lead to an instability of the solid-water interface and thus promote the occurrence of undesirable effects such as the penetration of water into the solid and/or the possible dissolution of the solid in water. It may be noted that, under the conditions of thermodynamic equilibrium, the solid-liquid interfacial free energy cannot attain negative values. However, under the nonequilibrium conditions of dynamic processes, if the traditional thermodynamic concept of a solid-fluid interfacial free energy is assumed to hold, then it is possible to arrive at negative values for this interfacial free energy in situations wherein the solid-fluid interface is unstable.

5.   CONCLUSIONS

Being endowed with a high degree of surface mobility in comparison with other types of solids, the surfaces of polymeric solids can undergo considerable structural alterations between different environments in order to lower their interfacial free energy with a surrounding medium. As a result of this, it is possible for these solids to reduce their interfacial free energy with an environment in which they have been equilibrated, to very low values. Such low interfacial free energy values can adversely affect the stability of the polymeric solid-fluid interface and thus give rise to undesirable effects such as the absorption of the fluid into the solid and/or the dissolution of the solid in the fluid.

The above considerations were investigated by a simple contact angle procedure, for the case of the polymeric solid-water interface, in this study. The results indicate that the polymeric solid-water interface, which typifies the interface of polymers with most biological environments, does indeed become unstable at very small solid-water interfacial free energy values (caused by the restructuring of the solid in the water environment in this case) and this instability promotes the penetration of water into the polymeric solid and the possible dissolution of the solid in water. The dynamics of polymer surface restructuring in the aqueous environment was followed by noting the changes in the solid-water-octane contact angles as a function of time on three different types of polymeric surfaces which gave rise to high, medium, and low interfacial free energies with water as a result of their surface restructuring in the aqueous environment. This was used to estimate, although only approximately, the polymeric solid-water interfacial free energy at the onset of instability of the solid- water interface (this instability was observed only in the case of the polymeric surface that gave rise to a very low solid-water interfacial free energy as a result of its surface restructuring under water). The contact angle results were well supported by the physicochemical characteristics of the different surfaces used in this study.

APPENDIX

Nomenclature

$\gamma_{s(o)}^{d'}$ equilibrium value of the solid's dispersion surface free energy component in an octane environment, dyn/cm

$\gamma_{s(o)}^{p'}$ equilibrium value of the solid's polar surface free energy component in an octane environment, dyn/cm

γ_o surface tension of octane, dyn/cm

γ_{so}' equilibrium value of solid-octane in terfacial free energy, dyn/cm

$\gamma_{s(w)}^{d}$ instantaneous value of the solid's dispersion surface free energy component in an aqueous environment, dyn/cm

$\gamma_{s(w)}^{d'}$ equilibrium value of the solid's dispersion surface free energy component in an aqueous environment, dyn/cm

$\gamma_{s(w)}^{p}$ instantaneous value of the solid's polar surface free energy component in an aqueous environment, dyn/cm

$\gamma_{s(w)}^{p'}$ equilibrium value of the solid's polar surface free energy component in an aqueous environment, dyn/cm

γ_w surface tension of water, dyn/cm

γ_{sw} instantaneous value of solid-water in terfacial free energy, dyn/cm

γ_{sw}' equilibrium value of solid-water interfacial free energy, dyn/cm

γ_{ow} octane-water interfacial tension, dyn/ cm

$\gamma_{s(air)}'$ equilibrium value of the solid's total surface free energy in an air environment, dyn/cm

$\gamma_{s(air)}^{d'}$ equilibrium value of the solid's dispersion surface free energy component in an air environment, dyn/cm

θ_1 instantaneous contact angle of water on the solid surface equilibrated in an octane environment

θ_1' equilibrium contact angle of water on the solid surface equilibrated in an octane environment

θ_2 instantaneous contact angle of octane on the solid surface equilibrated in the aqueous environment

θ_3 instantaneous contact angle of a nonpolar- and nonhydrogen-bonding liquid on the solid surface equilibrated in an air environment

ACKNOWLEDGMENT

This work was supported by the National Science Foundation.

REFERENCES

1. Ruckenstein, E., and Gourisankar, S. V., *J. Colloid Interface Sci.* **101**, 436 (1984) (Section 1.2 of this volume).
2. Yasuda, H., Sharma, A. K., and Yasuda, T., *J. Polym. Sci. Polym. Phys. Ed.* **19**(9), *1285* (1981).
3. Andrade, J. D., Gregonis, D. E., and Smith, L. M., *in* "Physicochemical Aspects of Polymer Surfaces" (K. L. Mittal, Ed.), Vol. 2. Plenum, New York, 1983.
4. Holly, J., and Refujo, M. F., *J. Biomed. Mater. Res.* **9**, 315 (1975).
5. Hamilton, W. C., *J. Colloid Interface Sci.* **40**, 219 (1972).
6. Ruckenstein, E., and Gourisankar, S. V., *J. Colloid Interface Sci.* **107**, 488 (1985) (Section 2.1 of this volume).
7. Kaelble, D. H., and Cirlin, E. M., *J. Polym. Sci.* A**2**(9), 363 (1971).
8. Dwight, D. W,, and Riggs, W. M., *J. Colloid Interface Sci.* **47**, 650 (1974).
9. Fowkes, F. M., *Ind. Eng. Chem.* **56**, 40 (1964).
10. Kloubek, J., *J. Adhesion* **6**, *293* (1974).
11. Schultz, J., Tsutsumi, K., and Donnet, J. B., *J. Colloid Interface Sci.* **59**, 277 (1977).
12. Bagnal, R. D., and Green, G. F., *J. Colloid Interface Sci.* **68**, 387 (1979).
13. "CRC Handbook of Chemistry and Physics," 62nd ed. CRC Press, Boca Raton, Fla., 1981-1982.
14. Gardon *in* "Encyclopedia of Polymer Science and Technology" (M. F. Mark, N. G. Gaylord, and N. M. Bikales, Eds.), Vol. 3. Wiley, New York, 1965.

2.3 Stability of Polymeric Surfaces Subjected to Ultraviolet Irradiation[*]

Sang Hwan Lee and Eli Ruckenstein
Department of Chemical Engineering, State University
of New York, Buffalo, New York 14260
Received June 1, 1986; accepted September 1, 1986

INTRODUCTION

Polymers are, in general, low surface free energy materials. Their polar surface free energy component being generally low in comparison to that of water, they maintain undesirably high values of the solid—water interfacial free energy when placed in contact with biological fluids (which are largely aqueous in nature). It was suggested previously (1) that for a biocompatible polymer the above interfacial free energy should be sufficiently low but not very low (of the order of a few dyn/cm), in order to fulfill the dual requirement of maintaining a low thermodynamic driving force for the adsorption of the fluid components on the solid surface as well as a mechanically stable solid—fluid interface. The interfacial free energy can however be varied by modifying the surface properties of the solid. Such surface modifications could be accomplished by plasma treatment (2, 3), surface fluorination (4), chemical etching techniques (5), and by ultraviolet irradiation. The last method is of interest in the present paper. Fox *et al.* (6) and Esumi *et al.* (7, 8) have previously employed ultraviolet irradiation to produce changes in the wettability of polymeric surfaces. From wetting angle measurements combined with the available computational methods, they have estimated the changes in the polar and dispersion components of the surface free energy of polymeric surfaces subjected to UV irradiation for various periods of time.

The results thus obtained show that the polar component of the solid surface free energy can be strongly affected by the irradiation treatment in the case of both low surface free energy (hydrophobic polymers) and medium surface free energy polymers. The ultraviolet treatment is believed to catalyze the photooxidation in air of the polymer surface and thereby inject polarity into the surface layers of the solid. However, this surface modification technique can also cause the degradation of the polymer chain which could lead to time-dependent effects on the wettability of the surface. It is, therefore, useful not only to measure the initial wetting characteristics of the irradiated polymer, but also to monitor the dynamics of its wettability in an environment of interest. Such investigations have been carried out in the present work.

EXPERIMENTAL AND METHODOLOGY

Polymethylmethacrylate (PMMA; Lucite, from DuPont) was radiofrequency (rf) sputtered onto single-crystal silicon wafers (Semimetals Inc. Westburg, NY). The polymer targets for sputtering were cut into 4-in. disks and were cleaned repeatedly with a detergent solution and distilled water. Finally, they were also sputter-cleaned in the sputtering chamber. Since this polymer is nonconducting,

---

[*] *Journal of Colloid and Interface Science*, Vol. 117, No. I, p. 172, May 1987. Republished with permission.

the rf sputtering, rather than the dc sputtering, had to be employed to prepare the thin film deposits. Argon was used as the sputtering gas, and the typical conditions adopted for sputtering were (a) rf power of 50 W, (b) pressure of 5 μm Hg in the chamber, (c) target-to-substrate spacing of 2.5 cm, and (d) sputtering time of 1.5 h. The scanning electron micrographs of the sputtered polymer films showed that the films were smooth and pinhole free. The deposited film thicknesses were measured by employing a Rudolph Research Thin Film Ellipsometer (Type 43603-200E) to which an NRC 8-mW He–Ne laser source was attached. Further, the rf-sputtered PMMA films on silicon substrates were exposed in air to the ultraviolet radiation of a mercury lamp (Oriel No. 6505, Stanford, CT). The 2537-Å wavelength radiation was filtered through an optical filter. The irradiated polymer films were then placed in a dark box for 1 day to allow the volatiles to escape and then they were washed repeatedly with distilled water to remove water-soluble oxidation products.

Three different sets of specimens were equilibrated, one in octane for 2 days, another in water for 2 to 3 days, and the third in air in a clean room for 2 days. The time allowed for equilibration was relatively long because the surfaces of the polymeric solids possess sufficient mobility to considerably alter their structure between different environments in order to minimize the free energy of the system. A polymeric surface will therefore tend to expose hydrophobic moieties to a hydrophobic environment and hydrophilic moieties to a hydrophilic environment. The contact angles of 0.8-to-1-mm-diameter sessile drops on the sample were measured using a NRL Contact Angle Goniometer (Rame–Hart, Inc. A-100, Mountain Lakes, NJ) and its accessories (especially the environmental chamber and the Polaroid camera attachment). The contact angles θ_1 of the water sessile drops in octane, for octane-equilibrated samples, the contact angles θ_2 of octane sessile drops under water, for water-equilibrated samples, as well as the contact angles θ_3 of methylene iodide, in air, for air-equilibrated samples, were measured as a function of time. The initial (instantaneous) wetting angles give information about the equilibrium values of the polar surface free energy of the solid in octane and water environments, as well as the equilibrium value of the dispersion component of the surface free energy in a nonpolar environment. The following expression can be used to calculate the equilibrium value of the polar surface free energy components of the solid in an octane environment (9):

$$\gamma_{s(o)}^{p'} = \left(\gamma_{ow}\cos\theta_{1i} + \gamma_w - \gamma_0\right)^2 / 4\gamma_w^p. \tag{1}$$

In this equation, $\gamma_{s(o)}^{p'}$ is the equilibrium value of the polar surface free energy component of the solid in an octane environment, γ_{ow} is the octane—water interfacial free energy (=50.8 dyn/cm), γ_w is the surface tension of water (=72.6 dyn/cm), γ_0 is the surface tension of water-saturated octane (=21.8 dyn/cm), γ_w^p is the polar component of the surface tension of water, θ_1 is the contact angle of a water drop on the solid surface equilibrated in an octane environment, and the additional subscript "i" indicates initial.

Similarly, the equilibrium polar component of the surface free energy in a water environment $\gamma_{s(w)}^{p'}$ was calculated using the equation (9)

$$\gamma_{s(w)}^{p'} = \left(-\gamma_{ow}\cos\theta_{2i} + \gamma_w - \gamma_0\right)^2 / 4\gamma_w^p, \tag{2}$$

where θ_{2i} is the contact angle of the octane drop under water.

In order to estimate the equilibrium dispersion component of the solid surface free energy in air, $\gamma_{s(air)}^{d'}$, methylene iodide ($\gamma_{mi} = 50.8$ dyn/cm) was employed as the nonpolar probe liquid for the surface equilibrated in air. The dispersion component is related to the wetting angle θ_3 via (9,10)

$$\gamma_{s(air)}^{d'} = \frac{\gamma_{mi}^2 \left(1 + \cos\theta_{3i}\right)^2}{4\gamma_{mi}^d}. \tag{3}$$

Stability of Polymeric Surfaces Subjected to Ultraviolet Irradiation

In this equation, $\gamma_{s(air)}^{d'}$ is the equilibrium value of the dispersion component of the surface free energy of the solid in air, γ_{mi} is the surface tension of methylene iodide, γ_{mi}^{d} is the dispersion component of the surface tension of methylene iodide ($=48.5$ dyn/cm), and θ_3 is the wetting angle. Since one can approximate

$$\gamma_{s(o)}^{p'} \simeq \gamma_{s(air)}^{p'}, \tag{4}$$

the total solid surface free energy, $\gamma'_{s(air)}$ can be estimated from

$$\gamma'_{s(air)} \simeq \gamma'_{s(o)} \simeq \gamma_{s(air)}^{d'} + \gamma_{s(o)}^{p'}. \tag{5}$$

Let us emphasize again that particularly Eqs. [1] and [2] are valid at time $t = 0$ only, since they involve the assumption that the surface has the same structure beneath both the environmental and probe liquids. As already noted, the area beneath the probe liquid can start to restructure as soon as the two come into contact. In Eq. [3], both the environmental and probe liquids are nonpolar. For this reason, one expects the angle θ_3 to remain almost unchanged in time. The more complete approach of Ruckenstein and Gourisankar (9) (which also involves the equilibrium wetting angle of a water drop on a solid surface equilibrated in an octane environment) could be employed to calculate the equilibrium values of both the polar and dispersion components of the surface free energy in both octane and water environments, if they are influenced only by the restructuring of the polymeric surface in different environments. However, in the present case, the photographs of a water drop on the solid surface in an octane environment, Figs. 3c–3e, indicate an increase in the turbidity of the drop with time. This evidently is a result of the extraction of some of the photooxidation degradation products of the polymer in the water drop, which, of course, changes the surface properties of both the liquid and the solid.

RESULTS

Results of contact angle measurement on rf-sputtered PMMA for both nonirradiated and UV-irradiated surfaces are given in Table I and plotted in Fig. 1. The accuracy of the contact angle measurement is, in general, $\pm 2°$. For the irradiated samples, the contact angles θ_1 decreased from θ_{1i} and reached the constant values θ_{1c} after 90 to 110 h. However, the constant values θ_{1c} were maintained only for 2 to 3 days, after which they started again to decrease with time. At this moment, it was observed that the water drops deformed and started to respond elastically to small external forces such as the touching of their surface with a needle. Finally, after about 240 h the

TABLE I

Contact Angles on the rf-Sputtered PMMA Surfaces as a Function of Irradiation Time

Time for UV irradiation (min)	θ_{1i}	θ_{1c}	θ_{2i}	θ_{3i}	$\tau(h)^a$
0	140	140	122	45.5	0
10	124	108.5	124	45	110
30	115.5	102	125.5	48.5	90
45	104	76	125.5	46.5	100
60	123.5	114	140	46.5	110

[a] τ is the time in which the water contact angle θ_1 on a surface reaches the constant value θ_{1c}.

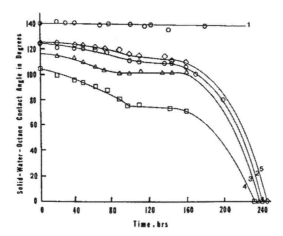

FIG. 1. The time dependence of the solid—water–octane angle on thin PMMA films for varying ultraviolet irradiation times. Curve 1 is for the nonirradiated sample, Curve 2 is for the sample irradiated for 10 min, Curve 3 for 30 min, Curve 4 for 45 min, and Curve 5 for 60 min.

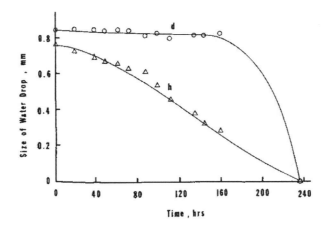

FIG. 2. The time dependence of the height and the base diameter of the water drop on the PMMA surface irradiated for 30 min.

drops disappeared leaving traces of the extractables on the surface. The size of the water drop decreased with time starting from the initial moment; the measured drop base diameter, d, and its height, h, are plotted in Fig. 2 as a function of time. For the nonirradiated samples the contact angle θ_1 and the dimensions d and h remained almost unchanged. The equilibrium values of the polar and dispersion components, and of the solid surface free energies, calculated by using Eqs. [1] to [5], are listed in Table II. In Tables I and II, the contact angles of methylene iodide in air on the polymeric surfaces, θ_3, and the corresponding values of the equilibrium dispersion surface free energy component of the solid surface free energies are also listed. It can be seen that these values are almost constant.

DISCUSSION

PMMA is a rigid polymer whose chain structure is not very mobile (11). The thin films of PMMA deposited by rf sputtering are tightly adherent to the substrate and probably even more cross-linked than the bulk PMMA used as the target. The contact angle of a water drop located on the

TABLE II

Estimated Values of Surface Free Energy Components of the rf-Sputtered PMMA as a Function of Irradiation Time (dyn/cm)

Time for UV irradiation (min)	$\gamma_{s(o)}^{p'}$	$\gamma_{s(w)}^{p'}$	$\gamma_{s(air)}^{d'}$	$\gamma_{s(air)}'$
0	0.7	29.7	38.5	39.2
10	2.5	30.9	38.8	41.2
30	4.1	31.7	36.8	40.9
45	7.3	31.7	37.9	45.2
60	2.6	39.6	37.9	40.5

nonirradiated sample remains unchanged with time because the rigid polymer would not allow surface restructuring which could expose its polar moieties to water. In contrast, the UV irradiation of the specimen results in an initial value of θ_{1i} lower than that for the nonirradiated surface and in a further decrease of θ_1 with time. This is a result of the following. The outer surface layers of the sputtered specimen are photooxidized by the UV irradiation. This injects polar groups into the surface and also generates oxidation products which degrade the polymer matrix. The more volatile oxidation products are removed from the surface layer while the sample is placed in the dark box for 1 day prior to the contact angle measurement. While some water-soluble oxidation products are expected to have been removed during the initial repeated washing with distilled water, other oxidation products may remain in the polymer matrix even after the washing. These polar products, which are probably covalently bonded to the surface, are extracted slowly by water, generating micropores, during the prolonged exposure of the polymer to water, thus contributing to the further degradation of the polymer surface and to a change in the contact angle.

The initial wetting angle θ_{1i} decreased from $140°$ for the nonirradiated sample to $104°$ for the sample irradiated for 45 min (Table I). Correspondingly, the equilibrium value of the polar surface free energy component in an octane environment, $\gamma_{s(o)}^{p'}$ changed from 0.7 dyn/cm for the nonirradiated sample to 7.3 dyn/cm for the sample irradiated for 45 min (Table II). For the sample irradiated for 60 min the initial value of θ_1 ($\theta_{1i} = 123.5°$) is greater than that for the samples irradiated for 30 and 45 min, but still lower than that for the nonirradiated sample. Correspondingly, the equilibrium value of the polar component of the surface free energy in octane, $\gamma_{s(o)}^{p'}$ is 2.6 dyn/cm (Table II). This indicates that the photooxidation effect increases rapidly for more than 45 min of irradiation time, thus injecting a higher polarity in the polymer which is, however, more easily water extractable during the initial washing with distilled water. On the other hand, the octane contact angle under water on the specimen equilibrated under water, θ_2, remained unchanged for samples irradiated for up to 45 min (Table I). In this case, the equilibrium value of the polar component of the solid surface free energy in an aqueous environment remains approximately the same, namely, 30–32 dyn/cm (Table II). This happens because the extractables have already been removed in the latter case during the 2 days of equilibration in water.

Of course, the contact of the solid surface with the water drop also stimulates the reorientation of the polar groups of the specimen toward water. This process is very slow for nonirradiated surfaces. For irradiated surfaces, the intrinsic polarity and the acquired polarity are probably cooperating in a more complex manner. The mobility of the polar groups will obviously be enhanced by the development of micropores that result from the extraction of the photooxidation products.

The size of the water drop decreased with time starting from the initial moment (Figs. 2 and 3). The water probably penetrates into the micropores developed by the extraction of the photooxidation products and by the degradation of the polymer matrix. Until the water contact angle reached the

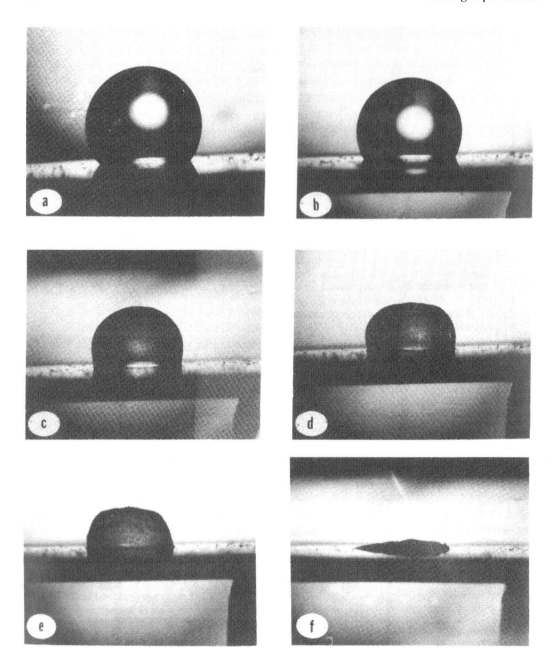

FIG. 3. Photographs of the water sessile drop, in an octane environment, on the 10-min UV-irradiated PMMA surface: (a) just after the placement of the drop ($t = 0$), (b) $t = 75$ h, (c) $t = 161$ h, (d) $t = 170$ h, (e) $t = 200$ h, and (f) $t = 240$ h.

constant value θ_{1c}, the initial shape of the original pure water drop remained unchanged. However, as soon as θ_1 reached the constant value θ_{1c}, the interface between water and octane lost its smoothness. After some time, probably when a critical concentration of the extractables was reached, the drop had an elastic response to small external forces such as the touching of its surface with a needle, and the upper portion of the interface between the drop and octane became flat and rough (Fig. 4). This elastic behavior of the probe liquid is due to the incorporation of the extractables.

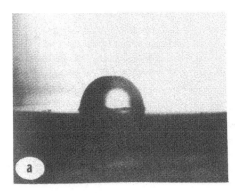 

FIG. 4. Photographs of a water sessile drop in an octane environment on the 45-min UV-irradiated PMMA surface showing the planar interface between water and octane: (a) $t = 92$ h after placement of the drop and (b) $t = 140$ h.

CONCLUSION

The polar component of the surface free energy of a polymeric surface can be affected significantly in some cases by ultraviolet irradiation. However, the ultraviolet treatment produces photooxidation products in the outermost layers of the solid surface and can also lead to the degradation of the polymer chains.

The above phenomena were investigated by the surface characterization of irradiated films of rf-sputtered polymethylmethacrylate (PMMA), by measuring the time-dependent change in the contact angle of a drop of water on the specimen equilibrated in an octane environment. The results show that the polymeric matrix is degraded by the UV irradiation and that the irradiation surface products are extracted slowly by water. The incorporation of the extractables in the probe water deforms the water drop, which at a given moment acquires a flat and rough shape in its top portion. The extraction develops micropores in the polymer matrix and promotes the penetration of water into the polymeric solid and the disappearance of the drop.

ACKNOWLEDGMENTS

This work was supported by the National Science Foundation. The authors are indebted to Dr. R. J. Good for allowing the use of his ellipsometer.

REFERENCES

1. Ruckenstein, E., and Gourisankar, S. V., *J. Colloid Interface Sci.* **101**, 436 (1984) (Section 1.2 of this volume).
2. Clark, D. T., Dilks, A., and Shuttleworth, D., in "Polymer Surfaces" (D. T. Clark and W. J. Feast, Eds.), p. 185. Wiley, New York, 1978.
3. Millard, M. M., and Pavlath, A. E., *J. Macromol. Sci. Chem. A* **10**, 579, 1976.
4. Corbin, G. A., Cohen, R. E., and Baddour, R. F., *Polymer* **23**, 1546 (1982).
5. Dwight, D. W., and Riggs, W. M., *J. Colloid Interface Sci.* **47**, 650 (1974).
6. Fox, R. B., Price, T. R., and Cain, D. S., *Adv. Chem. Ser.* **87**, American Chemical Society, Washington, DC, 1968.
7. Esumi, K., Meguro, K., Schwartz, A. M., and Zettlemoyer, A. C., *Bull. Chem. Soc. Japan* **55**, 1649 (1982).
8. Esumi, K., Schwartz, A. M., and Zettlemoyer, A. C., *J. Colloid Interface Sci.* **95**, 102 (1983).
9. Ruckenstein, E., and Gourisankar, S. V., *J. Colloid Interface Sci.* **107**, 488 (1985) (Section 2.1 of this volume).
10. Fowkes, F. M., *Ind. Eng. Chem.* **56**, 40 (1964).
11. Andrade, J. D., Gregonis, D. E., and Smith, L. M., in "Physicochemical Aspects of Polymer Surfaces" (K. L. Mittal, Ed.), Vol. 2, Plenum, New York, 1983.

2.4 Estimation of the Equilibrium Surface Free Energy Components of Restructuring Solid Surfaces[*]

Eli Ruckenstein and Sang Hwan Lee
Department of Chemical Engineering, State University
of New York, Buffalo, New York 14260

Received November 12, 1986; accepted January 29, 1987

INTRODUCTION

In the conventional methods of estimating the wetting characteristics of solids from contact angle measurements, the surface energetic properties of the solids were assumed to be the same in the environments of both the surrounding medium and the probe fluids. "The one-liquid method" (1, 2), in which the probe sessile drops are placed on the solid in air, and "the two-liquid method" (3–5), in which the sessile probe drops are placed on the solid in a liquid environment, have been widely used to estimate the surface free energy components of the solid, without taking into account that completely different surface energetic properties are possible beneath the surrounding medium and the probe liquid when one of them is hydrophilic and the other is hydrophobic.

Polymeric surfaces are, however, relatively mobile and adopt, in order to decrease the free energy of the system, different surface configurations in different environments (polar and nonpolar) (6–10). Two equilibrium quantities become therefore significant for such solids. One is the surface free energy of the solid in nonpolar environments, such as air and various hydrocarbons, $\gamma'_{S(L1)}$, and the other is that of the solid in polar environments such as water, $\gamma'_{S(L2)}$. The subscripts L1 and L2 are used for nonpolar and polar liquids, respectively, the symbol prime indicates the equilibrium value, and the symbol in parentheses identifies the equilibrating medium. If the total surface free energy of the solid is considered as composed of only dispersion and polar components, then

$$\gamma'_{S(L1)} = \gamma^{d'}_{S(L1)} + \gamma^{p'}_{S(L1)} \tag{1}$$

and

$$\gamma'_{S(L2)} = \gamma^{d'}_{S(L2)} + \gamma^{p'}_{S(L2)}, \tag{2}$$

where the superscripts d and p denote, respectively, the dispersion and polar components of the solid's surface free energy. The estimation of the values of these four unknown quantities, namely, $\gamma^{d'}_{S(L1)}$, $\gamma^{p'}_{S(L1)}$, $\gamma^{d'}_{S(L2)}$, and $\gamma^{p'}_{S(L2)}$ can be carried out by measuring a set of initial contact angles together with one final wetting angle, as suggested by Ruckenstein and Gourisankar (6). A different approach,

---

[*] *Journal of Colloid and Interface Science*, Vol. 120, No. 1, p.153, November 1987. Republished with permission.

Estimation of the Equilibrium Surface Free Energy Components

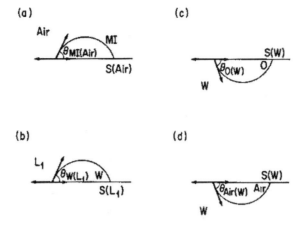

FIG. 1 Schematic diagrams of the four solid–probe–environment contact angles employed in the suggested surface free energy estimation procedure: (a) for Eq. [3], (b) for Eqs. [4]–[6], (c) for Eq. [8], and (d) for Eq. [9].

based on the measurements of initial wetting angles only, which is suggested here, is summarized in Fig. 1 and presented in the next section.

THE ESTIMATION OF THE SURFACE FREE ENERGY COMPONENTS

The following equation is used to estimate the equilibrium value $\gamma_{S(L1)}^{d'}$ of the dispersion component of the surface free energy of the polymeric surface equilibrated in octane (6, 7):

$$\gamma_{S(L1)}^{d'} \approx \gamma_{S(Air)}^{d'}$$

$$= \gamma_{MI}^2 \left(1 + \cos\theta_{MI(Air)}\right)^2 / 4\gamma_{MI}^d. \qquad [3]$$

In this equation, $\gamma_{S(Air)}^{d'}$ is the equilibrium value of the dispersion component of the surface free energy of the solid in air (which also provides a fair approximation of $\gamma_{S(L1)}^{d'}$), γ_{MI} is the surface tension of methylene iodide, γ_{MI}^d is the dispersion component of the surface tension of methylene iodide, and $\theta_{MI(Air)}$ is the initial contact angle of a methylene iodide drop located on a polymeric surface equilibrated in air.

The initial wetting angles (see under Discussion for the proper determination of their values) of water sessile drops located on the polymeric surface equilibrated in various hydrocarbons were used to estimate the equilibrium values of the polar component of the surface free energy in nonpolar environments. The following equations have been used in the calculations:

(Hexane)

$$\gamma_{S(H)}^{p'} = \left[\left(\gamma_{HW}\cos\theta_{W(H)} + \gamma_W - \gamma_H\right)/2\left(\gamma_W^p\right)^{1/2} - \left(\gamma_{S(H)}^{d'}\right)^{1/2}\left\{\left(\gamma_W^d\right)^{1/2} - \left(\gamma_H\right)^{1/2}\right\}/\left(\gamma_W^p\right)^{1/2} \right]^2, \qquad [4]$$

(Octane)

$$\gamma_{S(O)}^{p'} = \left(\gamma_{OW}\cos\theta_{W(O)} + \gamma_W - \gamma_O\right)^2 / 4\gamma_W^p \qquad [5]$$

and

(Cyclohexane)

$$\gamma_{S(C)}^{p'} = \left[\left(\gamma_{CW}\cos\theta_{W(C)} + \gamma_W - \gamma_C\right)/2\left(\gamma_W^p\right)^{1/2} - \left(\gamma_{S(C)}^{d'}\right)^{1/2}\left\{\left(\gamma_W^d\right)^{1/2} - \left(\gamma_C\right)^{1/2}\right\}/\left(\gamma_W^p\right)^{1/2}\right]^2. \qquad [6]$$

In Eq. [4], $\gamma_{S(H)}^{p'}$ is the equilibrium value of the polar component of the surface free energy in hexane, γ_{HW} is the hexane–water interfacial free energy, γ_W and γ_H are the surface tensions of water and hexane, respectively, γ_W^p is the polar component of the surface tension of water, and $\theta_{W(H)}$ is the initial contact angle of the water sessile drop in hexane. In Eqs. [5] and [6] for octane and cyclohexane environments, the subscript H from Eq. [4] is replaced by the subscripts O and C, respectively. The dispersion interaction terms do not appear in Eq. [5] because of the coincidental equality

$$\gamma_W^d = \gamma_O^d = \gamma_O = 21.8\,\text{dyn/cm} \qquad [7]$$

for octane. Therefore, the value of $\gamma_{S(O)}^{p'}$ can be determined from the measured value of $\theta_{W(O)}$. However, the dispersion interaction terms in the case of hexane and cyclohexane environments do appear. Therefore, the estimation of the values of $\gamma_{S(H)}^{p'}$ and $\gamma_{S(C)}^{p'}$ involves not only the measured values of $\theta_{W(L1)}$, but also the value of $\gamma_{S(L1)}^{d'}$ estimated by means of Eq. [3]. The arithmetic average of $\gamma_{S(H)}^{p'}$, $\gamma_{S(O)}^{p'}$, and $\gamma_{S(C)}^{p'}$ is further employed as $\gamma_{S(L1)}^{p'}$.

Similarly, the equilibrium value of the polar component of the surface free energy in a water environment $\gamma_{S(L1)}^{p'}$ was estimated by using the equation (6, 7)

$$\gamma_{S(W)}^{p'} = (-\gamma_{OW}\cos\theta_{O(W)} + \gamma_W - \gamma_O)^2/4\gamma_W^p, \qquad [8]$$

where $\theta_{O(W)}$ is the initial contact angle of an octane drop located on the polymeric surface equilibrated in water.

Finally, the equilibrium value of the dispersion component of the surface free energy in a water environment $\gamma_{S(W)}^{d'}$ was estimated by employing the equation (11)

$$\gamma_{S(W)}^{d'} = (\gamma_{OW}\cos\theta_{O(W)} - \gamma_W\cos\theta_{Air(W)} + \gamma_O)^2/4\gamma_O, \qquad [9]$$

where $\theta_{Air(W)}$ is the initial contact angle of an air bubble located on a polymeric surface equilibrated in water.

The following equations were used to estimate the interfacial free energies of both the solid equilibrated in a nonpolar environment–water, $\gamma_{S(L1)W}$, which is a nonequilibrium quantity, and the solid equilibrated in a polar environment–water, $\gamma'_{S(L2)W}$, which is an equilibrium quantity (2, 6):

(Nonpolar)

$$\gamma_{S(L1)W} = \left\{\left(\gamma_{S(L1)}^{d'}\right)^{1/2} - \left(\gamma_W^d\right)^{1/2}\right\}^2 + \gamma_{S(W)}^p \qquad [10]$$

(Polar)

$$\gamma'_{S(L2)W} = \left\{\left(\gamma_{S(L2)}^{d'}\right)^{1/2} - \left(\gamma_W^d\right)^{1/2}\right\}^2$$

$$+ \left\{\left(+\gamma_{S(L2)}^{p'}\right)^{1/2} - \left(\gamma_{(W)}^p\right)^{1/2}\right\}^2. \qquad [11]$$

Estimation of the Equilibrium Surface Free Energy Components

At this point, it is appropriate to summarize the assumptions involved in Eqs. [1]–[11]. They are (a) the total surface free energies of the solid and liquid phases can each be represented as the sum of their dispersion and polar components (1). (b) Both the dispersion and polar components of the work of adhesion can be expressed as the geometric averages (1)

$$W_{SL}^{d'} = 2\left(\gamma_S^{d'}\gamma_L^{d}\right)^{1/2} \tag{12}$$

and

$$W_{SL}^{p'} = 2\left(\gamma_S^{p'}\gamma_L^{p}\right)^{1/2}. \tag{13}$$

(c) The surface free energy components (particularly the dispersion component) of a solid in a non-polar environment is almost independent of the nature of the nonpolar fluid in which the surface is equilibrated, being almost the same in air and hydrocarbons.

It should be pointed out that, although the geometric mean approximation for dispersion interactions is well accepted, the geometric mean for polar interactions is not well accepted and, indeed, may be very inappropriate, particularly in situations involving strong hydrogen bond interactions. In addition, the specific acid–base interactions are ignored in the present treatment.

EXPERIMENTAL

Materials

Six solid surfaces of different hydrophilicity and surface rigidity were prepared from com merically available materials. Information concerning them is summarized in Table I. The siliconized glass surface (a) was manufactured by dipping a glass slide in a 10% by volume solution of a siliconizing fluid (Glassclad 6C, from Petrarch Systems, Inc.) in hexane. The surface was heat cured in an oven at 110°C for 1 h. Another siliconized glass surface (b) was prepared by dipping a glass slide in a 0.2% by volume aqueous silicone prepolymer (Glassclad 18, from Petrarch Systems, Inc.) solution. This silicone prepolymer is an alkylsilane dissolved in a mixture of t-butanol and diacetone alcohol, that reacts with water to form a silanol-rich prepolymer and an alcohol. This silanol-rich prepolymer reacts with the available hydroxy groups of the glass to generate a hydrophobic film. This surface was heat cured for 10 min at 110°C. A silicon elastomer (Silastic, medical grade sheet, vulcanized and reinforced, from Dow Coming Corp.) was deposited by ion beam sputtering onto single-crystal silicon wafers (Semimetals, Inc.), to manufacture specimen (c). Sputtering techniques have been already employed for the deposition of polymers as thin films, tightly adherent to a substrate. Ion beam sputtering offers some advantages over the other sputtering techniques, such as dc or rf sputtering, since the deposition can be achieved at a lower temperature and pressure. The rate of the process is, however, much smaller. The surface of the substrate was cleaned by ion beam sputtering just prior to deposition, in order to remove the surface contaminants and the adsorbed gases. This ensures the deposition of tightly adhering films. Argon ion beams of 1000 eV sputter etched the 4-in.-diameter polymeric target material and the sputter efflux was allowed to deposit on the surface of the substrate. The ion beam source to target spacing was 12 cm and the target to substrate spacing was 15 cm. The electric current density was 0.2 mA/cm². Twenty to thirty hours were necessary to obtain a 1000Å-thick polymeric film, whose thickness was measured with a profilometer. Since the sputter-depositing species are molecular scission fragments of the polymer, the chemical properties of the deposits may differ from those of the control polymer. In addition, sputtering injects into the film some of the oxygen atoms present in the sputtering chamber, thus generating some

TABLE I

Solid Surfaces of Different Hydrophilicity Prepared for the Contact Angle Measurements

Surfaces	Hydrophilicity	Description	Chemical composition
(a) Siliconized glass I	Hydrophobic	Prepared from Glassclad 6C (chlorine-terminated polydimethyl-siloxane telomer)	CH_3 groups on $Cl-Si-O-(Si-O)_n-Si-Cl$ with CH_3 substituents
(b) Siliconized glass II	Hydrophobic	Prepared from Glassclad 18 (silicone prepolymer formed an alkylsilane and water)	$C_{18}H_{37}-SiH_3 + H_2O$
(c) Silicone elastomer	Intermediate	Prepared from Silastic	$(-Si-O-)_n$ with CH_3 substituents
(d) Hydrogel I	Hydrophilic	Prepared from Hydron WD (bulk-polymerized poly(2-hydroxy-ethyl methacrylate), higher purity than Hydron N, high reduced viscosity (1.2–4.0 dl/g)	$(-CH_2-C-)_n$ with CH_3, $C=O$, $O-C_2H_4OH$
(e) Hydrogel II	Hydrophilic	Prepared from Hydron N (suspension-polymerized poly(2-hy-droxyethyl methacrylate), low reduced viscosity (0.5–0.8 dl/g)	$(-CH_2-C-)_n$ with CH_3, $C=O$, $O-C_2H_4OH$
(f) Glass	Hydrophilic	Microscope glass slide	

polarity. Another specimen (d), namely a hydrogel surface, was prepared by dipping a glass slide in 1.2 g/100 ml solution of poly(HEMA) (poly(2-hydroxyethyl methacrylate), Hydron WD, from Hydro Med Sciences, Inc.) in a mixed solvent (95% ethanol and 5% water). When the film-coated specimen was dried overnight at room temperature, a thin, hard, sterile film of optically clear polymer remained tightly bound to the glass surface. Hydron WD is the bulk-polymerized poly(HEMA) and contains 1.5 to 6% volatile impurities and less than 1000 ppm residuals. This polymer has a high reduced viscosity of 1.2 to 4 dl/g. Considering that the polymer contains such a low level of residual impurities and the manufactured film was equilibrated for 10 days in the environmental liquid, it is likely that if any extraction of constituents or contaminants took place, it occurred during equilibration. However, it is important to point out that poly(HEMA), as well as the HEMA monomer, is a highly surface-active material, and the release of a low concentration of this material can result in significant changes in surface and interfacial tensions. Another hydrogel surface (e) was similarly prepared, using, however, a different poly(HEMA) (Hydron N, from Hydro Med Sciences Inc.). The film was also hard, optically clear, and tightly adherent to the glass substrate surface. Hydron N is a suspension-polymerized poly(HEMA) and has a slightly higher level of impurities than Hydron WD. The volatiles are less than 5% and the residual monomer is less than 0.5%. The reduced viscosity of this material is low, being between 0.5 to 0.8 dl/g. In addition, a microscope glass slide

Estimation of the Equilibrium Surface Free Energy Components

(General Medical Scientific) was soaked in dichromate sulfuric acid cleaning solution overnight, rinsed in flowing distilled water, and dried at 150°C, to obtain solid (f).

Contact Angle Measurements

One specimen of each solid was equilibrated for 10 days in each of the following environments: hexane, octane, cyclohexane, and air in a clean space. Two other specimens of each of the solids (a)–(e) were equilibrated in double-distilled, deionized water for 10 days and for the rigid solid (f) for 1 month. The environments were changed daily, to eliminate the extractables as well as to avoid bacterialgrowth. The contact angles of 1–to 2–μl sessile drops on the samples were measured using a NRL contact angle goniometer (Rame-Hart, Inc., A-100, Mountain Lakes, NJ) and its accessories (especially the environmental chamber). The accuracy of the contact angle measurement was, in general, $\pm 2°$.

RESULTS

The results of contact angle measurements are listed in Table II. While the probe drops attained the initial contact angles $\theta_{O(W)}$, $\theta_{MI(Air)}$, and $\theta_{Air(W)}$ immediately after the placement of the drops on the solids, rapid decreases in the initial solid–water–hydrocarbon contact angles, $\theta_{W(H)}$, $\theta_{W(O)}$, and $\theta_{W(C)}$, plotted in Fig. 2, were observed for solids (c)–(f). The determination of the meaningful initial solid–water–hydrocarbon contact angles listed in the first three columns in Table II thus became a problem of concern (see under Discussion).

The values of the surface free energy components and the interfacial free energies of the solids equilibrated in both nonpolar and polar environments, estimated by using Eqs. [1]–[11], are given in Table III. Note that the values of $\gamma_{S(LI)}^{p'}$ in the first column of Table III were obtained as the arithmetic average of the three quantities, $\gamma_{S(H)}^{p'}$, $\gamma_{S(O)}^{p'}$, and $\gamma_{S(C)}^{p'}$, listed in Table IV. The values of the surface tensions and interfacial free energies of the liquids, used in the estimations of the surface free energy components of the solids, are given in Table V. It can be seen from Table III, that the order of hydrophilicity is as follows: siliconized glass I (most hydrophobic), siliconized glass II and silicone elastomer (intermediate), hydrogel I, hydrogel II, and glass (most hydrophilic). From the first two columns of Table III it can be seen that the equilibrium values of the polar components of the surface free energies $\gamma_S^{p'}$ are larger for the solids equilibrated in polar environments than for those equilibrated in nonpolar environments, as long as the solids have some polar moieties in their molecular structure. The siliconized glass surfaces (solids (a) and (b)) were both hydrophobic in nonpolar environments. However, solid (b), which contains some polar moieties that may be reoriented through the equilibration in a polar environment, showed an increase in the value of $\gamma_S^{p'}$ upon equilibration in a polar environment. In contrast, solid (a), which does not contain polar moieties, maintained its hydrophobicity. Solid (c), prepared by ion beam sputtering, exhibits some polarity,

---

TABLE II

Initial Equilibrium Contact Angles Measured on Solids Equilibrated in Both Nonpolar and Polar Environments

Surfaces	$\theta_{W(H)}$	$\theta_{W(O)}$	$\theta_{W(C)}$	$\theta_{O(W)}$	$\theta_{MI(Air)}$	$\theta_{Air(W)}$
(a) Siliconized glass I	154	149	155	30	62	81
(b) Siliconized glass II	148	144	151	55	54	93
(c) Silicone elastomer	121	114	115	119	72	116
(d) Hydrogel I	64	52	70	145	28	149
(e) Hydrogel II	48	45	60	152	34	158
(f) Glass	41	35	43	165	43	162

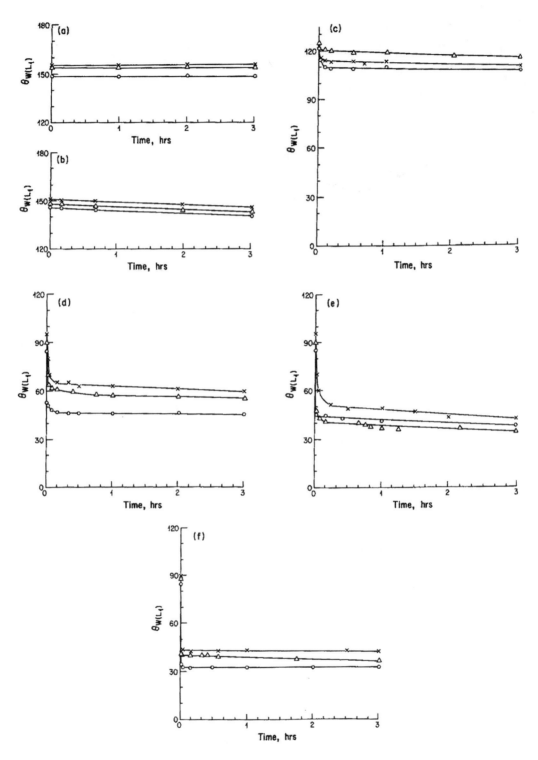

FIG. 2 Time dependence of the solid–water–hydrocarbon angle on solid surfaces of different hydrophilicity. (The symbols Δ, O, and × correspond respectively to the media hexane, octane, and cyclohexane.) (a) Siliconized glass I, (b) siliconized glass II, (c) silicon elastomer, (d) hydrogel I, (e) hydrogel II, and (f) glass.

TABLE III

Estimated Values of the Solid Surface Free Energies, Their Components, and Interfacial Free Energies for Solid Surfaces of Different Hydrophilicity, Equilibrated in Both Nonpolar and Polar Environments (dyn/cm)

Surfaces	$\gamma_{S(LI)}^{p'}$	$\gamma_{S(w)}^{p'}$	$\gamma_{S(LI)}^{d'}$	$\gamma_{S(w)}^{d'}$	$\gamma_{S(LI)}'$	$\gamma_{S(w)}'$	$\gamma_{S(LI)w}$	$\gamma_{S(w)w}'$
(a) Siliconized glass I	0.2	0.2	28.7	42.7	28.9	42.9	45.1	48.1
(b) Siliconized glass II	0.4	2.3	33.2	34.4	33.6	35.8	43.4	32.9
(c) Silicone elastomer	4.0	28.3	22.8	9.4	26.8	37.7	26.3	5.8
(d) Hydrogel I	27.5	42.0	47.0	20.3	74.5	62.3	8.3	0.4
(e) Hydrogel II	33.7	45.0	44.5	22.5	78.2	67.5	5.8	0.2
(f) Glass	39.7	49.0	40.2	20.1	79.9	69.1	3.5	0.1

TABLE IV

Values of the Initial Solid–Water–Hydrocarbon Contact Angles, $\theta_{W(LI)}$, and of the Polar Components of the Surface Free Energy of the Solids Equilibrated in Different Hydrocarbons. $\gamma_{S(LI)}^{p'}$ (dyn/cm), Estimated by Using Eqs. [4]–[6]

Surfaces	$\theta_{W(H)}$	$\theta_{W(O)}$	$\theta_{W(C)}$	$\gamma_{S(H)}^{p'}$	$\gamma_{S(O)}^{p'}$	$\gamma_{S(C)}^{p'}$	$\gamma_{S(LI)}^{p'}$
(a) Siliconized glass I	154	149	155	0.1	0.2	0.2	0.2
(b) Siliconized glass II	148	144	151	0.2	0.5	0.4	0.4
(c) Silicone elastomer	121	114	115	3.0	4.5	4.5	4.0
(d) Hydrogel I	64	52	70	25.2	33.2	24.1	27.5
(e) Hydrogel II	48	45	60	34.3	37.0	29.7	33.7
(f) Glass	41	35	43	38.2	42.0	38.8	39.7

TABLE V

Surface and Interfacial Tensions of the Liquids Used in the Contact Angle Experiments (dyn/cm)

Liquid	γ_{LV}	γ_{LV}^{d}	γ_{LV}^{p}	γ_{LW}
Hexane	18.4	18.4	0	51.1
Octane	21.8	21.8	0	50.8
Cyclohexcane	25.5	25.5	0	50.2
Methylene iodide	50.8	48.5	2.3	41.6
Water	72.6	21.8	50.8	—

while solid (a) is hydrophobic, even though the initial polymers employed in their preparation have the same chemical composition. This arises because sputtering injects oxygen atoms into the specimen (6). From the third and fourth columns one can see that the equilibrium values of the dispersion components of the surface free energies $\gamma_S^{d'}$ increased for very hydrophobic solids (especially solid (a)), but decreased for the solids whose $\gamma_S^{p'}$ had a large increase upon equilibration in a polar environment. One may note upon inspection of Eq. [11] that both the increase in the value of $\gamma_S^{p'}$ toward 50.8 dyn/cm (the value of γ_W^p) and the decrease in the value of $\gamma_S^{d'}$ toward 21.8 dyn/cm (the value of γ_W^d) contribute to the decrease in γ'_{SW} upon equilibration in water. One can also note that $\gamma_S^{p'}$ can have a dominant role in the calculation of γ'_{SW} values.

From the fifth and sixth columns one can see that the total surface free energy increases for hydrophobic solids, but decreases for hydrophilic solids, upon equilibration in water. From the seventh and eighth columns, one can conclude that the interfacial free energy between solid and water decreases significantly upon equilibration in water for all solids, except solid (a).

DISCUSSION

It is convenient to use Eqs. [3]–[9] which involve only initial wetting angles for the estimation of the four quantities, $\gamma_{S(L1)}^{p'}, \gamma_{S(W)}^{p'}, \gamma_{S(L1)}^{d'}$, and $\gamma_{S(W)}^{d'}$, because each of these quantities is calculated almost independently from a single equation. The estimation of $\gamma_{S(L1)}^{p'}$ entails, however, the identification of the meaningful initial contact angle value $\theta_{W(L1)}$. This is because of the initial rapid decrease in the angle with time. One may note that the initial decrease in the angle is particularly large for hydrophilic solid surfaces (Fig. 2). This initial rapid decrease in the contact angles can be attributed to the attainment of the equilibrium wetting angle from an initial nonequilibrium angle. Based on this consideration, the initial values of the equilibrium contact angles were taken as those attained after the initial sharp decrease (after a few seconds, though sometimes 1–2 min). The subsequent decrease in the contact angle is attributed here to surface restructuring.

A procedure very similar to the present one was recently suggested by Ruckenstein and Gourisankar (6) to estimate the four unknowns, $\gamma_{S(L1)}^{p'}, \gamma_{S(W)}^{p'}, \gamma_{S(L1)}^{d'}$ and $\gamma_{S(W)}^{d'}$. That procedure employed a set of initial contact angles together with one final wetting angle. The value of $\gamma_{S(W)}^{d'}$ was estimated from an equation (Eq. [11] in Ref. (6)) which involved all four unknown quantities.

Finally, let us note that (see Table III) the increase in polarity upon equilibration in water is significant. Various phenomena may be responsible for the surface modification of the solids upon equilibration in water, both individually and collectively. Some of them are (i) surface restructuring via the orientation of the polar moieties (6–10); (ii) penetration of water into the surface structure, resulting in surface swelling (12); and (iii) water structuring near the surface (13). The surface restructuring of the solids is probably the most prominent phenomenon associated with the equilibration of the solid in water (Fig. 3).

The decrease in the dispersion component of the surface free energy upon equilibration in water, while the surface polar component increases, is also worth noting (Table III). The restructured solid surface probably acquires a higher density of polar moieties in its outer surface layers. This decreases the dispersion contribution, if the dispersion interactions involving the polar moieties are weaker than those involving the nonpolar moieties.

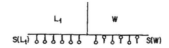

FIG. 3 Schematic diagram of a polymer surface layer equilibrated in nonpolar (L1) and water (W) environments. ○ denotes the polar head of the surface molecules.

Estimation of the Equilibrium Surface Free Energy Components

The unexpectedly small value (0.1 dyn/cm) of the water equilibrated glass–water interfacial free energy may be due to (1) water absorption in the surface pores of the glass and/or (2) the non-validity of the assumptions involved in the calculations (including the nature of the interactions) for a substrate such as glass.

CONCLUSION

A contact angle procedure, based on the measurement of four initial wetting angles, is suggested for the estimation of the surface characteristics of polymers in polar and nonpolar environments. This procedure was employed for the surface characterization of solids of different hydrophilicity. It was found that the equilibrium values of the polar components of the surface free energies, $\gamma_S^{p'}$ increased upon equilibration in a polar environment. On the other hand, the equilibrium values of the dispersion components of the surface free energies, $\gamma_S^{d'}$, increased for very hydrophobic solids, but decreased for the solids which had some polarity, upon equilibration in a polar environment. The solid–water interfacial free energies decreased upon equilibration in a polar environment. The differences in the surface energetics of the solids in the different environments are attributed here to the surface-restructuring capabilities of the solids.

ACKNOWLEDGMENT

This work was supported by the National Science Foundation.

REFERENCES

1. Fowkes, F. M., *Ind. Eng. Chem.* **56**, 40 (1964).
2. Kaelble, D. H., *J. Adhes.* **2**, 66 (1970).
3. Tamai, Y., Makuuchi, K., and Suzuki, M., *J. Phys. Chem.* **71**, 4176 (1967).
4. Matsunaga, T., *J. Appl. Polym. Sci.* **21**, 2847 (1977).
5. Schultz, J., Tsutsumi, K., and Donnet, J. B., *J. Colloid Interface Sci.* **59**, 272 (1977).
6. Ruckenstein, E., and Gourisankar, S. V., *J. Colloid Interface Sci.* **107**, 488 (1985) (Section 2.1 of this volume).
7. Lee, S. H., and Ruckenstein, E., *J. Colloid Interface Sci.* **117**, 172 (1987).
8. Holly, F. J., and Refojo, M. F., *J. Biomed. Mater. Res.* **9**, 315 (1975).
9. Andrade, J. D., Gregonis, D. E., and Smith, L. M., *in* "Physicochemical Aspects of Polymer Surfaces" (K. L. Mittal, Ed.), Vol. 2, p. 911. Plenum Press, New York, 1983.
10. Yasuda, H., Sharma, A. K,, and Yasuda, T., *J. Polym. Sci. Polym. Phys. Ed.* **19(9)**, 1285 (1981).
11. Andrade, J. D., Ma, S. M., King, R. N., and Gregonis, *D. E., J. Colloid Interface Sci.* **72**, 488 (1979).
12. Timmons, C. O., and Zisman, W. A., *J. Colloid Interface Sci.* **22**, 165 (1966).
13. Lee, H. B., Jhon, M. S., and Andrade, J. D., *J. Colloid Interface Sci.* **51**, 225 (1975).

2.5 Surface Restructuring of Polymers*

Sang Hwan Lee and Eli Ruckenstein
Department of Chemical Engineering, State University
of New York, Buffalo, New York 14260

Received December 15, 1986; accepted February 17, 1987

INTRODUCTION

Conventionally, the surface energetic properties of solids are assumed to be independent of the environment (polar and nonpolar) in which they are immersed. However, Holly and Refojo (1), taking contact angle hysteresis experiments into consideration, concluded that hydrogel surfaces changed their polarity according to the nature of the environment. Since then, such surface modification phenomena have been investigated by employing the ESCA method (2, 3), surface acidity titration (3), contact angle hysteresis (1, 4), and dynamic contact angle measurements (5, 6). Environmentally induced changes in the surface structure have been observed in those polymers which have relatively mobile surface layers and contain polar moieties. The polarity was in some cases incorporated by surface modification techniques, such as plasma treatment (7, 8), chemical oxidation (9), grafting hydrophilic polymers onto hydrophobic polymers (10), or sputtering (5, 6, 11). The objective of the present paper is to follow the surface modification behavior of polymers of different hydrophilicities and rigidities by dynamic contact angle measurements and to evaluate the time scale of their restructuring.

EXPERIMENTAL

MATERIALS

Thin films of polymeric materials deposited on glass substrates were employed. Glass microscope slides (General Medical Scientific) were cleaned by their dipping in a dichromate sulfuric acid solution overnight. This was followed by extensive rinsing in flowing distilled water and drying at 150°C. Six solid surfaces of different hydrophilicities and rigidities, which are listed in Table I, were prepared. (a) A cleaned glass slide was immersed in a 10% by volume solution of a siliconizing fluid (Glassclad 6C, from Petrarch Systems, Inc.) in hexane and the deposited film was heat cured at 110°C for 1 h to obtain a siliconized glass surface. (b) Another siliconized glass surface was prepared by dipping a cleaned glass slide in 0.2% by volume aqueous silicone prepolymer solution (Glassclad 18, from Petrarch Systems, Inc.) and the deposited film was heat cured at 110°C for 10 min. (c) A thin film of silicone elastomer (Silastic, medical grade sheet, from Dow Corning Corp.) was deposited by ion beam sputtering onto a single crystal silicone substrate (Semimetals Inc.). An electron-bombardment, ion thruster was employed as the ion beam source to generate an argon ion beam of energetic ions (1000 eV).

---

* *Journal of Colloid and Interface Science*, Vol. 120, No. 2, p. 529, December 1981. Republished with permission.

TABLE I

Solid Surfaces of Different Hydrophilicities

Surfaces	Description	Chemical composition
(a) Siliconized glass I	Prepared from Glassclad 6C (chlorine terminated polydimethylsiloxane telomer)	CH_3 $\quad$ CH_3 $\quad$ CH_3 $Cl-Si-O-(Si-O)_n-Si-Cl$ CH_3 $\quad$ CH_3 $\quad$ CH_3
(b) Siliconized glass II	Prepared from Glassclad 18 (silicone prepolymer forms an alkylsilane and water)	$C_{18}H_{37}-SiH_3 + H_2O$
(c) Silicone elastomer	Prepared from Silastic	CH_3 $(-Si-O-)_n$ CH_3
(d) Hydrogel I	Prepared from Hydron WD (bulk polymerized poly(2-hydroxyethyl methacrylate))	CH_3 $(-CH_2-C-)_n$ $C=O$ $O-C_2H_4OH$
(e) Hydrogel II	Prepared from Hydron N (suspension polymerized poly(2-hydroxyethyl methacrylate))	CH_3 $(-CH_2-C-)_n$ $C=O$ $O-C_2H_4OH$
(f) Glass	Microscope glass slide	

The surface of the target material was sputter cleaned just prior to deposition, in order to remove the surface contaminants. The ion beam sputter etched the target material which was placed 12 cm downstream of the ion source. The ejected surface atoms, molecules, or molecular fragments were deposited on the surface of the substrate, which was located 15 cm from the target. The composition of the deposited film was different from that of the control polymer. During the sputtering process, oxygen atoms present in the sputtering chamber were incorporated into the deposits (11). The process was relatively slow; 20 to 30 h was necessary to obtain a 1000-Å-thick polymeric film, (d) A cleaned slide was immersed in 1.2 g/100 ml solution of poly(HEMA), poly(2-hydroxyethyl methacrylate) (Hydron WD, from Hydro Med. Sciences Inc.), in a mixed solvent (95% by volume ethanol and 5% water) and the deposited layer was dried overnight at room temperature to prepare a hydrogel surface. (e) Another hydrogel surface was similarly prepared, using a different poly(HEMA) (Hydron N, from Hydro Med. Sciences Inc.). (f) The cleaned slide itself was employed as a sample.

Dynamic Contact Angle Measurements

One specimen of each of the solids was equilibrated in octane for 10 days while two specimens of each of the solids were equilibrated in doubly distilled, deionized, water for 10 days for solids (a)-(e) and for 1 month for solid (f), in order to ensure full equilibration of the more rigid surfaces. The equilibrating medium was replaced daily by a fresh one in order to avoid the growth of bacteria.

A drop of water was placed under octane on a surface which was previously equilibrated in octane and the change in time of the solid- water-octane contact angle $\theta_{w(0)}$ was measured. One of the water-equilibrated specimens was subsequently immersed in octane and the change in time of the contact angle of a water drop on the sample was determined. The water-equilibrated specimen was cleaned by flowing octane prior to its immersion in octane, in order to remove the adhering water. Another water-equilibrated specimen was immersed in octane for 1 day and the change in time of the contact angle of a water drop placed on the polymer surface was measured. The contact angles of 1-2 μl water drops placed on the specimens were measured using a NRL contact angle goniometer (Rome-Hart Inc. A-100) and its accessories (the environmental chamber). The accuracy of the contact angle measurement was $\pm$ 2°.

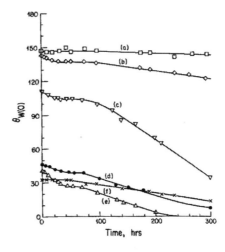

FIG. 1. The time dependence of the solid-water-octane angle on solid surfaces of different hydrophilicities. (a) Siliconized glass I, (b) siliconized glass II, (c) silicone elastomer, (d) hydrogel I, (e) hydrogel II, and (f) glass.

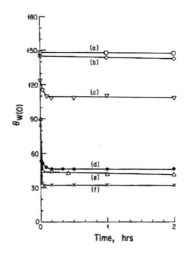

FIG. 2. The same as that in Fig. 1, but for a short time scale (2 h) to emphasize the rapid initial decrease in the contact angle.

RESULTS

The time evolution of the contact angle of water drops on the solids equilibrated in octane is plotted in Figs. 1 and 2. From Fig. 1, one can see that the order of hydrophilicity is siliconized glass I (most hydrophobic), siliconized glass II, silicone elastomer (intermediate), hydrogel I, hydrogel II, and glass (most hydrophilic).

For the most hydrophobic solid (a), the contact angle $\theta_{w(0)}$ remains constant for about 100 h and then starts to decrease slightly (Fig. 1). A somewhat different behavior is observed for the hydrophobic solid (b). The contact angle $\theta_{w(0)}$ decreases slowly for about 40 h and reaches a constant value, which is maintained for about 70 h. Afterward, the contact angle decreases again with time. For silicone elastomer (c), the contact angle decreases rapidly from 122 to 110° in about 5 min (Fig. 2) and then it further decreases at a slower rate for about 30 h to reach a constant value which is maintained for about 30 h (Fig. 1). The contact angle decreases at still greater times. For the very hydrophilic hydrogels (solids (d) and (e)), the behavior is similar to that of solid (c), but with a more rapid initial decrease from 90 to 46° (solid (d)) and from 90 to 44° (solid (e)) in the first few minutes (Fig. 2). Subsequently, the contact angle decreases at a slower rate for about 40-50 h to reach a brief constant value and then decreases again with time until the drop disappears (Fig. 1). However, for the glass surface (f) which is the most hydrophilic but has, in addition, a very rigid surface, the wetting angle shows an initial rapid decrease (in less than about 10 s) from 85 to 32°, followed by a relatively slow decrease for the remaining time span of the experiment (Fig. 1).

The changes with time in the size of the water drops (the drop base diameter, d, and the height, h) are plotted in Figs. 3a-3f along with the corresponding contact angles. While for solids (a)-(d) and (f) the drop diameter remains almost unchanged with time, the drop height decreases steadily. For solids (e), some widening of the drop base diameter (spreading) during the initial 30 h is observed.

Let us note that the phenomena, which were previously observed (6) for polymethylmethacrylate films irradiated with ultraviolet radiation such as the loss of smoothness of the interface between water and octane and the incorporation of the extractables in the probe water drop, were not detected with any of the present polymers.

The change of the contact angle with time for water drops under octane for polymers previously equilibrated in water are plotted in Figs. 4a-4f for various conditions. The measurements have been carried out: (i) just after immersion in octane, and (ii) after immersion in octane for 1 day.

The contact angle of the water drop measured on the most hydrophobic polymer (a) just after immersion in octane of the water- equilibrated sample has the same value as that for the octane equilibrated sample. Little change is therefore expected to occur in the surface structure of polymer (a) when the polar equilibrating environment is replaced by a nonpolar one. On the other hand, for polymers (b) and (c) somewhat lower initial contact angles of the water equilibrated sample than those of the octane-equilibrated specimen have been measured just after immersion in octane. In such cases some restructuring is expected to occur. On the hydrophilic solids (d), (e), and (f) equilibrated in water, the water drops spread completely as soon as the specimens were immersed in octane. This happens because the interfacial free energy between the solid and the octane is greater, and that between solid and probe liquid is smaller, than that on specimens equilibrated in octane. The spreading condition can be therefore satisfied at the moment of immersion.

When the contact angle of the water drop is measured on the water-equilibrated sample immersed in octane for 1 day, the initial contact angle $\theta_{w(0)}$ is comparable for all polymers to that on the octane-equilibrated sample. It is clear that the polymeric surface is subjected to "surface restructuring" not only in the polar environment but also in the nonpolar environment of octane. The contact angle decreases subsequently with time.

DISCUSSION

The following factors may be responsible for the dynamic behavior of the contact angle: (i) modification of the solid by the presence of the probe drop; (ii) gradual dissolution of the probe drop into the surrounding liquid phase; and (iii) the incorporation of the extractables from the solid surface into the probe liquid drop, thus altering the surface tension of the probe liquid and the interfacial free energy between drop and solid (6). However, for a water probe drop in an octane environment, the solubility of water in octane is about 100 ppm and, moreover, octane was saturated with water before use. Since the incorporation of extractable does not seem to be significant for the samples employed here, the changes in the structure of the surface of the solid are probably mainly responsible for the change of the contact angle with time.

Various phenomena can be responsible for the surface modification of the solid upon equilibration in water. They are (i) surface restructuring by reorientation of the polymer backbone, segments, and pending groups; (ii) penetration of water into the surface structure (absorption), sometimes resulting in surface swelling; and (iii) water structuring in the neighborhood of and/or on the polymeric surface (adsorption). Among these, the surface restructuring, which leads to the exposure of the polar moieties to a polar environment or of the hydrophobic moieties to a nonpolar medium, probably plays the main role. The surface modifications lead to different surface energetic characteristics of the polymeric surfaces equilibrated in polar and nonpolar environments. The values of the surface free energy components and the interfacial free energies of the solids equilibrated in both nonpolar and polar environments, estimated by a procedure suggested recently (11), are listed in Table II. As expected, the first two columns of this table show that the equilibrium value of the polar component of the surface free energy is larger for the solids equilibrated in polar environments than for those equilibrated in nonpolar environments (as long as the solid contains some polar moieties in its molecular structure). The difference in the dynamic behavior of the contact angle of hydrophobic siliconized glass surfaces (Figs. la and lb) can be explained by noting that solid (b) contains some polar moieties and exhibits, therefore, an increase in the value of the polar component of the surface free energy upon equilibration in water. The decrease in the angle during the first 40 h to a constant value (Fig. lb) can be attributed to the surface reorientation of the polar side groups of the polymer surface beneath the probe water drop. The subsequent decrease in the angle may perhaps be due mainly to the penetration of water into the polymer (Fig. 3b). This decreases the height of the drop and decreases the interfacial free energy between the solid and the probe liquid. As a result, the wetting angle decreases. The different values of the surface free energy components as well as the different dynamic behavior of the contact angle for the silicone surfaces (a) and (c) which were prepared from compounds of similar chemical composition can be explained by noting that the ion beam sputtering injects some polarity into the polymeric film. As a result, surface restructuring and penetration of water can occur for surface (c). For the more hydrophilic solid surfaces (solids (c)-(f)), an initial rapid decrease in the contact angle (which lasts a few seconds, though sometimes it may last 5 min or more) was observed (Fig. 2). This rapid initial decrease in the contact angle can be attributed in part to the attainment of the equilibrium wetting angle from an initial nonequilibrium angle. However, the surface restructuring may also play a role when the surface is highly mobile, as in the case of hydrogels (solids (d) and (e) (13)). The surface restructuring of hydrogel surfaces probably begins just after contact with water and lasts for 40-50 h (Fig. 1) until a constant value for the angle is achieved. The angle is maintained constant for 20-30 h and then decreases slowly until the drop disappears (after about 200-300 h). While in the initial stage of 40-50 h both restructuring and penetration of water play a role, in the final stage, penetration by water molecules probably plays the main part. Refojo has shown that the average pore diameter in the hydrogels is about 4 Å (14). After restructuring, the porosity can become greater and the water molecules can penetrate more easily into the polymer.

Surface Restructuring of Polymers

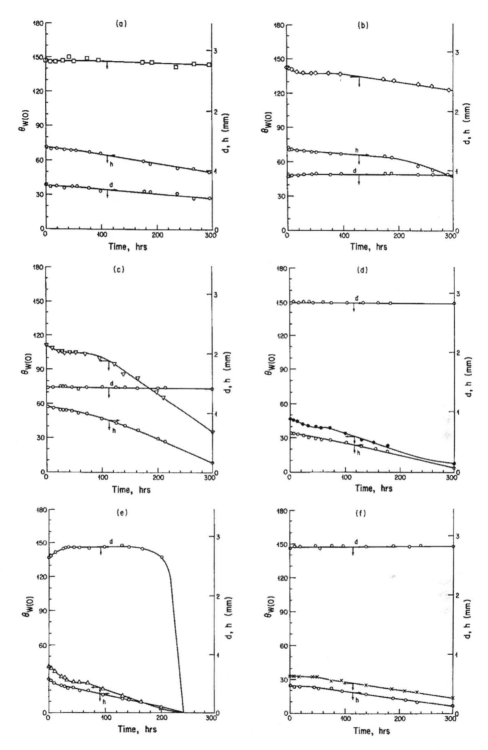

FIG. 3. The time dependence of the height h and the base diameter d of the water drop, along with the contact angle, on solid surfaces of different hydrophilicities. (a) Siliconized glass I, (b) siliconized glass II, (c) silicone elastomer, (d) hydrogel I, (e) hydrogel II, and (f) glass.

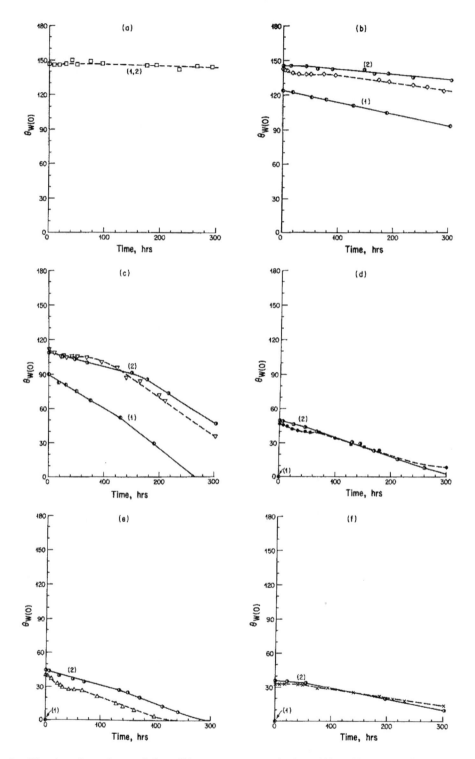

FIG. 4. The time dependence of the solid-water-octane angle for solids subjected to various conditions after equilibration in water. Curve 1, the measurements have been carried out immediately after immersion in octane; curve 2, after immersion in octane for 1 day. The dotted line is for solids equilibrated in octane. (a) Siliconized glass I, (b) siliconized glass II, (c) silicone elastomer, (d) hydrogel I, (e) hydrogel II, and (f) glass.

TABLE II

Estimated Values of the Solid Surface Free Energies, Their Polar and Dispersion Components, and Interfacial Free Energies for Solid Surfaces of Different Hydrophilicities, Equilibrated in Both Nonpolar and Polar Environments (dyn/cm) (Ref. (11))

Surfaces	$\gamma^{p'}_{s(0)}$	$\gamma^{p'}_{s(w)}$	$\gamma^{d'}_{s(0)}$	$\gamma^{d'}_{s(w)}$	$\gamma'_{s(0)}$	$\gamma'_{s(w)}$	$\gamma_{s(0)v}$	$\gamma'_{s(w)w}$
(a) Siliconized glass I	0.2	0.2	28.7	42.7	28.9	42.9	45.1	48.1
(b) Siliconized glass II	0.4	2.3	33.2	34.4	33.6	35.8	43.4	32.9
(c) Silicone elastomer	4.0	28.3	22.8	9.4	26.8	37.7	26.3	5.8
(d) Hydrogel I	27.5	42.0	47.0	20.3	74.5	62.3	8.3	0.4
(e) Hydrogel II	33.7	45.0	44.5	22.5	78.2	67.5	5.8	0.2
(f) Glass	39.7	49.0	40.2	20.1	79.9	69.1	3.5	0.1

Note: $\gamma_{s(0)}$ and $\gamma_{s(0)}$ are the surface free energies of the solid in the nonpolar and polar environments, respectively; the superscripts p and d refer to the polar and dispersion components; and the sign prime indicates an equilibrium value. $\gamma_{s(0)w}$ and $\gamma'_{s(w)w}$ are the interfacial free energies between the solid equilibrated in a nonpolar medium and water and between the solid equilibrated in water and water, respectively.

The relatively slow decrease in the contact angle for glass surfaces (Fig. 1f), which are obviously rigid, cannot be attributed to either surface restructuring or penetration of water into the solid. They probably arise because of the slow accumulation of impurities on the glass surface and/or the interface between the water droplet and the octane.

Immersion of the water-equilibrated solids in octane for different durations (Figs. 4) shows that surface restructuring also takes place in octane and increases the hydrophobicity of the polymeric surface. Even a very hydrophilic hydrogel surface can behave quite hydrophobically when exposed to a hydrophobic environment.

CONCLUSION

When probe water drops are placed on polymer surfaces (equilibrated in octane or water) of different hydrophilicities and rigidities under octane, a change with time of the contact angle is observed. The surface modification produced by the presence of the probe water drop is mainly responsible for the dynamic behavior of the contact angle. The height of the water drop decreases steadily, while, in general, the diameter of its base remains almost constant, indicating that water penetrates into the polymer. In addition, the reorientation of the buried polar groups takes place.

REFERENCES

1. Holly, F. J., and Refojo, M. F., *J. Biomed. Mater.Res.* **9**, 315 (1975).
2. Ratner, B. D., *in* "Biomaterials: Interfacial Phenomena and Applications" (Cooper and Peppas, Eds.), ACS Advances in Chemistry Series Vol. 199, p. 9. Amer. Chem. Soc., Washington, DC, 1982; Ratner, B.D., J. Appl. Polym. Sci. **22**, 643 (1978).
3. Lavielle, L., and Schultz, J., *J. Colloid Interface Sci.* **106**, 438 (1985); Schultz, J., Tsutsumi, K., and Donnet, J. B., *J. Colloid Interface Sci.* **59**, 272 (1977).
4. Andrade, J. D., Gregonis, D. E., and Smith, L. M., *in* "Physicochemical Aspects of Polymer Surfaces" (K. L. Mittal, Ed.), Vol. 2, p. 911. Plenum, New York, 1983.
5. Ruckenstein, E., and Gourisankar, S. V., *J. Colloid* Interface Sci. **109**, 557 (1985) (Section 2.2 of this volume).
6. Lee, S. H., and Ruckenstein, E., *J. Colloid Interface Sci.* **117**, 172 (1987) (Section 2.3 of this volume).
7. Clark, D. T., Dilka, A., and Shuttleworth, D., *in* "Polymer Surfaces" (D. T. Clark and W. J. Feast, Eds.), p. 185. Wiley, New York, 1978.

8. Yasuda, H., Marsh, H. C,, Brandt, S., and Reilley, C. N., *J. Polym. Sci. Polym. Chem. Ed.* **15**, 991 (1977).
9. Baszkin, A., Nishino, M., and Ter-Minassian-Saraga, L., *J. Colloid Interface Sci.* **59**, 516 (1977).
10. Ratner, B. D., *in* "Polymer Science and Technology" (Lee, Ed.), Vol. 12B. p.691. Plenum, New York, 1980.
11. Ruckenstein, E., and Lee, S. H., *J. Colloid Interface Sci.* **120**, 153 (1987) (Section 2.4 of this volume).
12. Lee, H. B., Jhon, M. S., and Andrade, J. D., *J. Colloid Interface Sci.* **51**, 225 (1975).
13. Yasuda, H., Sharma, A. K., and Yasuda, T., *J. Polym.* Sci. *Polym. Phys. Ed.* **19**, 1285 (1981).
14. Refojo, M. F., *J. Appl. Polym. Sci.* **9**, 3417 (1965).

2.6 Adsorption of Proteins onto Polymeric Surfaces of Different Hydrophilicities—A Case Study with Bovine Serum Albumin*

Sang Hwan Lee and Eli Ruckenstein

Department of Chemical Engineering, State University of New York, Buffalo, New York 14260

Received July 20, 1987; accepted December 1, 1987

INTRODUCTION

Protein adsorption from a biological fluid on a solid surface is a problem of interest in a variety of biological, medical, and technological processes (1–3). It occurs on all types of solid surfaces, even on surfaces which have the same kind of charge as the protein. The adsorbed protein layers serve as a conditioning film which determines the extent of subsequent cellular adhesion or thrombosis.

Even though the interactions between proteins and adsorbent surfaces have been extensively studied for decades (for reviews see Refs. (1–9)), there are still discrepancies in the reported results, and it is often difficult to compare the results obtained by different authors. This is partly because the experimental conditions, for instance the purity of the protein, the surface energetic properties of the adsorbent, and the rinsing scheme employed, have not been clearly indicated.

It has been shown that protein adsorption is entropically driven (8–18). For example, Norde et al. (10) concluded that the increase in entropy (ΔS_{ads}) upon adsorption dominates the enthalpic change (which in some cases is endothermic, $\Delta H_{ads} > 0$), resulting in a negative free energy change of adsorption ($\Delta G_{ads} = \Delta H_{ads} - T\Delta S_{ads}$). Considerable effort has been devoted to the understanding of the structure of the adsorbed proteins (8–26), but it is difficult to obtain direct structural information from most available experimental techniques. Nevertheless, Walton et al. (17,18) reported a decrease in the α-helical content of the desorbed albumin molecules and concluded that during adsorption the protein molecules undergo major conformational changes.

As expected, electrostatic interactions also strongly influence the patterns of protein adsorption (8–16, 27–41). For instance, Cosgrove et al. (34), among others, observed that with an increase in surface charge on positively charged polystyrene latex particles, the amount of proteins adsorbed decreased. They further concluded that "flat" configurations were attained by the highly charged polyelectrolyte upon adsorption on highly charged positive or negative surfaces.

---

* *Journal of Colloid and Interface Science*, Vol. 125, No. 2, p. 365, October 1988. Republished with permission.

Of course, one can control protein adsorption by changing the pH and ionic strength of the solution (8–16, 23, 34–41). For example, Norde et al. (8–16) have shown that protein adsorption is a strong function of pH, with a maximum in the amount adsorbed at pH values near the isoelectric point of the protein. They explained this result in terms of the effect which the charge of the protein molecule has on its configuration and in terms of the repulsion between the adsorbed molecules. Furthermore, they emphasized that the adsorption patterns of albumin were similar on adsorbent surfaces that differ both in their electrical charge and in their hydrophobicity. Consequently, they concluded that the adsorption behavior is determined mainly by the structural changes of the albumin molecule. On the other hand, Penners et al. (38), who investigated collagen adsorption, observed that in this case adsorption was mainly dependent on the surface energetic properties of the adsorbent and much less on pH and ionic strength.

It has been often noted that the affinity between a given protein and an adsorbent increases with the hydrophobicity of the surface and that proteins desorb more easily from hydrophilic than from hydrophobic surfaces (25, 37, 39–46). In contrast, Wojciechowski et al. (47) concluded that the affinity of the protein for an adsorbent cannot be correlated with the hydrophobicity of the latter. Baszkin (48) carried out experiments with albumin, γ-globulin, and fibrinogen on a series of relatively hydrophobic surfaces of various degrees of polarity. He concluded that the ratio of the polar and dispersion components of the work of adhesion W_A^p / W_A^d determines the degree of affinity of the protein for the adsorbent and that maximum affinity occurs when W_A^p / W_A^d approaches unity. Andrade (49) and Ruckenstein and Gourisankar (50) concluded that neither the total surface free energy of the solid nor the fractional contributions of the solid's component surface free energy should be considered indicators of its ability to adsorb proteins. Rather, the interfacial free energy between the solid and water is a more natural parameter which should be related to protein adsorption. The larger the latter, the greater the adsorption. On this basis a surface energetic criterion of biocompatibility of surfaces was suggested (50).

Protein adsorption has often been found to follow a Langmuir or modified Langmuir isotherm, having a finite initial slope and reaching a plateau after a critical protein concentration, in spite of the fact that the necessary conditions for the validity of this kind of isotherm are not satisfied (8–17, 39–40, 51–53). The plateau values were often found to be not too far from those of the close-packed monolayer of native molecules in a side-on or end-on orientation. On the other hand, multilayer adsorption has also been noted (38, 54–58). The amount adsorbed can vary significantly depending on whether adsorption is carried out under flow or static conditions, and on the scheme adopted for rinsing the unbound or loosely bound protein fractions. Walton (18) noted that protein layers adsorbed on a surface under flow conditions exhibit characteristics different from those formed under static conditions.

Regarding the reversibility of protein adsorption, it has been concluded that complete reversibility does not exist (5, 9), because the protein molecules attach to multiple sites on the adsorbent. Further, as suggested by Norde et al. (10), it is useful to distinguish among reversibilities toward (i) dilution of the solution, (ii) changes in pH and ionic strength, and (iii) exchange against a displacer protein which has higher affinity for the surface. Adsorbed proteins could often be released by changing the solution properties (such as pH, ionic strength, and addition of a displacer), while the dilution of the solution did not cause desorption. Some authors noted that protein adsorption is irreversible to dilution (5, 6, 31, 34, 40). Many others noted, however, a partial reversibility of the adsorbed protein molecules (2, 23, 56–64), indicating that some of the molecules are strongly adsorbed on the surface, while others are only loosely bound. Burghardt et al. (57) identified at least three desorbable fractions and attributed the observed partial reversibility to the multiplicity of the bound layers.

In summary, there are many variables which affect the adsorption of proteins onto synthetic surfaces. The results obtained in the literature regarding several aspects of protein adsorption are sometimes contradictory and confusing. The present investigation constitutes an attempt to reexamine protein adsorption in terms of somewhat better defined conditions. The role of the surface energetics in determining the amount of adsorbed protein and the role of the pH and ionic strength are emphasized. Adsorption equilibrium and adsorption kinetics have been investigated by employing a radiolabeling technique. Results indicate that loosely bound protein multilayers build up on the

EXPERIMENTAL

MATERIALS

The experimental details of the preparation of thin coatings of polymeric materials have been described previously (65). Table I summarizes some surface properties of the solids employed in the present study. The solids were equilibrated in buffers of given pH and salt concentration for a time sufficient to allow the equilibration of the adsorbent with the solvent, prior to the adsorption experiments.

Bovine serum albumin (BSA), essentially fatty acid free (Sigma Chemical Co., St. Louis, MO), was used without further purification or treatment. The protein was radiolabeled with ^{125}I, by the chloramine-T method, using $Na^{125}I$ of 100 mCi/ml (Amersham (code IMS.30), Arlington Heights, IL). Aliquots of 5 μl of 1 mg /ml BSA solution containing 0.01 M sodium phosphate buffer (pH 7.4) were introduced into a polystyrene tube to which, in addition, 20 μl of 0.05 M sodium phosphate buffer (pH 7.4) was added. The iodination was initiated by introducing 5 μl (0.5 mCi) of $Na^{125}I$ solution and 10 μl of 1 mg/ml chloramine-T solution to the protein solution. The reaction mixture was continuously agitated for 20 s with a stirrer, after which the iodination reaction was terminated by the addition of a *meta*-bisulfite solution that reacts with ^{125}I. Further, 100 μl of transfer buffer (10 mg/liter NaI and 100 mg/liter sucrose in 100 ml of 0.01 M sodium phosphate buffer) was introduced to the mixture. A small column (0.9 × 12 cm) of Sephadex G-75 was employed to separate the labeled protein from the unreacted iodide. Only the fraction of labeled proteins free of the unreacted iodide was subsequently employed. The gel filtration medium was equilibrated in the phosphate buffer and packed into the column. The column was saturated with protein before use, in order to minimize the adsorption of labeled protein. This saturation was accomplished by passing 5.5 ml of protein solution (20 mg/liter BSA solution) through the column. The iodination reaction mixture was transferred to the separation column and eluted with 0.01 M sodium phosphate buffer (pH 7.4). The labeled protein eluted first while the remaining reactants and unwanted by-products eluted later. The specific activity of the resulting labeled protein was about 50 mCi/mg. The labeled protein

TABLE I
Characteristics of Solid Surfaces Employed in This Study

		$\theta_{O(W)}$	Hydrophobicity/hydrophilicity	Description
(a)	Siliconized glass I	29	Hydrophobic	Prepared from Glassclad 6C (Petrarch Systems, Inc.)
(b)	Siliconized glass II	55	Hydrophobic	Prepared from Glassclad 18 (Petrarch Systems, Inc.)
(c)	PMMA	91	Intermediate	Prepared from Lucite (E.I. du Pont de Nemours & Co., Inc.)
(d)	Hydrogel I	145	Hydrophilic	Prepared from Hydron WD (Hydro Med Sciences, Inc.)
(e)	Hydrogel II	152	Hydrophilic	Prepared from Hydron N (Hydro Med Sciences, Inc.)
(f)	Glass	165	High surface free energy	Microscope glass slide, sodaline glass (General Medical Scientific)

was collected and stored in a freezer as a stock solution until use and used within 2 weeks. Sodium phosphate buffer (pH 7.4, ionic strength $= 0.01\ M$) was employed in the majority of the experiments. For other pH values, sodium phosphate buffer was utilized in the pH range 5.5–8.0, and acetic acid buffer in the pH range 3.8–5.5 (at an ionic strength of 0.0l M). The ionic strength was adjusted by adding sodium chloride to the sodium phosphate buffer.

Surface Characterization

Traditionally, polymeric surfaces have been considered immobile and their surface energetic properties have been assumed to be independent of the environment (polar or nonpolar) in which they were immersed. In reality, however, they rearrange their surface structure in response to their environment. Recently, contact angle procedures for the estimation of the equilibrium values of the polar $\left(\gamma_S^p\right)$ and dispersion $\left(\gamma_S^d\right)$ components of the surface free energies of polymeric surfaces equilibrated in both polar and nonpolar environments (66, 67) have been proposed. Since four quantities are of interest, namely $\gamma_{S(O)}^d$, $\gamma_{S(O)}^p$, $\gamma_{S(W)}^d$, and $\gamma_{S(W)}^p$, the characterization procedure involves the measurement of four initial wetting angles. The subscripts O and W indicate that the environment was a hydrocarbon (oil) or water. The surface energetic characteristics of the adsorbent surfaces thus obtained are summarized in Table II. The estimated polar components of the surface free energies show that the surfaces differ considerably in hydrophobicity and that the polar components are larger for the solids with polar moieties in their molecular structure equilibrated in polar environments. The interfacial free energies between the solids and water are also included in Table II. Since octane was employed as a probe fluid, it is clear from Ref. (66, Eq. [8]) that the wetting angle $\theta_{O(W)}$ acquires its maximum value of 180° when $\gamma_{S(W)}^p = 50.8\,\mathrm{dyn/cm}$. Consequently, polar components which are greater than 50.8 dyn/cm cannot be determined by this method. For this reason, it is difficult to characterize a highly polar material, such as glass (solid (f)) by this method. The angle of 165° measured in this case is too near the upper limit for the values listed in Table II for glass to be accurate. One may note, however, that this limitation could be overcome by using probe liquids which have very high surface tension and high interfacial free energy with water (such as mercury and gallium) (68). For reasons

---

TABLE II

Estimated Values of the Solid Surface Free Energies, Their Polar and Dispersion Components and Interfacial Free Energies for Solid Surfaces of Different Hydrophilicities, Equilibrated in Both Nonpolar and Polar Environments (dyn/cm)

	Surfaces	$\gamma_{S(O)}^p$	$\gamma_{S(W)}^p$	$\gamma_{S(O)}^d$	$\gamma_{S(W)}^d$	$\gamma_{S(O)}$	$\gamma_{S(W)}$	$\gamma_{S(O)W}$	$\gamma_{S(W)W}$
(a)	Siliconized glass 1	0.2	0.2	28.7	42.7	28.9	42.9	45.1	48.1
(b)	Siliconized glass II	0.4	2.3	33.2	34.4	33.6	35.8	43.4	32.9
(c)	PMMA	7.0	13.2	43.0	29.9	50.0	43.1	23.7	12.9
(d)	Hydrogel I	27.5	42.0	47.0	20.3	74.5	62.3	8.3	0.4
(e)	Hydrogel II	33.7	45.0	44.5	22.5	78.2	67.5	5.8	0.2
(f)	Glass[a]	39.7	49.0	40.2	20.1	79.9	69.1	3.5	0.1

Note: $\gamma_{S(O)}$ and $\gamma_{S(W)}$ are the surface free energies of the solid in the nonpolar and polar environments, respectively. The superscripts p and d refer to the polar and dispersion components. $\gamma_{S(O)W}$ and $\gamma_{S(W)W}$ are the interfacial free energies between the solid equilibrated in a nonpolar medium and water and between the solid equilibrated in water and water, respectively.

[a] See comments in the text.

Adsorption of Proteins onto Polymeric Surfaces of Different Hydrophilicities

already discussed in Ref. (68), the interfacial free energy between glass and water is expected to be relatively large.

Protein Adsorption Experiments

In order to perform equilibrium experiments, preequilibrated samples were first introduced into polystyrene tubes which contained 1 ml of sodium phosphate buffer (pH 7.4, ionic strength = 0.01 M), before being exposed to the protein solution. Any air bubbles which would adhere to the sample were removed by allowing the samples to cross the air/buffer interface several times. Small aliquots (30 μl) of the labeled BSA solution from the stock solution were added to several 1-ml unlabeled BSA solutions of known concentrations. These mixtures of labeled and unlabeled proteins were then introduced into the tubes. After the protein solution remained in contact with the samples for 20 h at room temperature, the adsorption was terminated by removing the solution from the tube by vacuum suction (in order to avoid contact of the sample with the protein solution/air interface where some denaturation can take place). Since some solution adhered to the samples, it was necessary to rinse them with at least 2 ml of buffer before counting the radioactivity of the total adsorbed proteins. The samples were further rinsed gently until the radioactivity of the surface remained constant. The amount of adsorbed proteins was determined by gamma radiation counting, using a gamma counter (Beckman Gamma 4000). Further desorption of protein molecules from the samples was also determined by allowing the preadsorbed protein films to desorb in 400 ml of fresh buffer in a new container until another constant radioactivity of the surface was measured.

In the adsorption kinetics experiments, samples were exposed to 1 mg/ml solution mixtures of labeled and unlabeled proteins for varying periods of time. Other experimental details are the same as those for the previously described equilibrium experiments.

In the experiments in which the pH was varied, the solids were equilibrated in the corresponding buffer before the addition of protein. Protein mixtures of 1 mg/ml concentration were made by introducing labeled and unlabeled proteins in the corresponding buffer containing the sample. The sample was further exposed to the protein solution of the desired pH for 20 h at room temperature. After 20 h, the adsorbed amounts were determined as described earlier.

The effect of ionic strength was investigated by exposing the samples to 1 mg/ml protein solutions of different ionic strengths at pH 7.4. The protein solutions of different ionic strength were prepared by adding labeled protein solutions to unlabeled protein solutions of sodium phosphate buffer (pH 7.4, ionic strength = 0.01 M) with NaCl concentrations varying between 0 and 0.5 M. After the protein molecules were allowed to adsorb for 20 h, the amount adsorbed was determined as in the previous experiments.

RESULTS

PROTEIN ADSORPTION ISOTHERM

The adsorption isotherms of albumin on various solid surfaces are presented in Figs, la–f. The results illustrate that the total amount of adsorbed protein (denoted by A) increases with increasing protein concentration and does not reach a plateau value even at concentrations of 10 mg/ml. The amounts that remained adsorbed (i) after rinsing with buffer until a constant value was reached (denoted by B) and (ii) after static desorption in 400 ml of fresh buffer (denoted by C) are also plotted (see the next section for a more detailed examination of desorption). The experimental error in the amounts adsorbed was less than 5%.

The figures also show that the greatest amount of adsorption occurred on the hydrophobic surfaces (solids (b) and (c)), and significantly less adsorption occurred on the hydrophilic hydrogel surfaces (solids (d) and (e)). However, the most hydrophobic surface (solid (a)) adsorbed a smaller amount than solids (b) and (c). Significant adsorption was also observed on glass (solid (f)), which is a high surface free energy material.

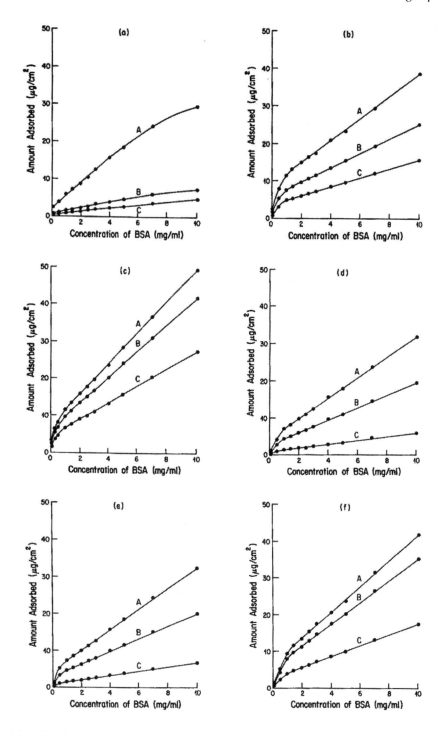

FIG.1. Adsorption isotherms for BSA at room temperature on different adsorbent surfaces in sodium phosphate buffer (pH 7.4, ionic strength = 0.01 M). Adsorption time = 20 h. (a) Siliconized glass I, (b) siliconized glass II, (c) PMMA, (d) hydrogel I, (e) hydrogel II, and (f) glass. The total amount of adsorbed protein is denoted A. The amounts of protein which remained adsorbed (i) after rinsing with buffer until a constant value is reached and (ii) after static desorption are denoted B and C, respectively.

PROTEIN DESORPTION

In Fig. 2, the amount of protein that remained adsorbed after rinsing with various amounts of buffer is plotted against the amount of rinsing buffer. Rinsing with at least 2 ml buffer was necessary to remove the radioactive protein solution that adhered to the solid (use of amounts less than 2 ml led to surface radioactivities which were one to two orders of magnitude larger). Rinsing with 10 to 30 ml of buffer removed the loosely bound proteins. There was no further decrease in the amount adsorbed with any subsequent rinsing. From Fig. 3, one can see that additional loosely bound protein molecules were desorbed when the solids were immersed, after the rinsing with 2 ml of buffer, in 400 ml of fresh buffer for up to 400 h. The amount which remained adsorbed reached another constant value after approximately 300 h.

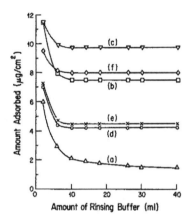

FIG. 2. The amount of BSA adsorbed on different adsorbent surfaces. The preadsorbed samples were rinsed mildly with different amounts of sodium phosphate buffer (pH 7.4, ionic strength = 0.01 M). (a) Siliconized glass I (△), (b) siliconized glass II (□), (c) PMMA (▽), (d) hydrogel I (○), (e) hydrogel II (×), and (f) glass (◊).

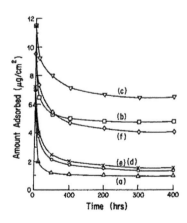

FIG. 3. Kinetics of static protein desorption from various surfaces. The preadsorbed samples were immersed in 400 ml of sodium phosphate buffer (pH 7.4, ionic strength = 0.01 M) for different periods of time, (a) Siliconized glass I (△), (b) siliconized glass II (□), (c) PMMA (▽), (d) hydrogel I (○), (e) hydrogel II (×), and (f) glass (◊).

Protein Adsorption Kinetics

The results regarding the kinetics of adsorption are presented in Figures 4a–f. Plateau values were reached after 10 to 20 h. In Fig. 5, the ratios of the adsorbed protein remaining after rinsing with flowing buffer (B) to the amount initially adsorbed (A) are plotted for the samples used in the investigation of adsorption kinetics (Fig. 4)

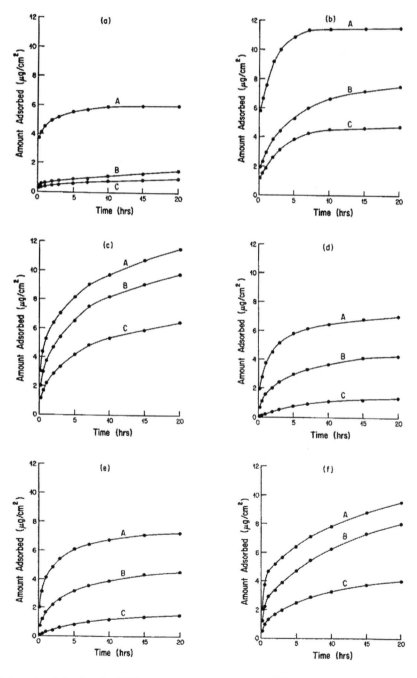

FIG. 4. Adsorption kinetics for BSA at room temperature on different adsorbent surfaces in sodium phosphate buffer (pH 7.4, ionic strength = 0.01 M). Concentration of BSA solution = 1 mg/ml. (a) Siliconized glass I, (b) siliconized glass II, (c) PMMA, (d) hydrogel I, (e) hydrogel II, and (f) glass.

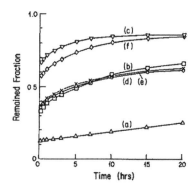

FIG.5. Release of adsorbed BSA from the solid surfaces utilized in the adsorption kinetics study (Fig. 4). The remaining fractions were expressed as *B/A*. (a) Siliconized glass I (Δ), (b) siliconized glass II (□), (c) PMMA (∇), (d) hydrogel I (○), (e) hydrogel II (×), and (f) glass (◊). The time on the abscissa is the adsorption time.

The Effect of pH

The effect of the pH of the solution is presented in Fig. 6, where the amount of protein adsorbed from a solution containing 1 mg/ml of protein after 20 h is plotted against pH. The curves exhibit a maximum between pH 4.5 and 5. One may note that the isoelectric point of BSA is approximately 4.8 and that the adsorption maximum does not coincide exactly with the isoelectric point of the protein. The ratio of the adsorbed molecules remaining after rinsing with flowing buffer to the amount initially adsorbed versus pH is plotted in Fig. 7. The fraction retained was slightly smaller at lower pH. Significant differences between the various solids indicate that adsorption depends on the properties of both the adsorbent and the protein.

THE EFFECT OF IONIC STRENGTH

The effect of the ionic strength on the adsorption from protein solutions in sodium phosphate buffer (pH 7.4) at different NaCl concentrations for a protein concentration of 1 mg/ml is examined

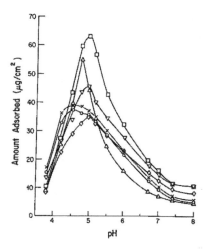

FIG.6. Adsorption of BSA at room temperature on different adsorbent surfaces as a function of pH in acetic acid buffer (pH 3.8 to 5.5, ionic strength = 0.01 M) and sodium phosphate buffer (pH 5.5 to 8.0, ionic strength = 0.01 M). Adsorption time = 20 h. Concentration of BSA solution = 1 mg/ml. Siliconized glass I(Δ), siliconized glass II (□), PMMA (∇), hydrogel I (○), hydrogel II (×), and glass (◊).

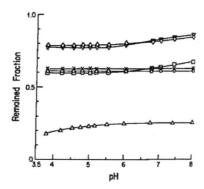

FIG.7. Release of adsorbed BSA from the solid surfaces utilized in the pH study (Fig. 6). The remaining fractions were expressed as *B/A*. Siliconized glass I (△), siliconized glass II (□), PMMA (▽), hydrogel I (○), hydrogel II (×) and glass (◊).

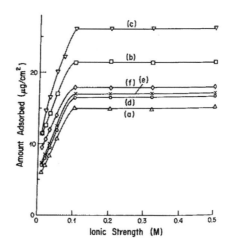

FIG.8. Adsorption of BSA at room temperature on different adsorbent surfaces as a function of ionic strength in sodium phosphate buffer (pH 7.4, ionic strength = 0.01 *M*). Adsorption time = 20 h. Concentration of BSA solution = 1 mg/ml. Ionic strengths were adjusted by addition of NaCl. (a) Siliconized glass I (△), (b) siliconized glass II (□), (c) PMMA (▽), (d) hydrogel I (○), (e) hydrogel II (×), and (f) glass (◊).

in Fig. 8. An increase in the ionic strength from 0.01 to 0.1 *M* resulted in increases in the amount adsorbed for all the solids. A further increase in ionic strength, however, did not have any additional effect on adsorption.

DISCUSSION

Protein Adsorption Isotherm

A protein molecule attaches to the surface of the solid via numerous segments. Therefore the free energy of adsorption is large and the protein molecules do not readily desorb when the system is diluted with solvent. Changes in pH or ionic strength, however, affect the interaction forces between protein and adsorbent and therefore can lead to desorption. Similarly, proteins which adsorb with larger free energy changes can displace the already adsorbed proteins.

The adsorption of proteins from aqueous solutions on surfaces is entropy driven (10). The increase in entropy is a result both of the freedom gained by the water molecules that result via the partial dehydration of the protein and surface and of the structural changes that occur in the less rigid,

Adsorption of Proteins onto Polymeric Surfaces of Different Hydrophilicities 123

partially dehydrated, protein molecules. The partially dehydrated proteins interact with the surface so as to optimize their hydrophobic and hydrophilic interactions. Only a fraction of the molecule is, however, attached to the surface; the remaining fraction dangles in the solution as loops and tails, interacting with the water molecules. The interactions with the surface generate a more extended molecule which because of the loops and tails and free bond rotation has more entropic freedom than the hydrated (more rigid) protein in solution. The increase in enthalpy produced by dehydration can be in some cases largely compensated by the decrease in enthalpy caused by the interactions with the surface. Consequently, the two positive entropic contributions, resulting from dehydration and from structural modifications produced by adsorption can indeed constitute the driving force for adsorption. Of course, the two enthalpic changes may not exactly compensate each other and therefore the enthalpy of adsorption can be endothermic. However, the entropy increase can be sufficiently large to compensate for the positive adsorption enthalpy. The adsorption process is also affected by the surface charges of the adsorbent and protein. Repulsive or attractive double-layer interactions are thus generated. In order to optimize the interactions, a protein with a net negative charge may expose positive patches to a negatively charged surface. However, and probably more importantly, the charge of the protein affects its configuration. A protein in solution is more extended when its net charge is higher and more globular when its net charge is smaller. The configuration has major effects on the amount of protein adsorbed, since a more globular molecule needs fewer sites for adsorption than a more extended molecule needs. Therefore the likelihood of adsorption is greater for a more globular configuration. In addition, there are repulsive interactions between the adsorbed charged molecules which hinder adsorption. The charge of the protein depends upon the difference between the pH and the isoelectric point of the protein. For pH values greater than the isoelectric point the net charge of the protein is negative, while for smaller values the net charge is positive. Thus, the amount adsorbed reaches a maximum at the isoelectric point because at this pH the configuration of the protein is more globular and there are no repulsive interactions between the adsorbed molecules.

One can visualize protein adsorption as follows. The first layer of protein is strongly adsorbed onto the adsorbent due to the above-mentioned thermodynamic driving force (largely entropic in nature). If the interfacial free energy between solid and water is high, the adsorbed proteins interact strongly with the solid surface. This adsorbed layer alters both the van der Waals and the electrostatic interactions between proteins and adsorbent. Additional layers are likely to adsorb more weakly and hence can desorb more easily than the first layer.

If the model of a close-packed monolayer is adopted, then the amount adsorbed should be between 0.21 (side-on adsorption) and 1.57 $\mu g/cm^2$ (end-on adsorption) (31). The amounts adsorbed for all the solids investigated were, however, greater than the above values. It is therefore clear that in the present experiments multilayer adsorption occurred. This is in contrast with some results in the literature which were explained in terms of monolayer adsorption (51–53). One may attribute this to the static adsorption and the mild rinsing scheme which were utilized in the present investigation.

In order to correlate the experimental results, the surface can be characterized by the interfacial free energy between the solid and the solvent (water), $\gamma_{S(w)w}$. For a given protein, the amount adsorbed is expected to be greater when the above interfacial free energy is greater. The interfacial free energy can be calculated by using the expression (66–69)

$$\gamma_{S(w)w} = \left\{ \left(\gamma_{S(w)}^d \right)^{1/2} - \left(\gamma_w^d \right)^{1/2} \right\}^2$$

$$+ \left\{ \left(\gamma_{S(w)}^p \right)^{1/2} - \left(\gamma_w^p \right)^{1/2} \right\}^2. \tag{1}$$

Equation [1] shows that, in order to obtain a low value for the solid—water interfacial free energy, the values of the individual surface free energy components (polar and nonpolar) of the solid should approach those of their respective counterparts of water. One should note that Eq. [1]

involves two simplifying assumptions: (i) the surface free energy can be decomposed in the sum of dispersion and polar components; and (ii) both the dispersion and the polar components of the work of adhesion can be expressed as geometric means. Even though other interactions such as specific acid—base interactions are not considered and the geometric mean approximation for the polar interactions may not be completely accurate, this approach will be used since a better one is not available.

The strength of the interactions between two bodies, a protein (p) and a solid (S) located in water (W), can be evaluated via the Hamaker constant A_{pWS},

$$A_{pWS} = -16\pi d_0^2 \Delta F_{pWS}(d_0),$$ [2]

where d_0 is the equilibrium distance between protein and surface (which is taken to be approximately 1.6 Å) and $\Delta F_{pWS}(d_0)$ is the change in free energy when the protein and solid located in water are brought from infinite to equilibrium distance. Hence $\Delta F_{pWS}(d_0)$ is given by the expression

$$\Delta F_{pWS}(d_0) = \gamma_{S(w)p(w)} - \gamma_{p(w)w} - \gamma_{S(w)w}.$$ [3]

(Note that the constant in Eq. [2] has the value 16 because the repulsion term from the Lenard–Jones potential was included in the calculations. If this term is not included, the value of the constant is 12.) The surface free energy components of the protein in its hydrated state were taken from the literature (9) ($\gamma_{p(w)}^p = 40.8$ dyn/cm, $\gamma_{p(w)}^d = 30.8$ dyn/cm, and $\gamma_{P(W)} = 71.6$ dyn/cm). These values are employed to calculate the corresponding interfacial free energies ($\gamma_{S(w)p(w)}$ and $\gamma_{p(w)w}$) by means of the expressions

$$\gamma_{S(w)p(w)} = \left\{ \left(\gamma_{S(w)}^d\right)^{1/2} - \left(\gamma_{p(w)}^d\right)^{1/2} \right\}^2$$
$$+ \left\{ \left(\gamma_{S(w)}^p\right)^{1/2} - \left(\gamma_{p(w)}^p\right)^{1/2} \right\}^2,$$ [4]

and

$$\gamma_{p(w)w} = \left\{ \left(\gamma_{p(w)}^d\right)^{1/2} - \left(\gamma_w^d\right)^{1/2} \right\}^2$$
$$+ \left\{ \left(\gamma_{p(w)}^p\right)^{1/2} - \left(\gamma_w^p\right)^{1/2} \right\}^2.$$ [5]

Utilizing the polar and dispersion components of the surface free energies of the polymeric solids from Table II as well as Eqs. [1]–[5], one can then calculate the interfacial free energies and the Hamaker constant. The obtained values are listed in Table III.

On the basis of Table III, the solids employed in the experiments can be divided into two classes. Solids (a), (b), and (c) have large Hamaker constants, while solids (d) and (e) have Hamaker constants which are one order of magnitude smaller. One would therefore expect that the amounts adsorbed will be larger for the first rather than for the second class. This indeed occurs for solids b and c but not for solid a. Perhaps the constant value taken for d_0 and/or the approximations involved in the evaluations of the components of the surface free energies and of the interfacial free energy are responsible for this anomaly.

One may also note that the ratio, B/A, between the amounts of protein that remain after rinsing and the initially adsorbed amounts (Figs. 1a–f), was generally greater for the solids with a higher initial adsorption. This indicates that stronger interactions occur between the protein and those solids.

TABLE III

Estimated Values of the Hamaker Constants, A_{pWS} (ergs), and of ΔF_{pWS} (ergs/cm²)

	Surfaces	$\gamma^p_{S(W)}$	$\gamma^d_{S(W)}$	$\gamma_{S(W)}$	$\gamma_{S(W)p}$	$\gamma_{S(W)w}$	ΔF_{pWS}	A_{pWS} [×10^{14}]
(a)	Siliconized glass I	0.2	42.7	42.9	36.3	48.1	−13.1	16.8
(b)	Siliconized glass II	2.3	34.4	35.8	23.8	32.9	−10.4	13.4
(c)	PMMA	13.2	29.9	43.1	7.6	12.9	−6.6	8.5
(d)	Hydrogel I	42.0	20.3	62.3	1.1	0.4	−0.6	0.8
(e)	Hydrogel II	45.0	22.5	67.5	0.8	0.2	−0.7	0.8
(f)	Glass	See text						

Note: The properties of the solid (subscript S) and of the protein (subscript p) are those obtained for surfaces equilibrated in a water environment. The subscript S(W) denotes that the solid was equilibrated in water environment. The superscripts p and d refer to the polar and dispersion components of the surface free energy. The surface and interfacial free energies are expressed in dynes/cm.

PROTEIN DESORPTION

Our experimental results indicate multilayer adsorption. The first layer is strongly bound, while the others are more loosely bound. Some layers could be removed by gentle rinsing and others by introducing the specimen in fresh buffer free of proteins. These layers are more loosely bound because their interactions with the protein-covered solid are weaker and because of the repulsive interactions between the strongly bound layer and the additional layers. From a surface energetic point of view, the adsorption of the first layer has already decreased the interfacial free energy between the solid and the water. The thermodynamic driving force for the adsorption of the additional layers is therefore decreased. One should note that the amount of protein which remains adsorbed after static desorption in fresh buffer (Fig.3) corresponds to multilayer (2, 5–3 layers) adsorption for the solids b, c, f (4–6.5 $\mu g/cm^2$) and to monolayer adsorption for the solids a, d, e (1–1.5 $\mu g/cm^2$). One should also note that the more tightly bound fractions could be further removed if a more severe desorption scheme (e.g., flowing buffer at a significant flow rate) is employed.

REARRANGEMENT OF THE ADSORBED PROTEIN

One may observe from Fig. 5 that the samples which were exposed to the BSA solution for longer periods of time had a stronger affinity for the BSA molecules. This suggests that the affinity of the adsorbed protein molecules for the surface increases in time. Hence the protein molecule as well as the polymeric surface rearrange after adsorption. This may also indicate that the water molecules that are intercalated between the adsorbed protein and the adsorbent are extruded in time, thus leading to strong interactions between proteins and solid.

THE EFFECT OF pH

At the isoelectric point of BSA (pH $\simeq$ 4.8), the protein molecules carry zero net charge. At pH values above the isoelectric point, the BSA molecules are negatively charged, while below the isoelectric point they are positively charged. The surface charge of the molecule generates electrostatic double-layer interaction and also affects the configuration of the proteins. The configuration is globular at the isoelectric point and because of intramolecular electrostatic repulsion becomes more extended for higher or lower values of pH. The decreased adsorption for pH values removed from the isoelectric point are due to the larger number of neighboring sites which are needed for the adsorption of more extended molecules as well as to the electrostatic double-layer repulsion between the charged adsorbed proteins. Indeed, the likelihood of adsorption is decreased when a larger number of neighboring sites are involved in adsorption. This is a surface exclusion effect of entropic origin. Even when the adsorbent is positively charged, the adsorption of the negatively charged BSA molecules (pH > isoelectric point) can invert the charge characteristics of the adsorbent thus changing the electrical characteristics of the surface (charge inversion). This leads to electrostatic double-layer repulsion for the additional adsorbed proteins.

Protein adsorption should not be interpreted solely on the basis of the nature of the protein. The differences in the values of the amount adsorbed as well as the position where the maximum is located for different adsorbent surfaces (Fig. 6) indicate that the charge and surface energetic characteristics of the adsorbent, in addition to the intrinsic properties of BSA, contribute to the adsorption process.

THE EFFECT OF IONIC STRENGTH

The results regarding the effect of the ionic strength on protein adsorption indicate that the surface charge of the proteins plays a role in adsorption. At low ionic strength (0.01 M), the charges of the protein molecules act fully and bring about both a greater contribution of the electrostatic forces to the total interaction and a more extended protein conformation (14, 34, 35). As noted earlier, this will decrease adsorption (Fig. 8). On the other hand, at higher ionic strengths, the surface charge of the protein molecules becomes increasingly shielded, resulting in reduced repulsive double-layer interaction as well as in a more globular configuration. Therefore, a greater amount of protein will adsorb on the surface. Since a further change in the amount adsorbed at ionic strengths above 0.1 M did not occur, charge does not appear to have a major effect for ionic strengths greater than 0.1 M. Both the shielding of the electric field and the binding of counterions may account for this observation.

CONCLUSIONS

Protein adsorption onto solid surfaces under static conditions leads to multilayer adsorption. However, only the first (or sometimes a few) layer(s) is strongly bound. The more loosely bound layers can be removed by rinsing or by desorption in fresh protein-free solutions. Three fractions of adsorbed proteins with different adsorption strengths have been measured experimentally. A more detailed explanation for the thermodynamic driving force of protein adsorption is proposed. One concludes that the entropic gains resulting from dehydration and from the configurational changes produced by adsorption are mainly responsible for protein adsorption. It is suggested that the interfacial free energy between the solid and the solvent (water) is an appropriate parameter to correlate protein adsorption data, and that higher adsorption should occur on a surface with a larger interfacial free energy.

The surface characteristics of six solids have been determined experimentally and used in the interpretation of the data. The Hamaker constants for the interactions between protein and solid have been evaluated with the aid of the surface characteristics. The maximum in protein adsorption that occurs in the amount adsorbed as a function of pH is explained as a result of the configurational changes induced by changes in pH. At the isoelectric point the molecule is more globular, but it becomes more extended for larger or smaller values of pH because of intramolecular electrostatic repulsion. A more extended molecule needs a larger number of neighboring free sites on the solid surface and therefore adsorbs with greater difficulty than a globular molecule does. In addition, the electrostatic repulsion between the adsorbed molecules hinders further adsorption of the more extended molecules. The results of a kinetic investigation indicate that time-dependent changes in the conformation of the adsorbed protein molecules and of the polymeric surface occur on the solid surface toward a higher affinity of the molecule for the solid surface.

REFERENCES

1. Andrade, J. D., *in* "Surface and Interfacial Aspects of Biomedical Polymers" (J. D. Andrade, Ed.), Vol. 2, p. 1. Plenum, New York, 1985.
2. Walton, A. G., and Koltiskio, B., *in* "Biomaterials: Interfacial Phenomena and Applications" (S. L. Cooper and N. A. Peppas, Eds.), ACS Advances in Chemistry Series, Vol. 199, p. 245. Amer. Chem. Soc., Washington, DC, 1982.
3. Horbett, T. A., *in* "Biomaterials: Interfacial Phenomena and Applications" (S. L. Cooper and N. A. Peppas, Eds.), ACS Advances in Chemistry Series, Vol. 199, p. 233. Amer. Chem. Soc., Washington, DC, 1982.
4. Brash, J. L., *in* "Interaction of the Blood with Natural and Artificial Surfaces" (E. W. Salzman, Ed.), p. 37. Dekker, New York, 1981.
5. MacRitchie, F., *Adv. Protein Chem.* **32**, 283 (1978).
6. Morrissey, B. W., *Ann. N. Y. Acad. Sci.* **283**, 50 (1977).
7. Lyklema, J., *Colloids Surf.* **10**, 33 (1984).
8. Norde, W., *Polym. Sci. Technol. B* **12**, 801 (1980).
9. Norde, W., *Adv. Colloid Interface Sci.* **25**, 267 (1986).
10. Norde, W., MacRitchie, F., Nowicka, G., and Lyklema, J., *J. Colloid Interface Sci.* **112**, 447 (1986).
11. Koutsoukos, P. G., Norde, W., and Lyklema, J., *J. Colloid Interface Sci.* **95**, 385 (1983).
12. Van Dulm, P., and Norde, W., *J. Colloid Interface Sci.* **91**, 248 (1983).
13. Van Dulm, P., Norde, W., and Lyklema, J., *J. Colloid Interface Sci.* **82**, 77 (1981).
14. Norde, W., *J. Colloid Interface Sci.* **66**, 257 (1978).
15. Norde, W., *J. Colloid Interface Sci.* **66**, 266 (1978).
16. Norde, W., *J. Colloid Interface Sci.* **66**, 295 (1978).
17. Soderquist, M. E., and Walton, A. G., *J. Colloid Interface Sci.* **75**, 386 (1980).
18. Walton, A. G. and Maenpa, F. C, *J. Colloid Interface Sci.* **72**, 265 (1979).
19. Park, K., Albrecht, R. M., Simmons. S. R., and Cooper, S. L., *J. Colloid Interface Sci.* **111**, 197 (1986).
20. Morrissey, B. W., and Han, C. C., *J. Colloid Interface Sci.* **65**, 423(1978).
21. Morrissey, B. W., *J. Colloid Interface Sci.* **46**, 152 (1974).
22. Gendreau, R. M., Leininger, R. I., Winters, S., and Jakobsen, R. B., *in* "Biomaterials: Interfacial Phenomena and Applications" (S. L. Cooper and N. A. Peppas, Eds.), ACS Advances in Chemistry Series, Vol. 199, p. 371. Amer. Chem. Soc., Washington, DC, 1982.
23. Watkins, R. W., and Robertson, C. R., *J Biomed. Mater. Res.* **11**, 915 (1977).
24. Kochwa, S., Brownell, M., Rosenfield, R. E., and Wasserman, L. R., *J. Immunol.* **99**, 981 (1967).
25. Jönsson, U., Lunström, I., and Rönnberg, I., *J. Colloid Interface Sci.* **117**, 127 (1987).
26. Ratner, B. D., *J. Colloid Interface Sci.* **83**, 630 (1981).
27. Hogt, A. H., Gregonis, D. E., Andrade, J. D., Kim, S. W., Danken, J., and Feijen, J., *J. Colloid Interface Sci.* **106**, 289 (1985).
28. Hattori, S., Andrade, J. D., Hibbs, J. B., Gregonis, D. E., and King, R. N., *J. Colloid Interface Sci.* **104**, 72(1985).
29. Van Wagenen, R. A., Coleman, D. L., King, R. N., Triolo, P., Brostrom, L., Smith, L. M., Gregonis, D. E., and Andrade, J. D., *J. Colloid Interface Sci.* **84**, 155 (1981).
30. Van Wagenen, R. A., and Andrade, J. D., *J. Colloid Interface Sci.* **76**, 305 (1980).
31. Schmitt, A., Varoqui, R., Uniyal, S., Brash, J. L., and Pusineri, C., *J. Colloid Interface Sci.* **92**, 25 (1983).
32. Ikada, Y., Iwada, H., Horii, F., Matsunaga, T., Taniguchi, M., Suzuki, M., Taki, W., Yamagata, S., Yonekawa, Y., and Handa, H., *J. Biomed. Mater. Res.* **15**, 697 (1981).
33. Falb, R. D., Leininger, R. I., and Crowley, J. P., *Ann. N.Y. Acad. Sci.* **283**, 396 (1977).
34. Cosgrove, T., Obey, T. M., and Vincent, B., *J. Colloid Interface Sci.* **111**, 409 (1986).
35. Zsom, R. L. J., *J. Colloid Interface Sci.* **111**, 434 (1986).
36. Bagchi, P., and Birnbaum, S. M., *J. Colloid Interface Sci.* **83**, 460 (1981).
37. MacRitchie, F., *J. Colloid Interface Sci.* **38**, 484 (1972).
38. Penners, G., Priel, Z., and Silberberg, A., *J. Colloid Interface Sci.* **80**, 437 (1982).
39. Shirahama, H., and Suzawa, T., *J. Colloid Interface Sci.* **104**, 416 (1985).
40. Suzawa, T., Shirahama, H., and Fujimoto, T., *J. Colloid Interface Sci.* **86**, 144 (1982).
41. Mizutani, T., *J. Colloid Interface Sci.* **82**, 162 (1981).
42. Jönsson, U., Ivarsson, B., Lundström, and Berghem, L., *J. Colloid Interface Sci.* **90**, 148 (1982).
43. Fletcher, M., and Pringle, J. H., *J. Colloid Interface Sci.* **104**, 5 (1985).
44. Cuypers, P. A., Hermens, W. T., and Hemeker, H. C, *Ann. N.Y. Acad. Sci.* **283**, 77 (1977).
45. Paul, L., and Sharma, C. P., *J Colloid Interface Sci.* **84**, 546 (1981).

46. Dillman, W. J., Jr., *J. Colloid Interface Sci.* **44**, 221 (1973).
47. Wojciechowski, P., Hove, P. T., and Brash, J. L., *J. Colloid Interface Sci.* **111**, 455 (1986).
48. Baszkin, A., and Lyman, D. J., *J. Biomed. Mater. Res.* **14**, 393 (1980).
49. Andrade, J. D., *J. Med. Instrum.* **7**, 110 (1973).
50. Ruckenstein, E., and Gourisankar, S. V., *J. Colloid Interface Sci.* **101**, 436 (1984) (Section 1.2 of this volume).
51. Lee, R. G., and Kim, S. W., *J. Biomed. Mater. Res.* **8**, 251 (1974).
52. Brash, J. L., and Samak, Q. M., *J. Colloid Interface Sci.* **65**, 495 (1978).
53. Lok, B. K., Cheng, Y. L., and Robertson, C. R., *J. Colloid Interface Sci.* **91**, 104 (1983).
54. Ito, Y., Sisido, M., and Imanishi, Y., *J. Biomed. Mater. Res.* **20**, 1157 (1986).
55. Nyilas, E., Chi, T. H., and Herzlinger, G. A., *Trans. Amer. Soc. Artif. Inter. Organs* **20**, 480 (1974).
56. Gendreau, R. M., and Jakobsen, R. J., *J. Biomed. Mater. Res.* **13**, 893 (1979).
57. Burghardt, T. P., and Axelrod, D., *Biophys. J.* **33**, 455 (1981).
58. Bagnall, R. D., *J. Biomed. Mater. Res.* **11**, 947 (1977).
59. Beissinger, R. L., and Leonard, E. F., *J. Colloid In terface Sci.* **85**, 521 (1982).
60. Beissinger, R. L., and Leonard, E. F., *Trans. Amer. Soc. Artif. Intern. Organs* **27**, 225 (1981).
61. Beissinger, R. L., and Leonard, E. F., *ASAIO J.* **3**, 160 (1980).
62. Royce, F. H., Ratner, B. D., and Horbett, T. A., *in* "Biomaterials: Interfacial Phenomena and Applications" (S: L. Cooper and N. A. Peppas, Eds.), ACS Advances in Chemistry Series, Vol. 199, p. 453. Amer. Chem. Soc., Washington, DC, 1982.
63. Stupp, S. I, Kauffman, J. W., and Carr, S. H., *J.Biomed. Mater. Res.* **11**, 237 (1977).
64. Eberhart, R. C., Lynch, M. E., Bilge, F. H., Wissinger, J. F., Munro, M. S., Ellsworth, S. R., and Quattrone, A., *in* "Biomaterials: Interfacial Phenomena and Applications" (S. L. Cooper and N. A. Peppas, Eds.), ACS Advances in Chemistry Series, Vol. 199, p. 293. Amer. Chem. Soc., Washington, DC, 1982.
65. Lee, S. H., and Ruckenstein, E., *J. Colloid Interface Sci.* **120**, 529 (1987) (Section 2.5 of this volume).
66. Ruckenstein, E., and Lee, S. H., *J. Colloid Interface Sci.* **120**, 153 (1987) (Section 2.4 of this volume).
67. Ruckenstein, E., and Gourisankar, S. V., *J. Colloid Interface Sci.* **107**, 488 (1985) (Section 2.1 of this volume).
68. Gourisankar, S. V., and Ruckenstein, E., *J. Colloid Interface Sci.* **109**, 591 (1986) (Section 1.3 of this volume).
69. Kaelble, D. H., *J. Adhesion.* **2**, 66 (1970).

3 Experiments on Wetting by Catalysts

Eli Ruckenstein and Gersh Berim

INTRODUCTION TO CHAPTER 3

Chapter 3 contains experimental studies of various catalysts supported on the surfaces of various materials. In particular, the following problems were discussed: (a) wetting of alumina films by platinum (Sec. 3.1, 3.2, 3.3), iron (Sec. 3.4), palladium (Sec. 3.5), and silver (Sec. 3.6) crystallites and the theory of such kinds of wetting (Sec. 3.7); (b) The rate of formation of the desired product and the selectivity for composite catalysts wherein small particles of zeolite are distributed non-uniformly within a large particle of the more porous silica-alumina (Sec. 3.8); (c) The behavior of nickel crystallites supported on titania (rutile) upon heating in H_2 and O_2 atmospheres (Sec. 3.9); (d) The physical and chemical changes that occurred in model catalysts formed of small crystallites of Ni, Co, or Fe supported on thin, electron transparent films of nonporous alumina due to heating in chemical atmospheres having compositions in the range encountered in the steam-reforming reaction (Sec. 3.10).

3.1 Redispersion of Platinum Crystallites Supported on Alumina–Role of Wetting[*]

Eli Ruckenstein and Y. F. Chu[†]

Faculty of Engineering and Applied Sciences, Slate University of New York at Buffalo, Buffalo, New York 14214.

Corresponding Author

[†] Y. F. Chu is associated with Mobil Research and Development Corporation, Paulsboro Laboratory, Paulsboro, New Jersey 08066.

Received November 6, 1978; revised February 26, 1979

INTRODUCTION

Supported metal catalysts used for high-temperature hydrocarbon processing or auto emission control generally deactivate with time. The deactivation is due among other causes to the agglomeration of the metal crystallites. Rejuvenation by redispersion of the crystallites of the aged catalyst has been reported (1 – 10). Kearby *et al.* (3) have shown that crystallites larger than 20 nm of the aged catalyst redisperse to crystallites smaller than 5 nm. The regeneration process generally involves heating the aged catalyst in an oxygen-containing atmosphere such as O_2 or steam, or heating the aged catalyst in an oxygen atmosphere followed by heating in a hydrogen atmosphere. The aged catalysts are sometimes pretreated with chlorine. The conditions and the mechanism or mechanisms by which redispersion of Pt crystallites occurs are not yet clear. Johnson and Keith (2) reported redispersion of Pt crystallites in a commercial Pt on alumina catalyst when heated in dry air around 510°C or in 1 atm O_2 around 580°C. Temperatures higher than 580°C caused sintering. Johnson and Keith believe that below a critical temperature a Pt-alumina complex existing in an oxidized state is stable and is responsible for redispersion. This critical temperature is near 510°C at 0.2 atm O_2 and near 580°C at 1 atm O_2. Ruckenstein and Pulvermacher (11) suggested that redispersion is sometimes caused by the loss of particles to the substrate. Flynn and Wanke (12) reported a redispersion in a commercial Pt on alumina catalyst when heated in O_2 around 600°C and 1 atm. They observed sintering at temperatures lower than 500°C or higher than 600°C. Flynn and Wanke explain redispersion as a result of the loss of metal oxide molecules from the crystallites to the support.

A direct examination of the changes in size, shape, and position of each crystallite on the support by transmission electron microscopy (TEM) during various stages of treatment can give some insight concerning redispersion. A model catalyst of Pt evaporated on thin films of alumina is suitable for this purpose. Ruckenstein and Malhotra (10) carried out such an experiment and observed that redispersion of crystallites occurs when aged specimens were heated at 500°C in air. The purpose of the present paper is to examine the behavior of Pt crystallites of the model catalyst during alternating heating in O_2 and H_2. The present experiments show that no appreciable sintering or redispersion occurs when the specimens are initially heated at 750°C either in pure O_2 or in pure H_2. However, after several cycles of alternating heating in O_2 and H_2 redispersion or sintering start to occur.

---

[*] *Journal of Catalysis* 59, 109–122 (1979). Republished with permission.

Redispersion of Platinum Crystallites Supported on Alumina

Thereafter, redispersion and sintering can be produced periodically by changing the chemical atmosphere from an oxidizing to a reducing one. The crystallites redisperse during the oxidizing step and sinter during the reducing step. Heating in wet N_2 has the same effect on redispersion as heating in O_2. When heating has occurred in wet H_2, some crystallites increased in size and others decreased in size, but the behavior of the Pt crystallites appeared to be independent upon size. Redispersion during heating in O_2 and sintering during heating in H_2 are explained below in terms of wetting.

EXPERIMENTAL

Preparation of alumina substrate. Thin films of alumina were prepared by anodization of aluminum foils as described previously (*10, 13*). The anodization solution contained 3% wt tartaric acid with a pH adjusted to 5.5. The voltage was kept at 20 V and the anodization lasted 1 min. This produced a nonporous layer of alumina of about 30 nm thickness on the aluminum foil. After the remaining aluminum foil was dissolved by amalgamation, the thin films of alumina thus obtained were transferred to a large amount of distilled water and picked up on gold grids. They were allowed to dry and were kept in an evacuated dessicator. The alumina films supported on the grid were then heated in a furnace in air at 850°C for 72 hr. During this treatment the amorphous nonporous alumina transformed to γ-Al_2O_3. The substrate did not change its morphology and structure upon further prolonged heating in O_2 or H_2. No reaction between Pt and alumina substrate has been detected by electron diffraction throughout the experiment.

Preparation of Pt supported on alumina substrate. Pt films 1 nm thick were deposited on the alumina substrate by evaporating the corresponding amount of Pt wire in a vacuum unit. Pt wire of 99.999% purity was purchased from Ventron Corporation. The pressure in the vacuum unit during the evaporation was always maintained below 10^{-6} Torr and the substrate was always at room temperature. Electron diffraction has shown that the component evaporated is Pt.

Heat treatment. The heating was carried out in a quartz tube located in a furnace. The temperature of the furnace was controlled to within ±5°C. The specimen introduced in the quartz tube was evacuated by a mechanical pump. After a pressure of 10^{-2} Torr was reached, the mechanical pump was switched off and nitrogen was allowed to flow through the tube at room temperature for 2 hr. The temperature was then raised to the chosen temperature in about 1 to 2.5 hr. As soon as the chosen temperature was reached, N_2 was replaced by H_2 or O_2. The experiments were performed at atmospheric pressure and the flow rate of the gas was 35 cm^3/ min. The gases used were all > 99.999% purity and were obtained from Linde Division, Union Carbide Corporation. Traces of water were removed by passing the gas through 5 A molecular sieves before entering the unit. In the experiments performed in wet N_2 or wet H_2, N_2 or H_2 were respectively passed through a bubbler tube containing H_2O at room temperature before entering the unit. After the desired period of heating, the gas was replaced by N_2 and the quartz tube containing the specimen was cooled to room temperature in about 1.5 hr.

Each specimen was subjected to heat treatment in the following succession:

(i) Thin films of alumina with Pt deposited on them were heated in the quartz tube in flowing H_2 at 750°C and 1 atm for 12 hr. During this treatment, the Pt film broke up and formed Pt islands. The resulting particle size distribution is considered as the initial size distribution.

(ii) The specimen was heated in O_2 at 750°C and 1 atm for 6 to 48 hr.

132   Wetting Experiments

(iii) The specimen was then heated in H_2 at 750°C and 1 atm for 6 to 48 hr.

(iv) The specimen was subjected to several cycles of alternating heating in oxygen and hydrogen until significant changes were observed in the sizes of the crystallites. Each cycle consisted of heating in O_2 at 750°C and 1 atm for 1 to 3 hr followed by heating in H_2 at the same temperature and pressure for 2 to 6 hr.

(v) The specimen was heated alternatively in O_2 and H_2 at 1 atm and a specified temperature.

By proper rotation, the same regions of the specimen were inspected by TEM after heating so that the change in size, shape, and position of each particle could be detected. The electron microscope has a resolution of about 1 nm. More than 2000 particles were measured on magnified electron micrographs of different regions of at least four specimens which had been heated similarly. The average crystallite size is calculated as $\sum n_i \tilde{D}_i^3 / \sum n_i \tilde{D}_i^2$, where n_i is the number of crystallites which have a diameter between D_i and $D_i + \Delta D_i$ and $\tilde{D}_i = D_i + (\Delta D_i / 2)$.

Similar final results were obtained when similar specimens were heated continuously for the same total time and exposed only once to air and electron beams for examination in TEM. The effect of contamination due to repetitive contacts with air or exposure to electron beams of the specimens during successive experiments was therefore minimal.

RESULTS

EFFECT OF O_2 AND H_2 ON THE REDISPERSION

The average size of Pt crystallites after each heating stage is given as a function of time in Table 1. It is clear from runs 1 through 4 in Table 1 that no significant changes in the size of the crystallites occur when the specimen is heated in O_2 for 48 hr followed by heating in H_2 for 48 hr. The average size of the crystallites increased from about 3 to 4 nm. This confirms a previous observation (13) that pure O_2 or pure H_2 does not appreciably stimulate sintering of Pt crystallites deposited by evaporation on thin films of γ-Al_2O_3. No redispersion of Pt crystallites occurred during this treatment. [It should be noted that, in contrast to the model catalyst used here, appreciable sintering of the crystallites occurs when industrial catalysts are heated in O_2 at high temperature (6, 8, 9, 12). The difference in the method of preparation and the nature of the substrate probably explains the different behaviors.]

The specimens were then heated alternatively in O_2 or H_2 for five cycles at 750°C and 1 atm (runs 5 through 14, Table 1). The five cycles of heating caused a small increase in the average size of the Pt crystallites from about 4 to 5 nm. The electron micrograph of Fig. 1a shows the morphology of Pt on the alumina substrate after run 14. The black particles are Pt and the substrate is γ-Al_2O_3. The grain boundaries of γ-Al_2O_3 are readily observed. After the specimen was heated in O_2 at 750°C and 1 atm for 3 hr (run 15, Table 1) the crystallites decreased in size (Fig. 1b). Furthermore, most of the crystallites from Fig. 1a can no longer be detected in Fig. 1b. Compare, for example, regions A through F and those crystallites marked by numbers or arrows of both Figs. 1a and b. The undetected crystallites did not evaporate from the specimen since, as will be shown later, Pt reappears during heating in H_2 (see Fig. 2a). The crystallites were not detected by TEM because the initial crystallites redispersed as particles with sizes smaller than the resolution of TEM. As explained below, platinum is probably oxidized to platinum oxide which better wets the substrate. Chu and Ruckenstein have provided evidence by electron diffraction that platinum is oxidized in an oxygen atmosphere (14).

The specimen of Fig. 1b was further heated in flowing H_2 at 850°C and 1 atm for 8 hr. The Pt crystallites reappeared and grew to an average size of about 15 nm (run 16, Table 1). They are relatively thin but have clear boundaries and irregular shapes on the substrate (see Fig. 2a).

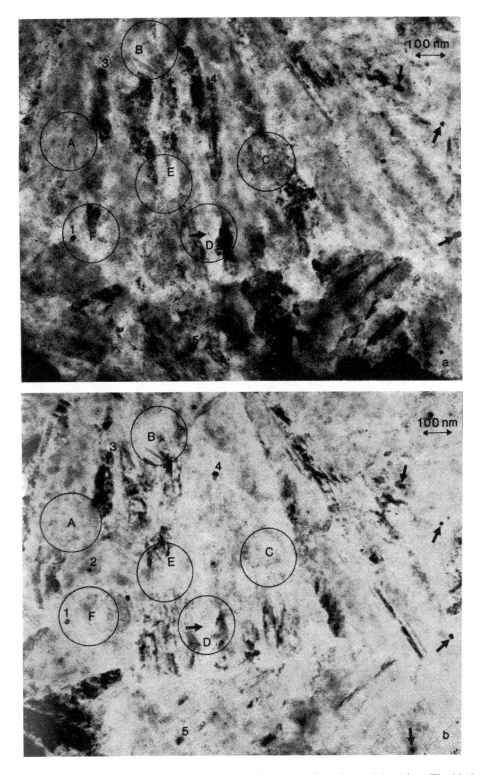

FIG. 1. Transmission electron micrographs showing the same region of a model catalyst. The black particles correspond to Pt and the substrate is γ-Al_2O_3. (a) After five cycles of alternating heating in O_2 and H_2 at 750°C and 1 atm, (b) an additional 3 hr heating in O_2 at 750°C and 1 atm after (a). Most of the Pt crystallites in Fig. 1b are not detected by TEM.

The dumbell-shape morphology of some Pt crystallites shown by arrows in Fig. 2a gives indication that migration and coalescence occurred.

After several cycles of alternating heating in O_2 and H_2 (runs 5 through 14, Table 1), redispersion or sintering could be produced periodically by heating in O_2 or H_2, respectively. Similar results were obtained on at least four specimens. The electron micrographs shown in the present paper are taken from the same region of one of the specimens.

TABLE 1
Effect of Experimental Conditions on the Average Size of Pt Crystallites Supported on γ-Al_2O_3

		Treatment conditions				
Run	Figure	Atmosphere (1 atm)	Temperature (°C)	Time (hr)	Average size (nm)	Comments
				a	3.3	
1		O_2	750	0–6	3.3	No change
2		O_2	750	6–48	3.3	No change
3		H_2	750	48–54	3.7	Slow sintering
4		H_2	750	54–96	3.9	Slow sintering
5		O_2	750	96–99	4.0	No change
0		H_2	750	99–105	4.2	Slow sintering
7		O_2	750	105–108	4.2	No change
8		H_2	750	108–114	4.3	Slow sintering
9		O_2	750	114–117	4.3	No change
10		H_2	750	117–123	4.5	Slow sintering
11		O_2	750	123–126	4.5	No change
12		H_2	750	126–132	4.6	Slow sintering
13		O_2	750	132–135	4.6	No change
14	la	H_2	750	135–141	4.6	No change
15	1b	O_2	750	141–144	b	Redispersion
16	2a	H_2	850	144–152	15.4	Sintering
17	2b	O_2	750	152–154	b	Redispersion
18		O_2	750	154–166	b	No change
19	2c	H_2	750	166–167	3.7	Sintering
20		H_2	850	167–168	10.6	Sintering
21		H_2	850	168–169	11.5	Sintering
22		O_2	750	169–170	b	Redispersion
23		H_2	550	170–180	b	No change
24		H_2	750	180–182	4.7	Sintering
25		H_2	850	182–183	5.7	Sintering
26		Wet H_2	750	183–186	b	Redispersion and sintering
27	3a	Wet H_2	750	186–192	b	Redispersion and sintering
28	3b	O_2	750	192–194	b	Redispersion
29	3c	H_2	750	194–196	4.2	Sintering
30		H_2	850	196–198	13.1	Sintering
31		Wet N_2	750	198–200	b	Redispersion
32		H_2	750	200–203	4.1	Sintering

[a] Fresh specimen after heating in H_2 at 750°C and 1 atm for 12 hr.

[b] Most Pt crystallites have sizes smaller than the resolution of TEM.

REDISPERSION AND SINTERING OF PT CRYSTALLITES

The specimen of Fig. 2a was further heated in O_2 at 750°C and 1 atm for 2 hr. Significant changes occur (see Fig. 2b). Comparing, for example, regions A to F in Figs. 2a and b, one can observe that most of the crystallites in Fig. 2a no longer appear in Fig. 2b. Again, Pt did not disappear from the specimen since it reappears as crystallites when heated in H_2 (Fig. 2c). The large crystallites 1 to 5 in Fig. 2a decrease in size during heating in O_2 (Fig. 2b). They also change their shapes. It may be noticed that crystallite 4 and 5 have smooth boundaries in Fig. 2a but sharp edges (indicated by arrows) in Fig. 2b. We intended to redisperse the remaining large crystallites further by additional heating in O_2 for 12 hr (run 18, Table 1). However, no significant change could be detected on the specimen. The small crystallites and/or the metal-oxide molecules undetected by TEM in Fig. 2b sintered and grew in size during heating in H_2 at 750°C and 1 atm for 1 hr (run 19, Table 1). A large number of small crystallites of about 4 nm appears (Fig. 2c). Regions A to F of Fig. 2 show the same regions of the specimen. It can be seen that most of the crystallites in Fig. 2c are new crystallites occupying positions where no crystallites were detected previously in Figs. 2a and b. Almost all the crystallites of Fig. 2c are of about 4 nm size, hence they are much smaller than the crystallites of Fig. 2a which are of about 15 nm size. The number of Pt crystallites observed in Fig. 2c is at least 25 times larger than that in Fig. 2a. The alternating heating in O_2 and H_2 was continued several times (Table 1). Redispersion occurred when the specimen was heated in O_2, and sintering occurred when the specimen was heated in H_2.

EFFECT OF WET H_2 OR WET N_2 ON REDISPERSION

After alternating heating in O_2 and H_2 the specimen was further heated in wet H_2 at 750°C and 1 atm for 3 hr (run 26, Table 1). Some crystallites increased in size and others decreased in size. However,

FIG. 2. Transmission electron micrograph showing the same region of the model catalyst. (a) An additional 8 hr of heating in H_2 at 850°C and 1 atm after 1b. Pt crystallites sinter and grow in size. (*Continued*)

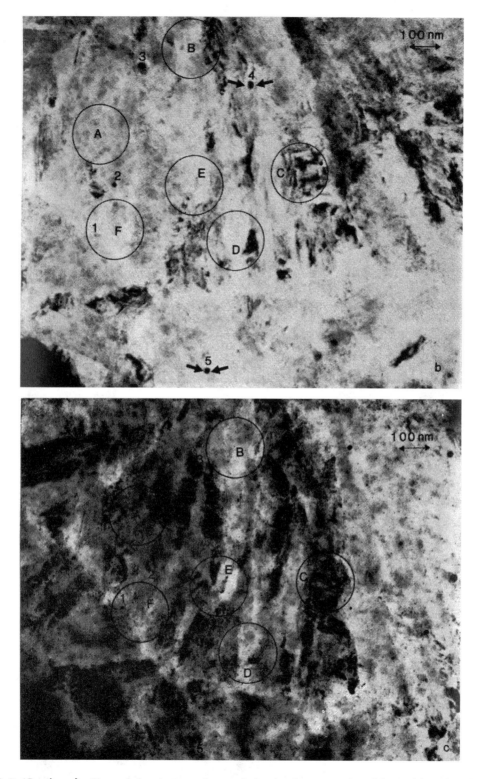

FIG. 2. (*Continued*) Transmission electron micrograph showing the same region of the model catalyst. (b) An additional 2 hr heating in O_2 at 750°C and 1 atm after 2a. Redispersion of Pt crystallites occurs. (c) An additional 12 hr of heating in O_2 and 1 hr heating in H_2 at 750°C and 1 atm after 2b. Pt crystallites sinter and reappear.

their behavior appeared to be independent of size. In addition it was found that redispersion of the Pt crystallites occurs during heating in wet N_2.

DISCUSSION

Two mechanisms appear to explain the redispersion observed in our experiments: (a) fracture of crystallites, and (b) spreading of platinum oxide (formed in the oxidizing atmosphere) over the surface of the substrate.

REDISPERSION BY FRACTURE

Crystallites 6, 7, and 8 of about 40 nm size in Fig. 3a are fractured into smaller particles in Fig. 3b. The crystallites resulting through fracture sintered after heating in H_2 (Fig. 3c). This fracture can be explained in terms of the stability theory developed by Ruckenstein and Dunn (*15*). Let us consider a thin film on a substrate and apply a small, spatially periodic displacement to its free interface. If the response of the system increases the effect of the perturbation, then the film may fracture. In the converse case the film is stable. The response of the system is determined by the surface diffusion of the atoms, which, in turn, is generated by a gradient of the chemical potential along the perturbed free interface of the film. The chemical potential is expressed in terms of the curvature of the interface, of the preexisting internal stresses in the film, and of the interaction forces between the metal atoms at the gas–solid interface and those of the film and substrate. The latter interactions have to be taken into account when the range of the interaction forces between one atom of the film and the entire substrate is larger than the thickness of the film. Surface diffusion can amplify,

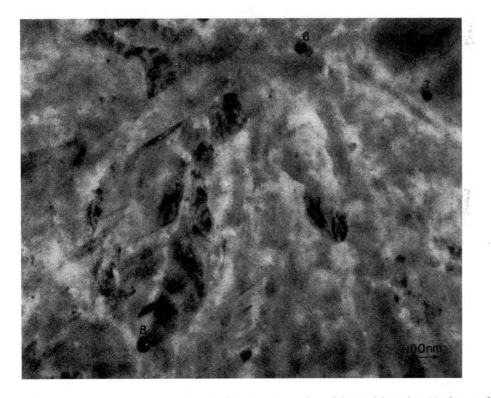

FIG. 3. Transmission electron micrographs showing the same region of the model catalyst (a) after run 27 (Table 1); *(Continued)*

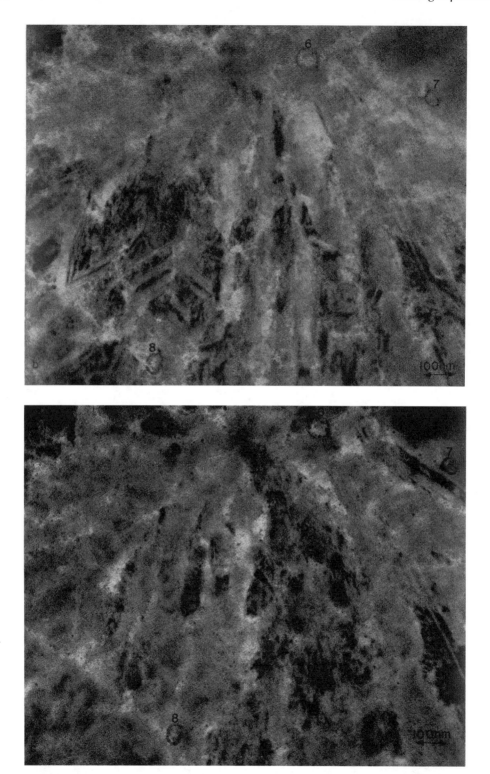

FIG. 3. (*Continued*) Transmission electron micrographs showing the same region of the model catalyst (b) an additional 2 hr of heating in O_2 at 750°C and 1 atm after 3a; (c) an additional 2 hr of heating in H_2 at 750°C and 1 atm after 3b.

under some conditions, the initial perturbation leading to fracture. The calculations predict a critical prestress required for fracture which decreases as the surface tension decreases. The theoretical considerations (*15*) are valid for thin films and can probably be extended only to sufficiently large crystallites. In the case of small crystallites, their shape and size are expected to affect stability.

Oxidation of platinum to platinum oxide generates an internal stress and decreases the film-gas surface tension. Fracture occurs when the internal stress is larger than the critical stress. Rapid cooling also generates internal stresses because of differences in the thermal expansion coefficients of the metal and substrate, and these can induce fracture. This kind of behavior was observed in references 10 and 14.

REDISPERSION BY SPREADING

Two kinds of spreading are distinguished here. In the first kind, no wetting angle can exist between crystallites and substrate; in the second kind a wetting angle is possible, but a two-dimensional fluid coexists with the crystallites. The thermodynamic origin of the two-dimensional fluid is examined at the end of this section.

For a crystallite in thermodynamic equilibrium on the substrate, the three surface tensions are related through Young's equation (Fig. 4)

$$\sigma_{gs} - \sigma_{ms} = \sigma_{mg} \cos\theta, \tag{1}$$

where σ is the surface tension (the subscripts gs, ms, and mg refer to the gas-substrate, metal-substrate, and metal-gas interfaces, respectively) and θ is the wetting angle. For very small crystallites the equilibrium problem is more complicated and has to be examined along the lines developed by Ruckenstein and Lee (*16*). For the sake of simplicity and because the present discussion is largely qualitative, Young's equation is used as the starting point.

In vacuum, the metals have high surface tensions. For instance, Pt has a surface tension of 2340 erg cm^{-2} at 1310°C; for Ni at 1250°C it is 1850 erg cm^{-2} (*17*). The oxides have a surface tension lower than that of the metals. [Since no values are available for platinum oxide, we mention that the surface tension of Ag at 932°C is a function of the partial pressure of oxygen, decreasing from about 1100 erg cm^{-2} at 10^{-6} atm to about 300 erg cm^{-2} at 1 atm (*18*)]. For this reason the metals do not wet the oxides. However, if the metal is oxidized, its surface tension decreases appreciably. If the entire crystallite is oxidized, the crystallite-substrate interfacial tension is also expected to decrease. In these conditions it is possible to have

$$\sigma_{gs} - \sigma_{ms} > \sigma_{mg}. \tag{2}$$

No wetting angle is, however, compatible with inequality (2), which leads to $\cos\theta > 1$, and therefore the crystallite has to spread over the entire surface available (*11*), ultimately as single molecules. When platinum oxide is reduced by heating in H$_2$, the crystallites form once again since platinum does not wet alumina and hence $|\sigma_{gs} - \sigma_{ms}| < \sigma_{mg}$.

The first five or six cycles of heating in O$_2$ and H$_2$ probably generate some porosity in the crystallites and, in this manner, enhance the oxidation of the smaller crystallites. They also produce some reconstruction of the substrate. Only after this initial process condition (2) is satisfied and can redispersion occur. The large crystallites are oxidized completely only near their leading edge where

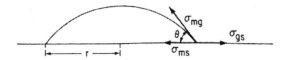

FIG. 4. Droplet on a substrate.

140 Wetting Experiments

they are sufficiently thin. Therefore, only that part of the larger crystallites spreads as molecules of platinum oxide over the substrate. The remaining part which does not spread is constituted of pure metal protected by an oxide layer. The sharp edges of the larger crystallites, observed in our experiments after redispersion during heating in O_2 (Fig. 2b), may be due to this process.

In the kind of spreading just discussed, no wetting angle can exist between crystallites and substrate. However, another kind of wetting can also explain the observed phenomena. In this case crystallites coexist with a thin layer of platinum oxide which can be a two-dimensional fluid, a bilayer, or a multilayer. In the present experiments one starts with platinum metal which, for reasons already mentioned, does not wet the substrate. A part of the platinum oxide formed during heating in oxygen will spread through the leading edge of the crystallite to wet the substrate as a two-dimensional fluid of molecules of platinum oxide. Reducing platinum oxide to Pt which does not wet the substrate, the two-dimensional fluid recondenses into metal crystallites, being either recaptured by the remaining crystallites or forming new ones.

It is interesting to observe that even in the second kind of wetting, platinum oxide crystallites smaller than a critical size have the tendency to wet the substrate completely. This is due to a Kelvin-type effect which generates in the smaller crystallites a higher dissolution surface pressure than in the larger ones. This dissolution surface pressure has to be added to σ_{gs} and, as long as the radius of the crystallite is small enough, can lead to the inversion of the inequality $\sigma_{gs} - \sigma_{ms} < \sigma_{mg}$ and hence to spreading (see also Appendix A).

A thermodynamic explanation of the origin of the two-dimensional fluid was provided in Ref. (*19*) using as a starting point the specific free energy of formation σ of a thin film on a substrate. If the film is thick, σ is given by the expression

$$\sigma\left(\infty\right) \equiv \sigma_{\infty} = \sigma_{mg} + \sigma_{ms} - \sigma_{gs}. \tag{3}$$

Spreading occurs if $\sigma_{\infty} < 0$. As soon as the range of the interaction forces between one atom of the film and the substrate becomes larger than the thickness of the film, the film can no longer be assumed macroscopic and the specific free energy of formation σ becomes a function of the thickness h of the film

$$\sigma\left(h\right) = \sigma_{\infty} + f\left(h\right), \tag{4}$$

where $f(h) \rightarrow 0$ when $h \rightarrow \infty$. If a Lenard-Jones potential is used for the interaction potential the following approximate expression is obtained for $f(h)$ (*19*):

$$f\left(h\right) = \frac{\alpha}{h^2} - \frac{\beta}{h^8}, \tag{5}$$

where α and β are constants. If $\sigma\,(h)$ is always negative, spreading occurs for all possible loadings of the substrate, whereas if $\sigma\,(h)$ is always positive no spreading occurs. When, however, $\sigma_{\infty} > 0$ but $\sigma\,(h) < 0$ for $h < h_0 \ll \infty$, then a thin film (probably a submonolayer) may coexist with the crystallites. A detailed examination of the problem with its implications for both sintering and redispersion is presented in Ref. (*19*).

CONCLUDING REMARKS

Several cycles of alternating heating in O_2 and H_2 at 750°C and 1 atm are needed before redispersion of Pt crystallites occurs. After these cycles of treatment, redispersion and sintering can be produced periodically by changing the chemical atmosphere from an oxidizing to a reducing one. Crystallites redisperse during the oxidation step and sinter during the reduction step. Wet N_2 has the same effect as O_2 as concerns the redispersion of Pt model catalyst. Redispersion is caused in

Redispersion of Platinum Crystallites Supported on Alumina

an oxygen atmosphere by spreading of platinum oxide over the surface of the substrate. Spreading occurs either because no wetting angle can exist between the oxidized crystallites and substrate or, most probably, because a two dimensional fluid of platinum oxide can coexist with the crystallites. There are conditions under which splitting of large crystallites can also occur. Sintering occurs during heating in H_2 because the oxide is reduced to metal and the metal does not wet the substrate. A thermodynamic explanation of the origin of the two-dimensional fluid is provided.

APPENDIX A

Following the procedure used to derive an equation for the effect of curvature on the vapor pressure, one obtains for the dissolution surface pressure p_s

$$p_s = p_{s\infty} \exp\frac{\delta}{r} \tag{A1}$$

where

$$\delta \equiv \sigma_e S_e / kT, \tag{A2}$$

$p_{s\infty}$ is the dissolution surface pressure for large crystallites, r is the radius of the crystallite–substrate interface, S_e is the surface area per molecule at the crystallite–substrate interface, k is Boltzmann's constant, T is the absolute temperature, and σ_e is the line tension in erg cm^{-1} at the leading edge of the crystallite. Further, denoting by $n_{s\infty}$ the number of molecules of platinum oxide per unit area of substrate in equilibrium with the large crystallites and assuming a two-dimensional ideal gas (although a two-dimensional van der Waals fluid is a more adequate assumption), one can write

$$p_{s\infty} = n_{s\infty} kT. \tag{A3}$$

Near the leading edge of the crystallite, on the substrate, there are at a given moment n molecules per cm^2. These molecules of platinum oxide generate a surface pressure nkT which opposes wetting.

Hence one can find a critical radius r_c below which spreading is possible from the equality

$$\sigma_{gs} + \left[n_{s\infty}\left(\exp\frac{\delta}{r_c} \right) - n \right] kT - \sigma_{ms} = \sigma_{mg}. \tag{A4}$$

Of course, only if the detachment of molecules during spreading maintains $r < r_c$ will spreading continue. Equation (A4) leads to

$$\frac{r_c}{\delta} = \left[\ln\frac{\sigma_{mg} + \sigma_{ms} - \sigma_{gs} + nkT}{n_{s\infty} kT} \right]^{-1}. \tag{A5}$$

Detachment of molecules will occur if the area concentration of molecules in equilibrium with a crystallite of radius r, $n_{s\infty} \exp(\delta/r)$, is larger than n. Because the surface tensions are of the order of a few hundred erg cm^{-2}, taking $\sigma_{mg} + \sigma_{ms} - \sigma_{gs} + nkT \approx 10^2$ erg cm^{-2} and $kT = 1.3 \times 10^{-13}$ erg, a relatively large value of $n_{s\infty} = 7 \times 10^{14}$ cm^{-2} is needed to obtain $r_c/\delta \approx 10$. No values are available for the line tension. Assuming $\delta \approx 10^{-7}$ cm results in a value of r_c of the order of 10^{-6} cm.

To be consistent the contribution σ_e/r_c of the line tension should be subtracted from the left hand side of Eq. (A4). This contribution is negligible for sufficiently large radii, but becomes important when the radius is sufficiently small. However, as already mentioned, Young's equation (corrected to include the contribution of the line tension) and, hence, Eq. (A4) are no longer valid for sufficiently small radii [16].

ACKNOWLEDGMENT

This work was supported by the National Science Foundation.

REFERENCES

1. Adler, S. F., and Keavney, J. J., *J. Phys. Chem.* **64**, 208 (1960).
2. Johnson, F. L., and Keith, C. D., *j. Phys. Chem.* **67**, 200 (1963).
3. Kearby, K. K,, Thorn, J. P., and Hinlieky, J. A., U.S. Patent, 3,134,732 (1964).
4. Brennan, H. M., Seelig, H. S., and Yander Harr. R. W., U.S. Patent, 3, 117,076 (1964).
5. Coe, R. H., and Randaltt, H. E., U.S. Patent. 3,278,419 (1965).
6. Jawarska-Galas, Z., and Wrzyszcz, J., *lut. Chem. Eng.* **6**, 604 (1966).
7. Kraft, M., and Spindler, H., *4th Int. Cong. Catal. Moscow,* 1252 (1968).
8. Schwarzenbek, E. F., and Turkevieh, J., U.S. Patent 3,400,073 (1968).
9. Weller, S. W., and Montagna, A. A., *J. Gated.* **20**, 394 (1971).
10. Ruckenstein, E., and Malhotra, M. L., *J. Catal.* **41**, 303 (1976).
11. Ruckenstein, E., and Pulvermaeher, B., *J. Catal* **29**, 224 (1973).
12. Flynn, P. C., and Wanke, S. E., *J. Catal.* **34**. 390 (1974).
13. Chu, Y. F., and Ruckenstein, E., *J. Catal.* **55**, 281 (1978).
14. Chu, Y. F., and Ruckenstein, E., *Surface Sci.* **67**, 517 (1977).
15. Ruckenstein, E., and Dunn, C., *Thin Solid Films* **51**, 43 (1978) (Section 5.2 of Volume I).
16. Ruckenstein, E., and Lee, P. S., *Surface Sci.* **52**, 298 (1975) (Section 3.1 of Volume I).
17. Blakely, J. M., and Maija, P. S., *in* "Surfaces and Interfaces I" (J. J. Burke *et al,* Eds.), p. 325. Syracuse University Press, 1967.
18. Butiner, F. H., Funk, E. R., and Udin, H., *J. Phys. Chem.* **56**, 657 (1952).
19. Ruckenstein, E. Proceedings of the Meeting on "Growth and Properties of Metal Clusters," Villeurbanne, France, 1979; *J. Cryst. Growth* (in press).

3.2 Redispersion of Pt/Alumina via Film Formation[*]

I. Sushumna and Eli Ruckenstein

Department of Chemical Engineering, State University of New York, Buffalo, New York 14260

Received April 7, 1987; revised June 9, 1987

INTRODUCTION

Pt/Al_2O_3 catalysts have been widely used in a number of industries. These supported catalysts, however, gradually lose their activity. This is attributed in part to the gradual decline in the exposed metal surface area due to sintering. Sintering of supported metal catalysts has been extensively investigated with both model and industrial catalysts (see (1–3) for reviews). The approach to the more significant problem of the regeneration of the sintered industrial catalyst has been more or less empirical. However, especially in the case of supported Pt catalysts, a number of investigations (1–17) regarding regeneration have been carried out and the mechanisms involved have been debated. Ruckenstein and Malhotra (4) suggested that the catalyst redispersion was a result of particle splitting due to a buildup of strain in the particles. Subsequently Stulga et al. (9) reported that (in the case of a similar system under similar conditions) they did not observe particle redispersion. Gollob and Dadyburjor (10), on the other hand, observed redispersion which they attributed to particle splitting. Ruckenstein and Pulvermacher (11) had suggested earlier that redispersion could occur by emission of particles, and Fiedorow and Wanke (5) attributed the observed increase in dispersion (measured by H_2 chemisorption) to a similar cause, namely emission of atoms or molecules. Dautzenberg and Wolters (12) suggested that an oxygen treatment could recover only the fraction of platinum complexed with alumina during prior high-temperature H_2 treatment and that it could not actually redisperse the sintered large platinum crystallites. Lieske et al. (13) also argued that oxygen by itself cannot lead to redispersion and that redispersion is possible only in the presence of chloride ions. Straguzzi et al. (14) and Boumonville and Martino (15) also suggested that chlorine plays a role in the redispersion of alumina-supported platinum catalysts. Ruckenstein and Chu (16) reported that in the case of Pt/Al_2O_3, following a few cycles of alternate heating in H_2 and O_2 at 750°C, sintering in H_2 and redispersion in O_2 were observed. They suggested that the redispersion in O_2 may be due to the coexistence of a two-dimensional phase of single atoms, or of a multilayer film, with the crystallites. The concept of the coexistence of single atoms and a particulate phase was used also by Yao et al. (17) to explain sintering or redispersion. Evidence for the redispersion of large crystallites via extension, film formation, and its breakup as well as by splitting of crystallites was reported recently in the case of Fe/Al_2O_3 (18, 19) and Ni/Al_2O_3 (20) model catalysts.

We report here results with model Pt/Al_2O_3 catalysts heated alternately in H_2 and O_3 at temperatures in the range 300–750°C, which show that there is an alternation in the size, namely an apparent decrease in the size of the particles on heating in O_2 and a recovery of more or less the original size on subsequent heating in H_2. This size alternation continued over a number of cycles and was observed at temperatures between 500 and 700°C. The results are discussed in terms of the redispersion of the crystallites via the formation of disconnected individual films around the particles, which is a result of the strong interactions between crystallites and support.

---

[*] *Journal of Catalysis 108*, 77-96 (1987). Republished with permission.

EXPERIMENTAL

The method of preparation of model Pt/Al_2O_3 catalysts and the treatment procedure are presented in detail in Refs. (16, 21) and they are therefore not elaborated here. The term hydrogen refers to the as-received ultra-high-purity hydrogen (≤ 1 ppm O_2 and ≤ 3 ppm moisture) from Linde Division, Union Carbide Corp., and "purified hydrogen" refers to the above hydrogen purified further by passing it through a Deoxo unit (Engelhard Industries), followed by a column of silica gel, a bed of 15 wt% MnO on silica, and finally, through a column of 13x and a column of 5-A molecular sieves immersed in liquid nitrogen.

RESULTS

The results presented here have been obtained with the same samples for which the sintering results were described in Ref. (21). The experiments have been carried out with a number of specimens. The results obtained with five of them are presented in some detail.

1. A 2-nm initial film thickness sample was heated at 500°C for 8 h in purified hydrogen; this generated small particles of about 1.6 nm average size. In order to more easily identify the processes that occur, the size of the particles was increased further by heating the sample in H_2 at 600°C for 4 h followed by heating at 700°C for 5 h, at 800°C for 2 h, and finally again at 600°C for 2 h. This heat treatment led to relatively large particles of about 7 nm average size. On subsequent heating of the sample in O_2 at 300°C, there was a decrease in the size of all the particles. Additional heating in O_2 at progressively higher temperatures in steps of 50°C up to 600°C (2-h duration at each temperature) did not result in significant changes in either the size distribution or the shape of the particles. However, when the sample was subsequently heated in H_2 at 500°C, all the particles, small and large, increased significantly in size. Few particles disappeared, indicating that neither ripening nor migration and coalescence involving the detectable particles were responsible for the growth. Further heating in O_2 at 500°C resulted again in a decrease in the size of all the particles. Subsequently, when the sample was heated in H_2 at 500°C, the particles again increased in size and regained more or less the size they had prior to the oxygen treatment. Alternate heating in H_2 and O_2 at 500°C was carried out over six cycles and the above alternation in size was observed in each cycle (Fig. 1). A decrease in size on heating in O_2 and an increase in size on heating in H_2 were also observed when the same specimen was subsequently heated at 700°C.

2. To verify whether the observed alternations in size were caused by the gradual and prolonged heat treatment in O_2 between 300 and 600°C prior to the cycling in H_2 and O_2 at 500°C, another sample with a similar prehistory, but without the extended oxygen treatment between the initial size distribution and the cycling in H_2 and O_2 at 500°C, was investigated. During the first two cycles, the particles decreased in size only marginally on heating in O_2 and increased in size on heating in H_2. However, during the subsequent four cycles, the changes in size were more pronounced, even though the extent of size reduction was smaller than that for Sample 1. Perhaps the gradual heating in O_2 at progressively higher temperatures prior to the cycling in H_2 and O_2 enhanced the wetting of the substrate by the particles in the case of Sample 1. It is also to be noted that following the heating in O_2 at 500°C, the smallest of the detectable particles were still present on the substrate and did not appear to have disintegrated via molecular emission. In addition, following the heating in O_2, detectable films in the form of a halo around each particle could be seen (Fig. 2a). Subsequent to the above six cycles, the treatment in oxygen was carried out at higher temperatures. On heating in O_2 at 600°C few particles disappeared and the films around the particles could be seen even better than at 500°C (particles marked with

large arrows in Fig. 2c). Even the smallest particles remained (particles marked with small arrows in Figs. 2a–2d) and regained more or less their previous sizes on subsequent heating in H_2 at 500°C. A large number of crystallites considerably extended to form thin, light patches on the substrate on subsequent heat treatment in O_2 at 700°C (Regions 1,2, and 3 in Fig. 2e, for example). Some particles are no longer detectable. Not all the particles present before the last O_2 treatment were regained on subsequent heating in H_2 at 500°C. Smaller particles were generated in regions where the particles had extended considerably before (Regions 1–3 in Figs. 2e and 2f). Films around some particles could still be seen (particles marked with large arrows). A decrease in the number of detectable particles on the substrate surface during heating in O_2 may suggest the possibility of material loss via evaporation. Evaporation of Pt as PtO_2 in oxygen atmospheres at temperatures greater than 700°C has been suggested before (22–24). However, in light of the observations indicating the presence of detectable films around the particles on heating in O_2 and the generation of smaller particles in some regions where the particles had considerably extended before, it is more likely that a part of the material (Pt or PtO_x) diffused into or spread over the surface of the substrate. These probably happen due to the strong interactions between particles and substrate. Further, particles 0.8 nm in size or even smaller remained without disappearing through the various cycles of alternate heating in H_2 and O_2 at 500°C and also following the high-temperature (up to 700°C) O_2 treatments (particles marked with small arrows in Figs. 2a–2f), suggesting that molecular emission and/or evaporation played a minor role, if any.

3. A third sample of 2 nm initial film thickness and from the same batch as the preceding sample was heated in a slightly different sequence. Particles were generated as before by heating the Pt/alumina sample in purified H_2 at 500°C for 12 h and subsequently at 600°C for 5 h. Before the alternate heating in H_2 and O_2, Sample 1 described above was heated in O_2 between 300 and 600°C in steps of 50°C, without hydrogen treatments in between. If the particles were already oxidized at the lower temperature of 300°C, subsequent heat treatment of the oxidized sample in O_2 at higher temperatures would cause little change

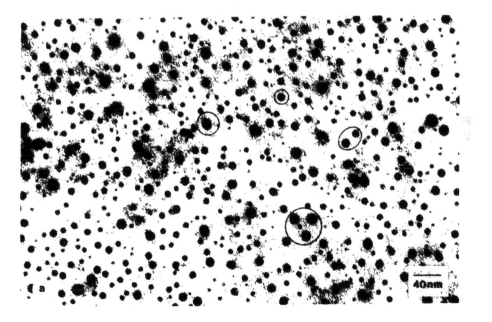

FIG. 1. Changes in particle size of a 2 nm initial film thickness Pt/Al_2O_3 sample on alternate heating in H_2 and O_2 at 500°C. The sample has had a long heating history prior to Fig. 1a (see text), (a) H_2, 500°C, 2 h.

(*Continued*)

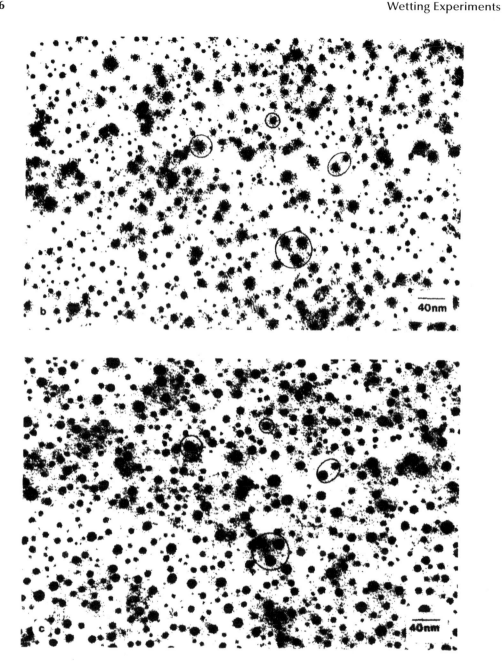

FIG. 1. (Continued) Changes in particle size of a 2 nm initial film thickness Pt/Al$_2$O$_3$ sample on alternate heating in H$_2$ and O$_2$ at 500°C. The sample has had a long heating history prior to Fig. 1a (see text), (b) O$_2$, 500°c, 2 h (or 16 h); (c) H$_2$, 500°C, 4 h.

as observed up to 600°C with Sample 1. In order to verify whether there is a threshold temperature beyond which film formation is detected following heating in O$_2$, it is appropriate to start with a reduced sample prior to the oxidation step at each temperature. Therefore, the present sample was also heated in O$_2$ in the range 300–600°C in steps of 50°C, but unlike in Sample 1, in between each oxidation step, this sample was heated in H$_2$ at 500°C in order to reduce the oxide to metal prior to the next oxidation. As reported earlier (21), except for some marginal sintering, mostly by the coalescence of neighboring particles, there was very little change in the size or shape of the particles up to the

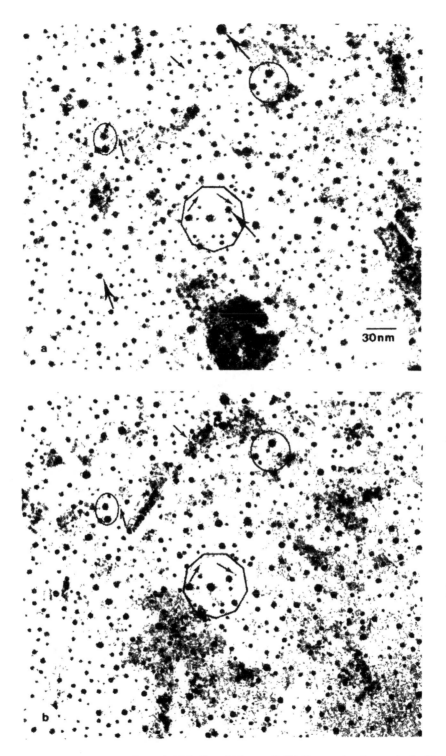

FIG. 2. Changes on heating a 2 nm initial film thickness Pt/Al$_2$O$_3$ sample alternately in H$_2$ and O$_2$. Micrographs for only a few cycles of alternate heating are shown. Prior to the following, the sample had undergone a few cycles of alternate heating in H$_2$ and O$_2$ (see text), (a) O$_2$, 500°C, 5 h; (b) H$_2$, 500°C, 2 h.

(*Continued*)

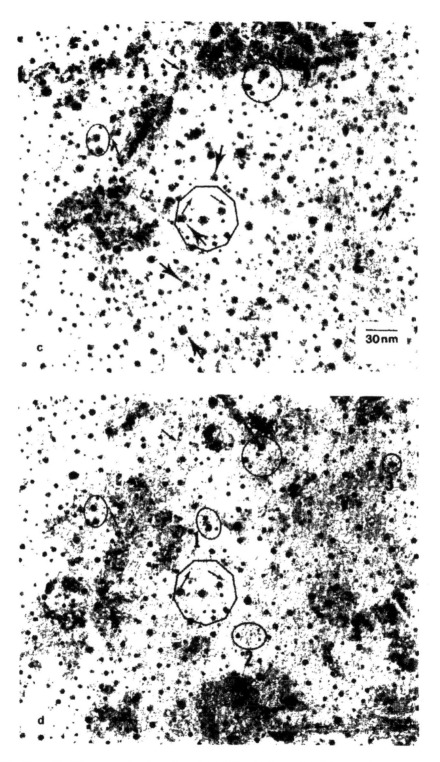

FIG. 2. (Continued) Changes on heating a 2 nm initial film thickness Pt/Al$_2$O$_3$ sample alternately in H$_2$ and O$_2$. Micrographs for only a few cycles of alternate heating are shown. Prior to the following, the sample had undergone a few cycles of alternate heating in H$_2$ and O$_2$ (see text), (c) O$_2$, 600°C, 5 h; (d) H$_2$, 500°C, 2 h.

(Continued)

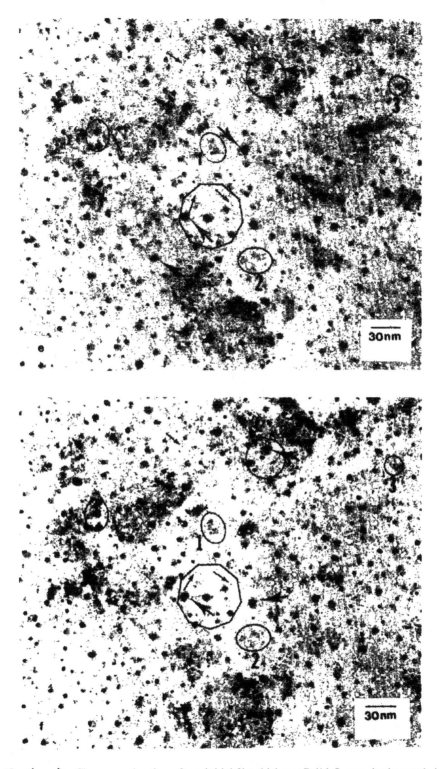

FIG. 2. (Continued) Changes on heating a 2 nm initial film thickness Pt/Al$_2$O$_3$ sample alternately in H$_2$ and O$_2$. Micrographs for only a few cycles of alternate heating are shown. Prior to the following, the sample had undergone a few cycles of alternate heating in H$_2$ and O$_2$ (see text), (e) O, 700°C, 5 h; (f) H$_2$, 500°C, 2 h.

500°C O_2 treatment. No detectable decrease in the particle size or the formation of films around the particles could be detected following the 300 to 450°C O_2 treatments. Particles as small as 0.8 nm in size or even smaller remained essentially unaffected following the above treatments in H_2 and O_2 as observed also with Sample 2 on cycling at 500°C. This indicates that even though an oxide might have formed at or below 500°C, no significant molecular emission was involved. (It is to be noted that oxidizing atmospheres are more likely to induce molecular/atomic emission.) At 500°C, changes regarding particle size and film formation begin to be noticed. As for Sample 2, the decrease in the size of the particles following heating in O_2 at 500°C was marginal during the first two cycles of heating but more pronounced during the subsequent cycles; the extent of size reduction was again smaller than that for Sample 1.

The behavior of this sample on subsequent heating in O_2 at higher temperatures was slightly different. On heating in O_2 at 600°C a number of particles disappeared, as observed before (Fig. 3). Some large particles have decreased in size considerably or disappeared, while nearby smaller particles have only decreased marginally in size (Regions 1–5 in Fig. 3). Subsequent heating in O_2 at 700°C resulted in an almost complete disappearance of the existing particles on the substrate. Instead, a number of very large, thin, irregular hexagonal patches could be observed. These patches can be a result of either a redistribution of the particles (due to the strong interactions between particles and substrate) or contamination. These patches remain, however, and even increase in size, on heating in O_2 at 800°C. One would expect that they would disappear on heating in O_2 if they were hydrocarbon or carbon impurities. The patches have sharp edges and are facetted in irregular, hexagonal shapes as seen in Figs. 3d and 3e. (Fig. 3e shows a larger area of the sample.) Some patches overlap (particles marked (a) and (b)). The patches remain essentially unchanged on subsequent heating in H_2 at 500 or 700°C. From the micrographs, it appears that the patches represent large extended particles, which are formed as a result of the strong interactions between particles and support (which in turn are enhanced by the high-temperature, 700 or 800°C, oxygen treatment). The interactions are probably so strong that little change in their morphology occurs on subsequent heat treatment in H_2 or O_2. It is to be noted that a thin, hexagonal, pill box morphology of Pt particles observed on heating in H_2 at $\geq$500°C was previously attributed to the so-called "SMSI," which was reported to be restricted to particles on reducible oxide supports, such as TiO_2 (28).

4. In order to verify whether the observed alternation in size was also exhibited by samples with smaller average sizes (since the average particle size in industrial Pt/Al_2O_3 catalysts is usually small), another sample with a smaller loading (0.75 nm initial film thickness) was heated in H_2 and O_2 at 500°C. Since this specimen was not heated in H_2 or O_2 at >500°C at any time during either the pretreatment or the cycling in H_2 and O_2, the particles were small. Any marginal decrease in size of such small particles on heating in O_2 will not be easily detectable unless the particles spread completely. Consequently, very little change, other than that due to sintering, was observed. After heating the sample for a total of 80 h, which incidentally generated sufficiently large particles, clearly detectable alternations in size were subsequently observed on cyclic heating in H_2 and O_2 at 500°C. These results indicate that the films around the particles, or an apparent decrease in the particle size as a result of the film formation, become detectable only when the particles are sufficiently large. The inability to detect such changes by electron microscopy does not imply, however, that they do not occur for small particles as well.

5. Whether the alternation in size on cyclic heating in H_2 and O_2 and the formation of films around particles in O_2 were restricted to temperatures below the decomposition temperature of PtO_2, namely $\simeq$ 580°C, or it is possible also at higher temperatures was verified by heating another sample at 750°C. The formation of films around the particles following heating in O_2 was in fact more obviously exhibited by a sample (of 1.5 nm initial film thickness)

heated at 750°C. After Pt deposition, the sample was heated in as-received ultra-high-purity H$_2$ at 750°C for 15 h. This generated very large particles in the range 5–15 nm (Fig. 4). When this specimen was subsequently heated in O$_2$ at 750°C, a large number of small and large particles disappeared without any kind of preference or pattern (Figs. 4a and 4b). All the remaining particles, including the small ones decreased in size. It could be surmised that because of the higher temperature, emission of atoms/molecules to the substrate or material loss via evaporation was involved. However, on subsequent heating in H$_2$ at 750°C for 5 h, the remaining particles considerably increased in size while some grew even larger than their size prior to the O$_2$ treatment. There was very little decrease in the number of detectable particles. The increase in average particle size on switching to H$_2$ from O$_2$ was greater than could possibly be explained by the very small decrease in the number of detectable particles (Figs. 4b and 4c). This indicates that the decrease in size during prior heating in O$_2$ could not have been solely due to the loss of material by evaporation. Two of the possibilities are (i) the material was present on the substrate surface as single atoms/molecules

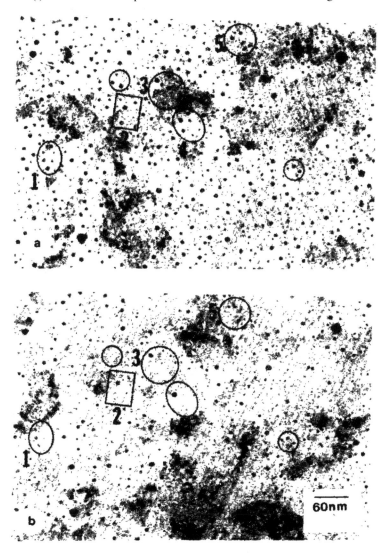

FIG. 3. Micrographs showing changes in a 2 nm initial film thickness Pt/Al$_2$O$_3$ sample subsequent to a few cycles of alternate heating in H$_2$ and O$_2$ (see text), (a) H$_2$, 500°C, 2 h; (b) O$_2$, 600°C, 5 h.   (*Continued*)

FIG. 3. (Continued) Micrographs showing changes in a 2 nm initial film thickness Pt/Al$_2$O$_3$ sample subsequent to a few cycles of alternate heating in H$_2$ and O$_2$ (see text), (c) (O$_2$, 700°C, 5 h) + H$_2$, 500°C, 2 h; (d) O$_2$, 800°C, 3 h; (e) same as (d), a larger area shown at a lower magnification.

following the O_2 treatment, which were subsequently captured by the particles on H_2 treatment, and (ii) following the O_2 treatment the material was present as individual unconnected films around each particle which subsequently contracted and merged with the respective particle on heating in H_2, to result in the observed increase in size.

A subsequent prolonged heat treatment of this sample for 12 h in H_2 at 750°C led to the appearance of films around the particles (Fig. 4d). This behavior is reminiscent of a similar observation in the case of Fe/Al_2O_3 on prolonged heat treatment in as-received H_2 (18). There it was suggested that the extension of the particles was a result of the oxidation of iron by the residual O_2 and H_2O present in the H_2 and the subsequent strong interaction between the iron oxide and the alumina. In the present case even though a stoichiometric platinum oxide (PtO_2) is not stable at 750°C, transient phases of platinum oxide might have formed which might have reacted with alumina. Even otherwise, because of the high temperature, an interaction compound between Pt and Al_2O_3 might have formed leading to the extension of the particles. In this regard, it is noted in the next section that the presence of an interaction compound is suggested in the electron diffraction patterns in both H_2 and O_2. Subsequently, an apparent decrease in size of the solid core and the appearance of films around the particles on heating in O_2 and an apparent increase in size on heating in H_2 were observed following three more cycles of alternate heating of this sample in H_2 and O_2 at 750°C (Fig. 5).

It follows from the above experimental results that an apparent decrease in size on heating in O_2 and a recovery in size on heating in H_2 is exhibited by model Pt/Al_2O_3 catalysts over a range of temperatures (500–750°C used in the present experiments) and initial metal loadings (0.75–2 nm initial film thickness) even though there are some differences perhaps due to the differences in the history of the samples.

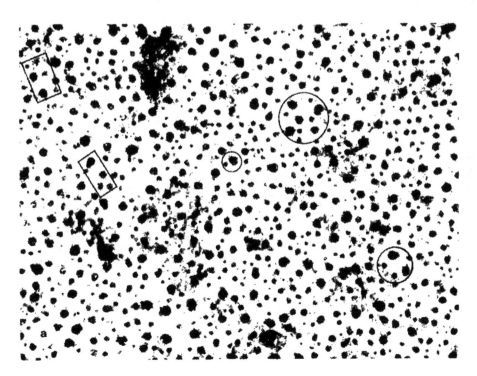

FIG. 4. Electron micrographs showing sequence of changes in a 1.5 nm initial film thickness Pt/Al_2O_3 specimen heated at 750°C. (a) Initial (H_2, 750°C, 15 h); *(Continued)*

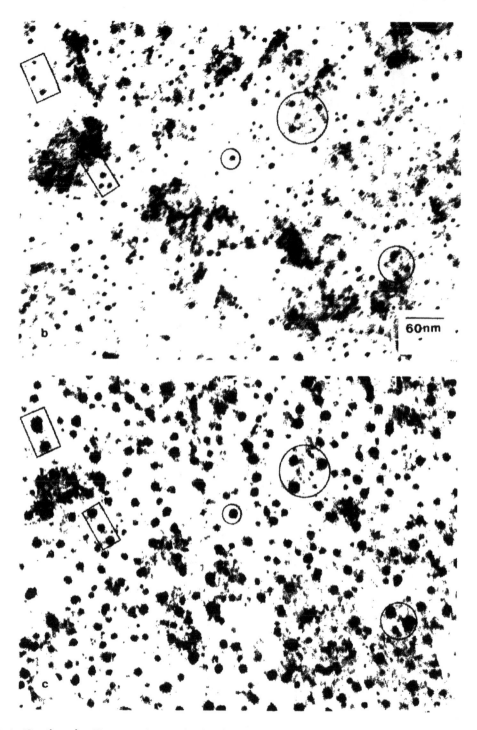

FIG. 4. (Continued) Electron micrographs showing sequence of changes in a 1.5 nm initial film thickness Pt/Al$_2$O$_3$ specimen heated at 750°C. (b) O$_2$, 750°C, 5 h; *(Continued)*

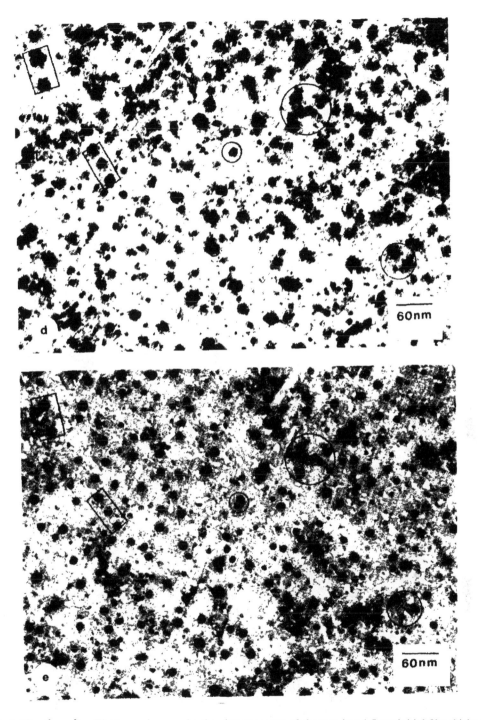

FIG. 4. (Continued) Electron micrographs showing sequence of changes in a 1.5 nm initial film thickness Pt/Al$_2$O$_3$ specimen heated at 750°C. (d) H$_2$, 750°C, + 12 h (17 h); (e) H$_2$, 750°C, + 6 h (23 h).

DISCUSSION

A few mechanisms have been suggested in the literature for the redispersion of sintered supported metal catalysts in an oxidizing atmosphere. The sintered crystallites can emit atoms or molecules to the substrate which can then coexist with the crystallites as a two-dimensional dispersed phase, thereby increasing the dispersion (*5, 16,17*). Such a molecular emission mechanism is expected to be favored at high temperatures and in an oxidizing atmosphere rather than in a reducing one, since in oxidizing atmospheres the possibility of a decrease in the free energy of the system via the strong interactions between the platinum oxide molecules and the alumina substrate creates a driving force for the emission of molecules from the particles to the substrate. On subsequent heating in H_2, the dispersed oxide molecules and the small crystallites are reduced (to metal), the interactions with the support become weaker, and subsequently two possibilities can arise. (i) Either new small particles are formed (not necessarily in the same locations where the particles might have disintegrated by emission of atoms or molecules to the substrate), and/or (ii) the atoms (reduced molecules) are captured by the existing three-dimensional crystallites. Since new particles have not been generated in our experiments either following heating in O_2 or on subsequent heating in H_2, one could suppose that on heating in O_2 dispersed single molecules were present on the substrate and that they were captured by the existing crystallites on subsequent heating in H_2. However, the extent of capture of atoms by various crystallites will be different because of the differences in their radii of curvature. Since their rate of emission is smaller, the large crystallites are expected to have a greater growth than the smaller ones (*1, 3*). Consequently, a size distribution different from the one before is expected. However, as noted under Results, especially at 500°C, all the detectable particles increase in size and attain more or less the size they had prior to the O_2 treatment. Therefore, it appears that a mechanism based on emission and capture of single atoms from a two-dimensional phase of single atoms located on the substrate surface is unlikely to be responsible for the observed behavior.

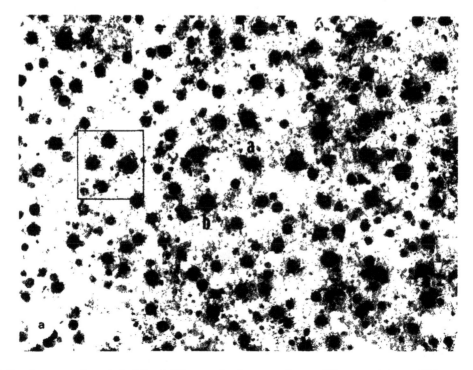

FIG. 5. Same sample as that of Fig. 4. Micrographs shown are from a different region, at a different magnification, and for treatments subsequent to those in Fig. 4e. (a) H_2, 750°C, 32 h. *(Continued)*

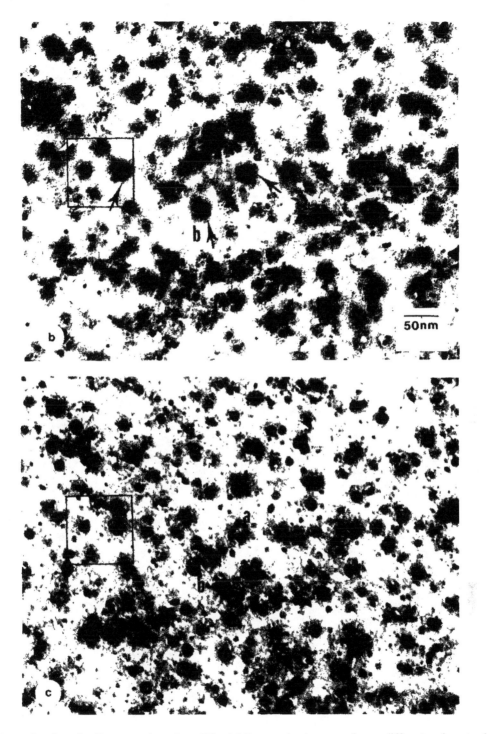

FIG. 5. (Continued) Same sample as that of Fig. 4. Micrographs shown are from a different region, at a different magnification, and for treatments subsequent to those in Fig. 4e. (b) O_2, 750°C, 12 h; (c) H_2, 750°C, 12 h.

158 Wetting Experiments

Another possible, and in the present case more likely, mechanism for redispersion in an oxygen atmosphere is by the spreading of a film in the form of a halo around each particle. As discussed in detail in (18, 25), the surface free energy (γ_{cg}) of a metal is larger than the surface free energy of the corresponding oxide. Also, since, in general, a metal oxide (platinum oxide in the present case) interacts more strongly with the substrate (which is another oxide) than a metal does, the interfacial free energy γ_{cs} between the crystallite and substrate is also smaller in an oxidizing than in a reducing atmosphere. Since γ_{cg} and γ_{cs} are both smaller in an oxygen atmosphere, the contact angle θ is also smaller as can be seen from Young's equation shown below and the particle would be more extended in an O_2 atmosphere than in a H_2 atmosphere:

$$\cos\theta = \frac{\gamma_{sg} - \gamma_{cs}}{\gamma_{cg}}.$$

The interfacial free energy γ_{cs} is related to the net interaction energy U_{cs} between crystallite and substrate via the expression

$$\gamma_{cs} = \gamma_{cg} + \gamma_{sg} - \left(U_{int} - U_{str}\right)$$

$$= \gamma_{cg} + \gamma_{sg} - U_{cs},$$

where U_{int} is the interaction energy per unit contact area between substrate and supported material and U_{str} is the strain energy per unit contact area due to the mismatch of their lattices. The net interaction energy is larger when the support and the supported material are similar in structure (two oxides, for instance, rather than an oxide and a metal), or when there are covalent interactions between the two. In the present case, if the interactions between the platinum oxide and the alumina are strong enough, the small particles (which will be more completely oxidized than the larger ones) could spread out completely (because of the large driving force for spreading when U_{cs} is very large or γ_{cs} is very small) while the larger particles would only extend. Therefore, the small particles would no longer be detected in the micrographs following heating in O_2, while the larger particles would appear smaller if the films extending from the leading edge of the particles are so thin as not to be detectable by electron microscopy. If the above interactions are not very strong (i.e., if the driving force for spreading is not very large), or if the kinetics of wetting is slow, even the smallest particles would only extend and could therefore still be detected in the micrographs (of course, with a smaller apparent size and an undetectable film around). In some cases, for example when the particle size is large, the extended film around the particles may be thick enough to be detected in the micrographs (particles marked with large arrows in Figs. 2 and 5). In these cases, the particles do not spread completely, but only acquire detectable or undetectable films around them. These films would contract and merge with the respective crystallites on subsequent heating in H_2 and, depending upon the reconstruction that occurs, the particles can regain the sizes they had prior to the oxygen treatment. This scenario appears to explain the observed results better than the two-dimensional phase of single atoms or molecules that covers the entire substrate can.

It is to be noted that at 500°C, the films around the particles are not likely to be interconnected to form a contiguous patch as observed in the case of Fe/Al$_2$O$_3$ at high temperatures (19). If the films were to be interconnected, the particles would be pulled closer together and sinter when the interconnecting film contracts on subsequent heating in H_2. Since little sintering is observed, it appears that the film constitutes an atmosphere around the particle (in the form of a halo), extended only in its immediate vicinity, so that on subsequent heating in H_2 it contracts and merges entirely with the parent particle. However, it is likely that there exists a critical metal loading beyond which the films would extend farther away from the particles, interconnect with those from nearby particles,

Redispersion of Pt/Alumina via Film Formation

and form a contiguous film covering the substrate surface. Similarly, when two or more particles are close to one another, the films around them may overlap and become interconnected.

Another alternative could also be envisioned. In a H_2 atmosphere, the Pt crystallites can catalyze the partial reduction of alumina. Similar suggestions have already been made (26). The partially reduced substrate is relatively more reactive. Also, because of its nonstoichiometric nature, the surface free energy γ_{sg} of the partially reduced substrate is larger than that of Al_2O_3. In addition, as suggested for Pt/TiO_2 (27, 29), the reduced substrate species may migrate onto the crystallites and because of their interaction with the crystallites they may decrease γ_{cg}. Similarly, because of its reactive nature, the net interaction energy U_{cs} between the substrate and the crystallite at the base of the crystallites is increased and therefore γ_{cs} may be decreased. As a result of the increase in γ_{sg} and decreases in γ_{cg} and γ_{cs}, the contact angle will decrease and the crystallite would extend on the substrate in a reducing atmosphere, as in the case of Pt/TiO_2 (28). That is, in the present case, the larger size of the crystallite, observed following heating in H_2, is because of its extended form. On subsequent heating in O_2, the partially reduced substrate can be reoxidized to Al_2O_3 and γ_{sg} will therefore decrease. If the interactions between platinum oxide and Al_2O_3 are weaker than those between Pt and the reduced alumina, γ_{cs} may increase compared to its value in the reducing atmosphere. Also, because now the interactions between the particles and the partially reduced substrate species migrated onto them are absent, γ_{cg} may also increase although it may be smaller compared to the initial of the metal. As a result of the increases in γ_{cg} and γ_{cs} and the decrease in γ_{sg}, the contact angle will increase and the particles would appear contracted and smaller. That is, the smaller size of the particles, observed following heating in O_2, is a result of their contraction to a larger wetting angle. However, this mechanism is not likely to be responsible for the observed behavior especially since in a few micrographs a kind of film is detected around the particles following heating in O_2. It is more likely that stronger interactions and extension occur in an oxygen atmosphere rather than in a reducing one.

Apart from the indirect evidence, via film formation, for strong interactions between the crystallites and support, there is also some direct evidence for such interactions from electron diffraction results, even though the compounds formed could not be identified with certainty. Electron diffraction patterns were recorded for all the samples following each heat treatment. Following heating in oxygen, especially above 500°C, additional rings, a few of which corresponded to Pt_3O_4 and Al_2Pt, appeared in the diffraction patterns. The rings corresponding to the compound between Pt and Al did not disappear on subsequent heating in H_2, only their d values changed marginally. This indicates that once an interaction compound is formed it is not easily decomposed. This compound formation may involve the diffusion of Pt species into the bulk of the alumina substrate and it may be difficult to be reversed. However, at the interface between the particle and the substrate, the compound may be relatively easily reduced (even if only partly) and lead to the observed contraction of the particles. It is to be noted that especially on heating in O_2 at temperatures greater than 600°C a large number of particles disappeared, suggesting possible loss of material by evaporation. In this regard, Chen and Schmidt (24) have reported that there is considerable loss of Pt as PtO_2 when Pt/SiO_2 is heated in air or O_2 at ≥ 700°C. However Leitz et al. (30) have noted that only in the case of chlorine-containing samples and at ≥ 800°C was some loss of Pt observed. It was pointed out in the previous section that only a small fraction, if any at all, of the material must have been lost via evaporation, since substantial recovery of the material in the form of crystallites occurred on subsequent heating in H_2 (Figs. 4a–4c). As suggested before, the apparent "loss" of material might have been due to considerable extension of the particles (formation of thin undetectable films on the substrate) as well as to the diffusion of material into the substrate to form an interaction compound with alumina.

Even though the identity of the compounds formed could not be ascertained, the appearance of additional rings in the electron diffraction pattern (some of which correspond to some known Pt–Al compounds) nonetheless suggests compound formation between PtO_x and Al_2O_3 to yield $Pt_pAl_qO_r$, where r can also be zero. In this regard it is to be noted that den Otter and Dautzenberg (31) proposed the formation of Pt_3Al to explain the suppression of room temperature H_2 adsorption activity following high-temperature (≥ 500°C) reduction. Lagarde et al. (32) suggested that a Pt–O bonding between crystallites and Al_2O_3 is more likely than a Pt–Al bonding. But they were inclined to the view that chlorine atoms are also present in the complex. As pointed out in the Introduction, Leiske *et al.* (*13*) argued that oxidative redispersion is possible only in the presence of chlorine or chloride ions, and that oxygen alone does not lead to redispersion. Even though the presence of chlorine or chloride ions in our samples is very unlikely, it cannot be ruled out completely. It is worth noting that in the procedure followed to obtain electron transparent alumina films, subsequent to the anodic oxidation of an aluminum foil, the alumina films are stripped of their aluminum backing by dissolving the latter in mercuric chloride solution. Even though the alumina films are washed repeatedly in distilled water, there may be strongly bound chloride ions left behind and, therefore, we cannot rule out its absence or confirm its presence. A number of other authors have postulated complex formation between crystallites and support in Pt/Al_2O_3 samples (*6, 32–36*) and the exact nature and mechanism of formation of the complex have been debated. The present results have shown, both via electron diffraction and via film formation, that there are strong interactions between the crystallites and support in Pt/Al_2O_3 model catalysts and that the oxidative redispersion (regeneration), routinely achieved in industries by oxidative treatments between 500 and 600°C, occurs more via spreading and film formation than by emission of molecules to a two-dimensional phase of single atoms or molecules. It can be further noted that while Pt/Al_2O_3 is differentiated as being a "non-SMSI" system, the results indicate that there are strong interactions in this system as well, similar to those exhibited by the SMSI systems.

CONCLUSION

Pt/Al_2O_3 model catalysts are observed to undergo alternate changes in particle size when heated alternately in H_2 and O_2 in the range 500–700°C. When a sample heated previously in H_2 at $T \geq 500$°C is heated in O_2 at $500 \leq T \leq 700$°C, there is an apparent decrease in the particle size. On subsequent heating in H_2, approximately the previous sizes of the particles are recovered. Even the smallest of the particles exhibit this behavior and (particles down to about 0.8 nm are observed to) remain on the substrate through the various cycles of heating, suggesting that the molecular emission-capture mechanism is unlikely to be the cause of the alternations in size. In some cases, especially at 600 or 700°C, on heating in O_2, a kind of film, in the form of a halo, is observed around each particle and some particles extend out considerably into detectable or undetectable patches. In light of these observations, the alternation in size at 500°C is explained on the basis of the formation in O_2 of a thin, undetectable film around each particle which results in the appearance of a smaller size of the particles since only the solid core is detected. It is surmised that the film is in the immediate vicinity of and in contact with the parent particle and that the films around the particles do not interconnect to form a contiguous patch. On subsequent heating in H_2, the films contract and merge with the respective parent particle. Depending upon the reconstruction that follows, the particles can recover approximately their previous sizes. Such extension in O_2 and contraction in H_2 are enhanced by prolonged pretreatment in O_2 and/or by a few initial cycles in H_2 and O_2 which perhaps improve the wetting characteristics. It is suggested that extension and film formation are possible and likely mechanisms for the routinely observed oxidative redispersion of supported platinum catalysts.

REFERENCES

1. Wynblatt, P., and Gjostein, N. A., *Prog. Solid State Chem.* **9**, 21 (1975).
2. Burch, R., *in* "Catalysis," (G. C. Bond and G. Webb, Senior Reporters), Vol. 7 (a specialist periodical report). The Royal Society of Chemistry, London, 1985.
3. Ruckenstein, E., and Dadyburjor, D. B., *Rev. Chem. Eng.* **1**(3), 251 (1983).
4. Ruckenstein, E., and Malhotra, M. L., *J. Catal.* **41**, 303 (1976).
5. Fiedorow, R. M. J., and Wanke, S. E., *J. Catal.* **43**, 34 (1976).
6. Johnson, F. L., and Keith, C. D., *J. Phys. Chem.* **67**, 200 (1963).

7. Kraft, M., and Spindler, H., "Proceedings, 4th International Congress on Catalysis, Moscow, 1968" (B. A. Kazansky, Ed.), p. 1252. Adler, New York, 1968.
8. Yates, D. J. C., U.S. Patent 1433864 (1976).
9. Stulga, J. E., Wynblatt, P., and Tien, J. K., *J. Catal.* **62**, 59 (1980).
10. Gollob, R., and Dadyburjor, D. B., *J. Catal.* **68**, 473 (1981).
11. Ruckenstein, E., and Pulvermacher, B., *J. Catal.* **29**, 224 (1973).
12. Dautzenberg, F. M., and Wolters, H. B. M., *J. Catal.* **51**, 26 (1978).
13. Lieske, H., Leitz, G., Spindler, H., and Volter, J., *J. Catal.* **81**, 8 (1983).
14. Straguzzi, G. I., Aduriz, H. R., and Gigola, C. E., *J. Catal.* **66**, 171 (1980).
15. Boumonville, J. P., and Martino, G., *in* "Studies in Surface Science and Catalysis" (B. Delmon and G.Froment, Eds.), Vol. 6, p. 159. Elsevier, Amsterdam, 1980.
16. Ruckenstein, E., and Chu, Y. F., *J. Catal,* **59**, 109 (1979). (Section 3.1 of this volume).
17. Yao, H. C., Sieg, M., and Plummer, H. K., Jr., *J. Catal.* **59**, 365 (1979).
18. Sushumna, I., and Ruckenstein, E., *J. Catal.* **94**, 239 (1985).
19. Ruckenstein, E., and Sushumna, I., *J. Catal.,* **97**, 1 (1986).
20. Ruckenstein, E., and Lee, S. H., *J. Catal.* **86**, 457 (1984).
21. Sushumna, I., and Ruckenstein, E., *J. Catal.* **109**, 433 (1988) (Section 3.3 of this volume).
22. Nowak, E. J., *Chem. Eng. Sci.* **21**, 19 (1966).
23. Schmidt, L. D., and Luss, D., *J. Catal.* **22**, 269 (1971).
24. Chen, M., and Schmidt, L. D., *J. Catal.* **55**, 348 (1978); **56**, 198 (1979).
25. Ruckenstein, E., and Sushumna, I., *in* "Hydrogen Effects in Catalysis" (Z. Paal and P. G. Menon, Eds.). Dekker, New York, 1987, in press.
26. Weller, S. W., and Montagna, A. A., *J. Catal.* **21**, 303 (1971).
27. Mériaudeau, D., Dutel, J. F., Dufaux, M., and Naccache, C., *Stud. Surf. Sci. Catal.* **11**, 95 (1982).
28. Baker, R. T. K., *J. Catal.* **63**, 523 (1980).
29. Ruckenstein, E., *in* "Strong Metal Support Interactions" (R. T. K. Baker, S. J. Tauster, and J. A. Dumesic, Eds.). ACS Symp. Ser. No. 292, Washington, DC, 1986.
30. Leitz, G., Lieske, H., Spindler, H., Hanke, W., and Walter, J., *J. Catal.* **81**, 17 (1983).
31. den Otter, G. J., and Dautzenberg, F. M., *J. Catal.* **53**, 116 (1978).
32. Lagarde, P., Murata T., Vlaic, G., Freund, E., Dexpert, H., and Boumonville, J. P., *J. Catal.* **84**, 333 (1983).
33. McHenry, K. W., Bertolacini, R. J., Brennan, G. M., Wilson, J. L., and Seelig, H. S., *in* "Proceedings, 2nd International Congress on Catalysis, Paris, 1960," Paper 117 (1960), Vol. **2**, p. 2295, Technip, Paris, 1961.
34. Ushakov, V. A., Moroz, E. M., Zhdan, P. A., Boronin, A. I., Bursian, N. R., Kogan, S. B., and Levitskii, E. A., *Kinet. Katal.* **19**, 744 (1978).
35. Kunimori, K., Ideda, Y., Soma, M., and Uchijima, T., *J. Catal.* **79**, 185 (1983).
36. Mansour, A. N., Cook, J. W., Jr., Sayers, D. E., Enrich, R. J., and Katzer, J. R., *J. Catal.* **89**, 462 (1984).
37. Margitfalvi, J., Kem-Talas, E., and Szedlacsek, P., *J. Catal.* **92**, 193 (1985).

3.3 Events Observed and Evidence for Crystallite Migration in Pt/Al₂O₃ Catalysts*

I. Sushumna and Eli Ruckenstein

Department of Chemical Engineering, State University of New York at Buffalo, Buffalo, New York 14260

Received February 3, 1987; revised July 28, 1987

INTRODUCTION

Alumina—supported Pt catalysts are relatively thermostable in comparison to supported Ag, Fe, Ni, etc. Nevertheless, Pt/ Al₂O₃ catalysts still lose their activity gradually on use in reducing atmospheres (such as in catalytic reforming). A part of the activity loss is attributed to the sintering of the metal particles (1–3). The supported catalysts lose exposed surface area also during heating in oxidizing atmospheres (such as in automotive exhaust oxidation and during coke burn off) (1, 4). That is, coarsening of the supported crystallites occurs in both reducing and oxidizing atmospheres, though its extent may vary, depending on the catalyst system and the treatment conditions employed. In specific cases, such as Pt/Al₂O₃ heated in O₂ between 500 and 580°C, redispersion of the catalysts has also been observed (5–10). Basically, two physical models have frequently been employed to explain the growth kinetics of the supported crystallites (11). One, known as the crystallite migration model, considers that the growth occurs via random migration, collision, and coalescence of crystallites (12, 13). The other, known as the atomic migration model or Ostwald ripening, assumes that the growth occurs via emission of single atoms or molecules by the small crystallites and their capture by the larger crystallites (14–16). For single—atom emission and capture, two possibilities have been noted. In one of them, which is global, the small crystallites lose atoms to a surface phase of single atoms dispersed over the substrate, while the large ones capture atoms from this phase. In the other case, called direct ripening (17), atoms released by a small particle move directly to a neighboring large crystallite.

A number of investigations with supported catalysts including Pt/Al₂O₃ have been carried out in an attempt to identify the mechanisms involved in, and the factors responsible for, the growth of the crystallites, as well as for the regeneration of the sintered catalysts (3, 11, 18–41). Both model and industrial—type catalysts have been investigated, predominantly by selective chemisorption of gases, transmission electron microscopy, and X–ray line broadening (for reviews, see Refs. (11, 14, 42–44). As a result of the anomalous chemisorption behavior (such as the suppression of room—temperature H_2 chemisorption following high—temperature (500°C) H_2 treatment) exhibited by certain catalyst systems such as Pt/Al₂O₃ (45, 46) and Pt/TiO₂ (47–49), it is now recognized that chemisorption of gases may require the availability of adsorption sites on the crystallites (which may be covered and hence prevent chemisorption, even though the dispersion is high). Consequently, there is not necessarily a direct correlation between the extent of chemisorption

---

* *Journal of Catalysis* 109, 433–462 (1988). Republished with permission.

Events Observed and Evidence for Crystallite Migration in Pt/Al$_2$O$_3$ Catalysts

and the catalyst dispersion. Caution is needed therefore in drawing conclusions on the basis of dispersion values obtained from chemisorption measurements alone (*50*). A number of investigations of the mechanism of sintering of supported metal catalysts have relied on such chemisorption measurements. Selective chemisorption of gases is, in general, indeed a very valuable method of measuring catalyst dispersion. However, especially in cases where ambiguities and anomalies of the kind mentioned above are involved, it is appropriate to use a complementary technique such as electron microscopy. As is well known, transmission electron microscopy (TEM), though tedious, provides direct information on the crystallite shape, size, size distribution, and, to a large extent, on the mechanisms involved in sintering and redispersion. Therefore, TEM has been employed by a number of researchers to investigate the mechanism of sintering of supported metal catalysts.

The results have been discussed in terms of both Ostwald ripening (*8, 11, 14–18, 26–30, 37*) and crystallite migration models (*32–40*). On the basis of their electron microscopy or chemisorption results, a few authors (*26, 28, 29*) reported that they had never observed crystallite migration or dumbbell—shaped particles on the supports with the catalysts they had investigated and that atomic or molecular migration is the only mechanism of sintering and redispersion. Recent results, from this laboratory as well as from others, of direct TEM observation of changes in the same region of the samples following successive heat treatments have shown, however, that crystallites can migrate and that crystallite migration and coalescence contribute significantly to the growth of supported crystallites. Besides the early electron microscopy observations of supported crystallite migration by Bassett (*51*) in the case of copper and silver islands on carbon, graphite, or molybdenite substrates and by Skofronick *et al.* (*52*) in the case of gold islands on carbon and silicon substrates, there are more recent observations of supported crystallite migration and coalescence. Chu and Ruckenstein (*24*) observed significant migration of crystallites larger than 20 nm in the case of Pt/C heated in O$_2$ atmospheres. They observed migration of particles larger than 10 nm also in the case of Pt/Al$_2$O$_3$ at 750°C (*25*). Heinemann and Poppa (*55*) presented direct *in situ* electron microscopic evidence for the simultaneous occurrence of short—distance crystallite migration and coalescence as well as ripening in the sintering of Ag/graphite system in the temperature range 25–450°C. They noted that though slow ripening occurred over the entire temperature range, the overriding surface transport mechanism was short—distance (<10 nm) cluster mobility. Chen and Ruckenstein (36) in the case of Pd/Al$_2$O$_3$, Ruckenstein and Lee (*40*) in the case of Ni/Al$_2$O$_3$, and Sushumna and Ruckenstein (*39*) in the case of Fe/Al$_2$O$_3$ observed crystallite migration among a large number of other events when the samples were heated in H$_2$ and O$_2$. Arai *et al.* (*37*) reported *in situ* TEM results of sintering of Ni, Pt, and Ag on amorphous carbon, SiO$_2$, and Al$_2$O$_3$ and noted that in the case of Pt/Al$_2$O$_3$ and above 600°C, only particles larger than 10 nm grew through abrupt surface movement and subsequent coalescence of adjacent pairs. Richardson and Crump (*32*) in the case of Ni/SiO$_2$ concluded, on the basis of magnetic measurements of the particle size distributions, that especially above 450°C particle migration and coalescence were responsible for the sintering in a He atmosphere. Granquist and Buhrman (*54, 55*), analyzing various previous experimental results on sintering, argued on the basis of the shape of the particle size distributions that crystallite (migration and) coalescence is more likely than ripening to be responsible for the coarsening of the supported particles. Abundance of experimental results notwithstanding, the mechanism(s) of sintering has yet to be clarified; in general, some authors support one mechanism, while others support the other mechanism. However, recent results (*36, 39*) seem to suggest that the phenomena of sintering and regeneration involve complex interrelated processes that cannot be restricted to one or even to both of the mechanisms noted above.

In the past, investigations of supported metal catalysts, including Pt/Al$_2$O$_3$ have been based in general on either short (1 or 2 h) (*26–30, 41*) or long (24 or 48 h) durations of heating (*9, 25, 29*). However, no systematic investigations of short intervals of heating, especially of a fresh catalyst, over extended time periods, at various temperatures, have been carried out. Especially on the basis of 1 or 2 h of heating, a few authors have reported that they never observed crystallite migration on the supports with the catalysts they had investigated (*26–30*). Evidence for migration and coalescence with Pd/Al$_2$O$_3$ (*36*), Pt/Al$_2$O$_3$ (*25*), Ni/Al$_2$O$_3$ (*40*), and especially Fe/Al$_2$O$_3$ (*39*) have been

164 Wetting Experiments

presented from this laboratory among others in the past. Here we present additional transmission electron microscopic evidence for crystallite migration among other phenomena in Pt/Al_2O_3 model catalysts heated at different temperatures in H_2 and O_2.

EXPERIMENTAL

PREPARATION OF ALUMINA SUPPORTS

The method of preparation of electron—transparent films of $\gamma-Al_2O_3$ has been described in detail previously (25, 39). Thin aluminum foils (99.999% pure, Alfa Products, Inc.) were cleaned thoroughly by chemical polishing and were subsequently washed in distilled water. An amorphous aluminum oxide layer about 30 nm thick was built up on the cleaned aluminum foil by anodic oxidation at 20 V in a 3 wt% tartaric acid solution. The oxidized foil was subsequently cleaned thoroughly in distilled water. The oxide layer was separated from aluminum by floating small pieces of the oxidized foil in mercuric chloride solution, in which the unoxidized aluminum dissolved. The aluminum oxide films were subsequently washed repeatedly in distilled water and eventually picked up on gold electron microscope grids. The oxide films on the grids were dried and subsequently heated at 800°C in laboratory air for about 72 h and then slowly cooled to the room temperature. The prolonged heat treatment is carried out to convert the amorphous films to crystalline $\gamma-Al_2O_3$ and to ensure that no further changes would occur on subsequent heat treatments. It is to be noted that Elfter the above heat treatment and prior to or after the deposition of the metal, these alumina samples were not brought into contact with any solution or liquid and therefore no artifacts of the kind discussed by Glassl et al. (56) were likely to have been generated.

Platinum films 0.75, 1.5, and 2 nm thick were deposited onto the alumina films in three different batches, by evaporating the corresponding amounts of Pt wire (99.999% pure, Alfa Products, Inc.) from a tungsten filament in an Edwards vacuum evaporator. The vacuum in the chamber was better than 10^{-6} Torr. The substrate was maintained at room temperature during deposition.

It should be pointed out that the industrial supported Pt catalysts are relatively thermostable. The very low metal loading of Pt/Al_2O_3 usually used in industry contributes to the slow growth rate. The model catalysts in general have a relatively high loading, which enhances sintering. However, even though the metal loadings of the model catalysts may not correspond to those of the industrial catalysts, the model systems provide at least a qualitative indication of the processes that occur in supported catalyst systems. In this regard it is worth noting that Smith et al. (41) and Baker et al. (27) have noted previously that the behaviors of the model catalysts of the kind described here are in general similar to those of the industrial catalysts.

The samples on which metal had been deposited were heated in a quartz boat inside a quartz tube. Ultrapure hydrogen, both as—received (99.999% pure with < 1 ppm O_2 and < 3 ppm moisture) and further purified by being passed through a Deoxo unit (Engelhard Industries) followed by a column of 3–Å molecular sieves immersed in liquid nitrogen, were used in the experiments. Helium, 99.999% pure with < 3 ppm moisture, was also passed through the above column of molecular sieves immersed in liquid nitrogen. Ultrahigh—purity– grade oxygen was used as received. All the gases were supplied by Linde Division, Union Carbide Corporation. For each heat treatment, the following procedure was adopted. With the sample inside, the tube was flushed with He for at least 5 minutes before the power to the furnace was turned on. The sample was heated in He until the preset temperature was reached, at which point He was turned off and the desired gas was let in. The sample was then heated for the predetermined length of time, at the end of which the furnace and the gas were turned off and He was allowed to flow through again. The sample was cooled slowly to the room temperature in He before it was exposed to the atmosphere.

The samples were observed in a JEOL 100U transmission electron microscope operated at 80 kV. Each sample was scanned thoroughly over the entire specimen area to check for uniformity in behavior, and pictures were taken at a few regions. Following each heat treatment, the same

Events Observed and Evidence for Crystallite Migration in Pt/Al$_2$O$_3$ Catalysts

regions were photographed at the same magnification (in the 60–100–K range). Electron diffraction patterns were also recorded following each heat treatment. A liquid nitrogen trap was used to reduce specimen contamination in the microscope. Also, the intensity of the electron beam and the specimen's exposure to the beam were kept low to minimize the contamination of the specimen. The alumina film moved very little on the supporting grid, and, especially following the initial one or two observations, the same regions could be located relatively easily after successive heat treatments. The particle migrations shown in the next section were followed with reference to certain fixed grain boundaries or other "landmarks" on the support.

RESULTS

Table 1 lists some of the events observed on Pt/Al$_2$O$_3$ model catalysts heated in H$_2$ and O$_2$. Figures 1, 2, and 3 show some micrographs of the time sequence of a sample with an initial film thickness of 1.5 nm heated in as—received ultrapure hydrogen. The letters on all the micrographs correspond to the respective events listed in Table 1. When the metal—deposited sample was heated in H$_2$ for 6 h at 500°C, detectable particles in the range 0.8–5 nm formed, with the average around 2 nm. The sample with this initial distribution was heated subsequently in the following cumulative sequence: 1, 2, 3, 4, 6, 8, 12, 17, 29, 43, and 65 h. Following the initial 1 h heating, a number of neighboring particles, especially the larger ones, coalesced, decreasing the particle number density (B, C, F in Figs, 1a-1d). A number of small particles 0.8–1.5 nm in size remained unaffected while a number of other small particles decreased in size or disappeared (regions 1 and H in Figs, 1a–1d and 2a–2d, for example). All the large particles also appear to have decreased slightly in size (I in Figs, la and lb, for example). Since, as just mentioned, a number of small particles on the substrate remained unaffected, the decrease in the size of the large particles may be a result of emission of atoms to the substrate, due not to a global ripening mechanism, but more likely to local variations in curvature along the periphery of the large particles. Or, it may be a result of extension of an undetectable film from around the particles or of diffusion of material into the substrate. It may also be a result of mere reconstruction with time to a more compact shape. During the next three 1–h intervals of heating, the number of particles decreased further. Both the decrease and disappearance of smaller particles (H, J) and the coalescence of nearby particles in the vicinity of unaffected or decreasing smaller particles were observed (C, F in Figs, 1b–1d). Also, a number of nearby small particles appear to have coalesced to form larger particles (K); this growth occurs either by migration and coalescence of the crystallites or by direct ripening between nearby crystallites, more likely by the former, as the particles are of about the same size and small, ~2 nm (K in Figs, 1b, 1c and 2b, 2c). In fact, in the region marked "a" in Figs, la and lb, a small particle (about 0.8 nm) appears to have grown to about 2.5 nm, very likely as a result of the migration and coalescence with it of a larger particle from nearby. A similar event is visible at "b" in Figs, 1b and 1c. The subsequent heat treatment for up to a total of 8 h brought very little change (not all the micrographs for the entire heating sequence are shown here). At the end of a total of 12 h of heating only a few additional pairs of nearby particles had coalesced. However, there was a marginal increase in the size of all the particles, both small and large. Small particles, as small as 1.5 to 2 nm still remained on the substrate. It is possible that all the particles over the entire observable size range could have increased in size marginally at the expense of particles much smaller than about 0.8 nm, either by their coalescence with the migrating crystallites or by a global ripening mechanism. On the other hand, the increase in size may also be a result of extension and/or reconstruction of the particles. Following 17, 29, and 43 h total heating, a number of pairs of large particles coalesced. Very few small particles disappeared. Particles were formed in some places where no particles had been apparent previously (region 1 in Figs. 3a, 3b). After heating for additional 22 h (to 65 h of total heating), a number of large crystallites (up to about 8 nm) migrated large distances (up to about 30 nm). Some of these crystallites coalesced with other particles (B in Figs, 2e, 2f and in 3b, 3c), while some stopped short of colliding (not shown in figures). A number of other particles near each other coalesced, most likely by migration and

TABLE 1

Events Observed on Pt/Al$_2$O$_3$ Model Catalysts Heated in H$_2$ and O$_2$

Type	Event	Schematics	Figure Nos.
A	Migration, without coalescence, of small or large particles toward or away from another particle		1, 2, 4, 5, 8, 10, 11
B	Migration, collision, and coalescence; gradual migration and coalescence		1, 2, 3, 4, 5, 8, 10, 11, 12
C	Coalescence of nearby particles		1, 2, 3, 4, 7, 8
D	Migration followed by a decrease in size or vice versa		9, 10
E	Dumbbell-shaped particles, particles in fusion		1, 2, 4, 5, 6, 8, 9
F	Coalescence of nearby larger particles adjacent to unaffected/decreasing smaller particles		1, 2, 8, 9
G	Transfer of atoms between two nearby particles via a neck or narrow whiskerlike bridge; formation and breakup of a contact between two particles via a narrow whiskerlike bridge		1, 2, 8
H	Decrease and/or disappearance of a small particle near a larger particle that grows in size or appears unaffected (ripening); decrease in size and/or disappearance of one particle amid particles of equal size		1, 2, 9
I	Decrease of a large particle nearby smaller particles		9, 10, 11
J	Disappearance of large and small particles		9, 10, 11
K	Growth of smaller particles or coalescence of small particles to yield larger particles		1, 2, 3, 10, 11
L	Collision, coalescence, and subsequent separation of two particles; separation of particles in contact		1, 2, 4, 12
M	Appearance of two particles in place of one (splitting, or wetting and reconstruction)		12
N	Overlapping particles		12

Events Observed and Evidence for Crystallite Migration in Pt/Al$_2$O$_3$ Catalysts

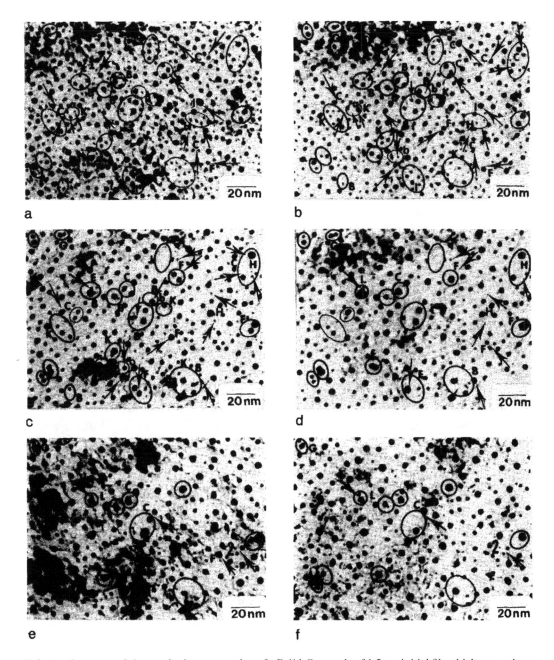

FIG. 1. Sequence of changes in the same region of a Pt/Al$_2$O$_3$ sample of 1.5 nm initial film thickness on heating in as—supplied hydrogen at 500°C. The following durations of heating, starting with (b), are cumulative. (a) Initial (6 h H$_2$); (b) 1 h; (c) 2 h; (d) 4 h; (e) 17 h; (f) (65 h H$_2$ +) 16 h O$_2$.

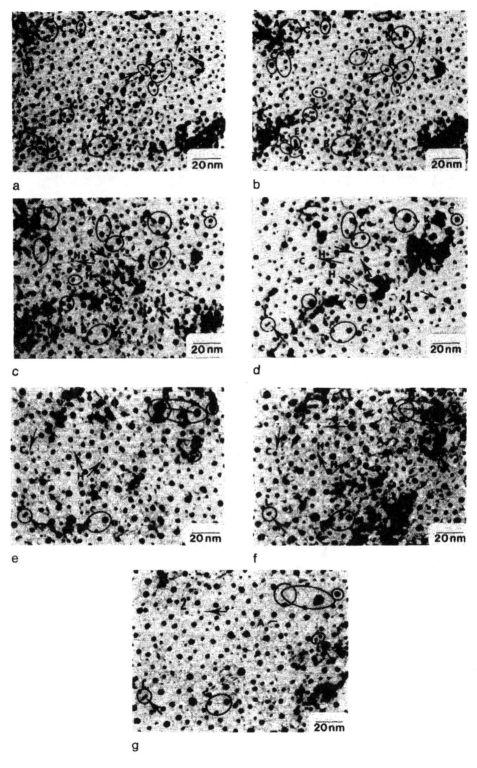

FIG. 2. Same sample as in Fig. 1. The micrographs are of a different region, (a) Initial (6 h H$_2$); (b) 1 h; (c) 2 h; (d) 12 h; (e) 29 h; (f) 65 h; (g) 16 h O$_2$.

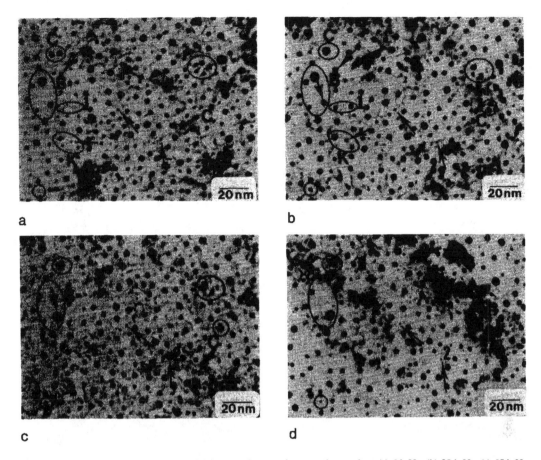

FIG. 3. Same sample as in Fig. 1. The micrographs are of yet another region, (a) 6 h H_2; (b) 29 h H_2; (c) 65 h H_2; (d) 16 h O_2.

collision. It is to be noted that long durations of heating lead to significant sintering, especially when the loading is high, though it is not as rapid and pronounced as in the initial stages. During this sequence of heating it was also observed that sometimes two nearby particles made contact via a faint, narrow neck or a whiskerlike bridge. The particles then either sintered via a transfer of atoms along the bridge or separated again as the interconnecting bridge was broken (G in Figs, 1b– 1f and 2a, 2b). In some cases the two particles more or less completely coalesced and then tended to separate again (L in Figs. 1d–1f and 2b–2g). Additional events observed are marked on the micrographs by various letters corresponding to the events listed in Table 1.

When the above sample was subsequently heated in O_2 at 500°C for up to 28 h, no particle migration or detectable ripening of particles was observed. However, most of the particles, small and large, decreased in size (region 2 in Figs. 1f, 2g, and 3d) as a result of either emission of atoms/molecules to the substrate or of the extension and formation of films undetectable by electron microscopy around the particles. This phenomenon is discussed further in Refs. (57, 58). Two additional cycles of heating in H_2 and O_2 for a total of 50 h resulted in marginal particle coarsening mostly via coalescence of neighboring particles (C in Figs. 4a–4b). Particle migration without coalescence (A in Figs. 4a, 4b) and elongation of some particles (O in 4a, 4b) could also be observed. Almost all the detectable particles of about 0.8 nm and even less remained (some of them are marked with arrows in Figs. 4a, 4b). Following subsequent cycles of alternate heating in H_2 and O_2, a large number of events of migration and coalescence of particles (A, B in Figs. 4c–4f and 5a–5e) and dumbbell—shaped particles (E in Figs. 4e–4f and 5a–5e) could be observed in both O_2 and H_2, as shown in Figs. 4 and 5.

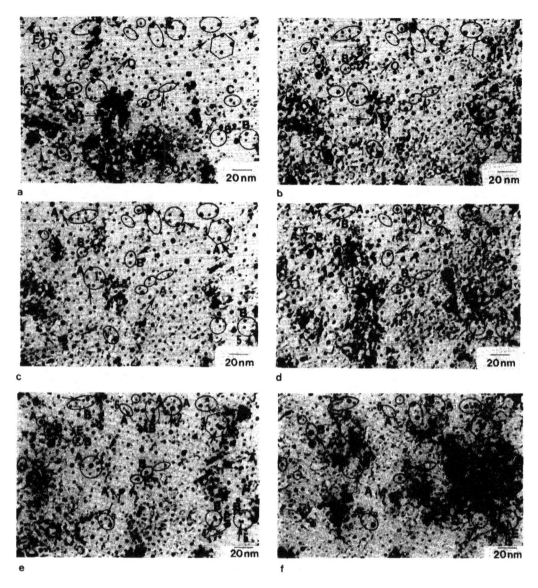

FIG. 4. Same sample as in **Fig.** 1 heated subsequently in H_2 and O_2 alternately at 500°C. (a) Initial (65 h H_2 + 28 h O_2 + 18 h H_2 + 14 h O_2); (b) 6 h H_2; (c) 12 h O_2; (d) 14 h H_2; (e) 12 h O_2; (f) 10 h H_2 + 10 h O_2.

As seen on the micrographs (Figs. 4 and 5), there is, in addition, a marginal decrease in the size of all the particles on heating in oxygen and an increase in size on heating in hydrogen (Figs. 4c and 4d), which is discussed in detail in Refs. (57, 58). It can be seen that while there is very little ripening, especially of the global kind, large particles (regions marked 1–5 in Figs. 4c–4f and regions 1 and 4 in Figs. 5a–5e), as large as 8 nm (region 1 in Figs. 5a–5e), migrate gradually over large distances (25 nm) and coalesce with other particles. At region 2 in Figs. 4e and 4f, the large particle appears to have migrated over a smaller particle. The smaller particle is probably trapped in a valley on the substrate surface and does not hinder the migration of the larger particle over it. On the other hand, at region 1 in Fig. 5, for example, the large particle incorporates into itself the smaller particles it encounters while migrating. The numerous events marked A, B, and E in the micrographs provide evidence for more than isolated incidents of crystallite migration in supported model catalysts. Also, dumbbell—shaped particles (E in Figs. 4 and 5) remain over extended periods of heat treatment,

Events Observed and Evidence for Crystallite Migration in Pt/Al$_2$O$_3$ Catalysts

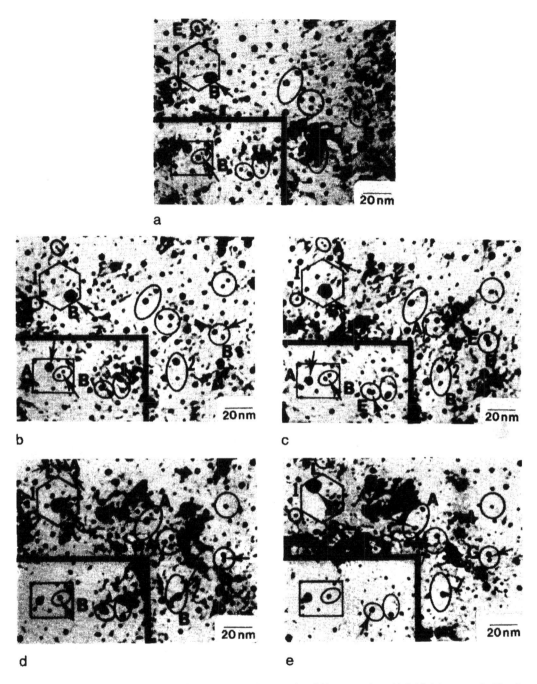

FIG. 5. Same sample as in Fig. 4. The micrographs are of a different region, (a) Initial (same as in Fig. 4); (b) 6 h H$_2$; (c) 12 h O$_2$ + 14 h H$_2$; (d) 12 h O$_2$; (e) 10 h H$_2$ + 10 h O$_2$. The regions bounded by the black margins are from a different area.

suggesting that coalescence can be a rate—limiting step. This is in contrast with the suggestion that coalescence must be fast (*14*). In addition, as noted before, sometimes two neighboring particles collide, partially coalesce, and then separate into two again (L in Figs. 4e, 4f and 5d, 5e).

It is clear that migration and coalescence play a role in the coarsening of Pt/Al_2O_3 catalysts not merely during the initial stages of sintering of a fresh sample but also after prolonged heat treatment of the sample. In contrast to the opinion that especially in an oxidizing atmosphere only atomic/molecular migration takes place, the present results show that migration and coalescence of crystallites also occur.

From the above results one can note that when an already sintered sample is heated further, changes begin to occur after two or three cycles of alternate heating in H_2 and O_2. Very little change with regard to sintering is noticed on extended heating in the same atmosphere. A change in exposure, either of an oxidized sample to a reducing atmosphere or of reduced particles to O_2, brings about particle motion, coalescence, etc. This is perhaps due to the heat of reaction involved, as suggested also by Glassl *et al.* (*33*). The above results show that migration of particles in Pt/Al_2O_3 catalysts occurs at the relatively low temperature of 500°C, both with a fresh sample in H_2 and a "sintered" sample in both O_2 and H_2.

A different sample of 2-nm initial film thickness was heated in additionally purified H_2 at 500°C for 12 h and subsequently at 600°C for 5 h to generate particles. Following heating at 500°C, a large number of particles of about 1.6 nm average size were formed. The specimen employed in the experiments described in the previous paragraphs, which had a lower metal loading of 1.5-nm initial film thickness, yielded slightly larger particles following only 6 h of heating at 500°C. The difference in the sintering behaviors of these two specimens points to the role of the type of hydrogen used. Sintering is faster and relatively more pronounced in the as—received hydrogen (which contains traces of water) than in the further purified H_2. Similar trends were also observed by Smith *et al.* (*41*). The additional heating at 600°C for 5 h caused severe sintering. Dumbbell—shaped particles could be observed following both the 500 and the 600°C heat treatments (particles marked with arrows in Figs. 6a and 6b). Even though the sintering mechanism could not be followed during these heat treatments because of the severity of sintering, the presence of dumbbell—shaped particles suggests that at both temperatures crystallite migration and coalescence were involved. This sample was subsequently heated alternately in O_2 and purified H_2. A few select micrographs are shown in Figs. 7 and 8. The heat treatment in O_2 was carried out from 300 to 500°C, with a 50°C increment in successive cycles. But the heat treatment in H_2 in each cycle was carried out at 500°C. (The incremental heating in O_2 was carried out in an attempt to identify the lowest temperature at which film formation around the particles, or a decrease in size of all the particles, occurs. Film formation around particles when Pt/Al_2O_3 model catalysts are heated in O_2 is discussed in Refs. (*57,58*).) Throughout this treatment and especially up to the 500°C O_2/H_2 cycling, the growth was slow and gradual and the marginal growth occurred mostly by coalescence of nearby particles (C in Figs. 7'b–7'f). However, at 500°C, as with the other sample reported in the previous paragraphs, significant migration and coalescence were observed on alternate heating in H_2 and O_2. A number of particles as small as 1 nm remained with very little change throughout the various cycles of heating for up to a total of about 100 h (some are marked with arrows in Figs. 7a–7f and 8a–8h), while a few others disappeared. This suggests that at low temperatures very little global ripening, involving a large number of particles, takes place (in spite of the alternating oxidizing and reducing treatments). Particle migration (A), migration and coalescence (B and C in Figs. 7 and 8), and direct ripening between a small and a larger particle via a narrow whiskerlike bridge (G, region 1 in Fig. 8) were all observed. In other cases nearby particles made and broke contact a few times without any change in the size of the particles (region 2 in Fig. 8). These events indicate that in some cases the coalescence is retarded or inhibited altogether and that coalescence may also be a rate—limiting step, as stated before.

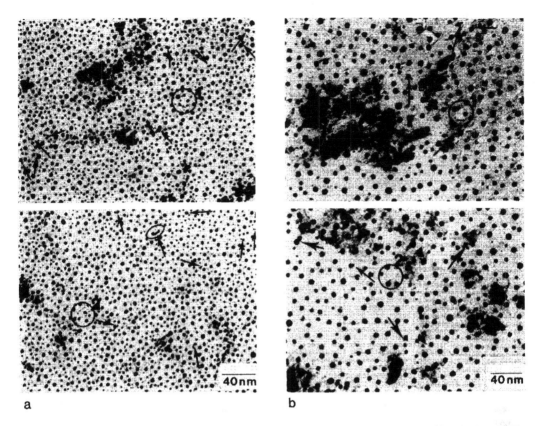

FIG. 6. Changes on heating a Pt/Al$_2$O$_3$ sample of 2 nm initial film thickness in purified hydrogen. The micrographs are of the same regions, (a) 12 h H$_2$, 500°C; (b) 5 h H$_2$, 600°C.

Another sample from the same batch as the preceding one with a 2-nm initial film thickness of Pt was heated similarly in purified H$_2$ at 500°C for 12 h and subsequently at 600°C for 5 h to obtain an initial crystallite distribution (Fig. 9a). Heating this sample further in purified H$_2$ at 700°C for only 1/2 h led to severe sintering (Fig. 9b). A large number of small and relatively large particles disappeared, other particles of the same size and even smaller remained unaffected, while some of the smaller particles also decreased in size (particles marked with arrows in Figs. 9a, 9b). In addition, a large number of neighboring pairs of crystallites coalesced, as indicated by the elongated and/or dumbbell—shaped particles in the micrographs (C, E in Figs. 9a, 9b). In one location, out of the two small particles adjacent to a much larger particle, the seemingly larger of the two decreased in size while the smaller grew or remained unaffected (regions 1, 2 in Figs. 9a–9d). In another place, two large particles of about the same size coalesced into one, while the nearby very small particles decreased only slightly in size (F in Figs. 9a–9d). This indicates that both local ripening and short—distance migration and coalescence occur adjacent to each other. Probably because of the high temperature, most of the particles were faceted and rectangular. Subsequent heat treatment

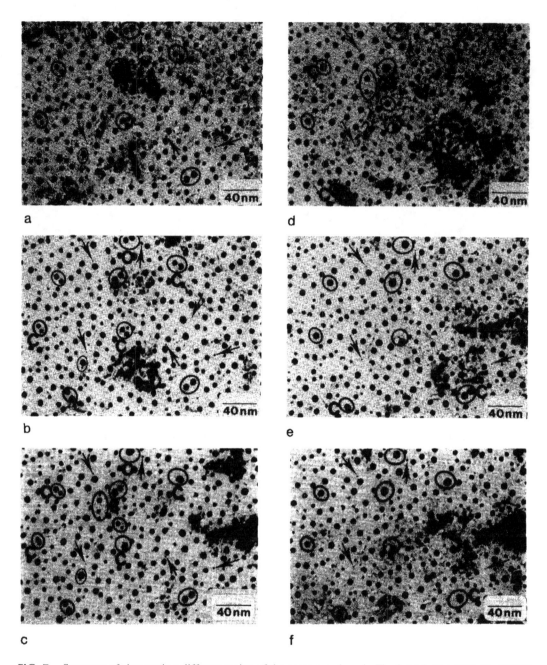

FIG. 7. Sequence of changes in a different region of the same sample as in Fig. 3. (a) Initial (12 h H_2, 500°C + 5 h H_2, 600°C); (b) 4 h O_2, 300°C; (c) 2 h H_2, 500°C; (d) 4 h O_2, 350°C + 4 h H_2, 500°C; (e) 3 h O_2, 400°C + 4 h H_2, 500°C; (f) 5 h O_2, 450°C.

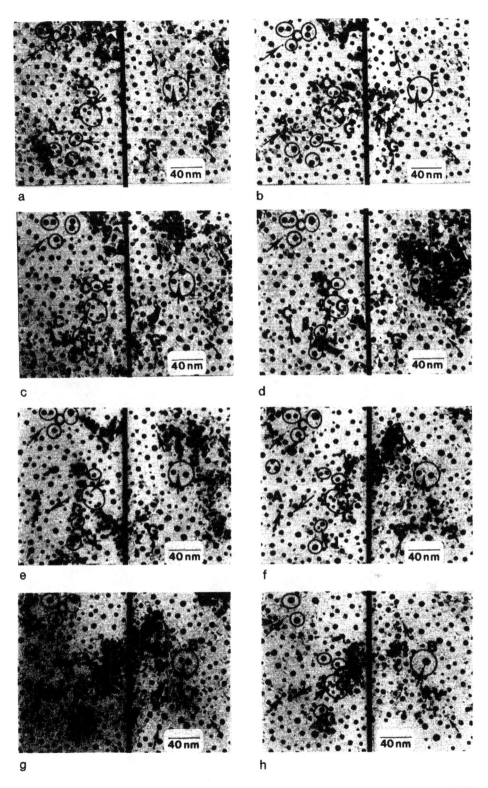

FIG. 8. Same sample as in Fig. 6. The micrographs are of a different region, (a) Initial (12 h H$_2$, 500°C + 5 h H$_2$(600°C); (b) 4 h O$_2$, 300°C + 2 h H$_2$, 500°C; (c) 4 h O$_2$, 350°C; (d) 4 h H$_2$, 500°C; (e) 3 h O$_2$, 400°C; (f) 4 h H$_2$, 500°C; (g) 4 h O$_2$, 450°C; (h) 2 h H$_2$, 500°C.

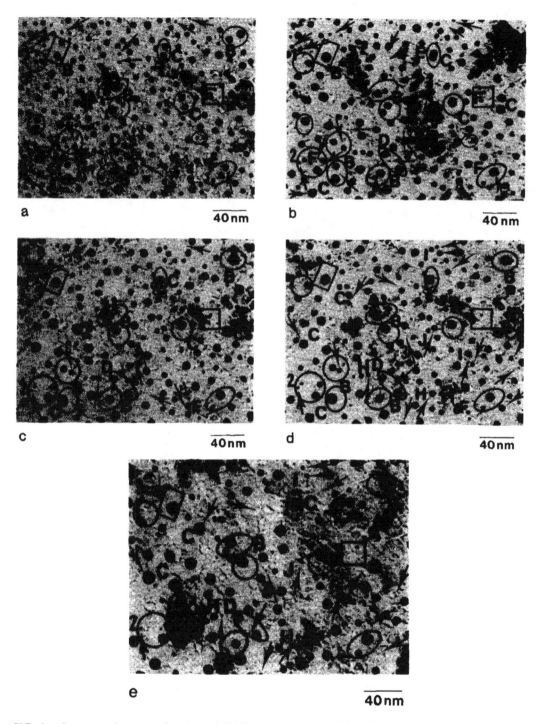

FIG. 9. Sequence changes on heating a Pt/Al$_2$O$_3$ sample of 2 nm initial film thickness in purified H$_2$. (a) Initial (12 h H$_2$, 500°C + 5 h H$_2$, 600°C); (b) 1/2 h H$_2$, 700°C; (c) 2 h H$_2$, 700°C; (d) 5 h H$_2$, 700°C; (e) 2 h H$_2$, 800°C.

Events Observed and Evidence for Crystallite Migration in Pt/Al$_2$O$_3$ Catalysts 177

at the same temperature for up to a total of 5 h resulted in gradual sintering, and similar events such as coalescence of neighboring particles (C), decrease in size of small particles (H), decrease in size and migration (D), gradual migration and coalescence (B), and disappearance of small particles (particles marked with arrows) could be observed (Figs. 9b–9d). Also, the particles became relatively circular. Heating this sample further in H$_2$ at 800°C for 2 h led to a drastic decrease in the number of particles and an increase in the average particle size. A large number of both small and large particles disappeared (J and particles marked with arrows in Figs. 9d, 9e). Some relatively large particles decreased in size while considerably smaller particles nearby remained unchanged (I). The sintering was very fast and it can only be inferred that both short—distance migration and subsequent coalescence of particles (B, C) and localized ripening (H) and migration of smaller particles, which decreased in size as they migrated (regions 1, 3 in Figs. 9d, 9e), contributed to the growth of the particles.

Another sample of 1.5-nm initial film thickness was heated in purified H$_2$ at 750°C for 1 h to generate crystallites. Only 1 h of this high—temperature treatment of the fresh sample with a lower loading (1.5 nm) yielded particles considerably larger than those obtained with a sample of higher loading (2 nm) after 12 h of heating at 500°C. As expected, the sintering is faster and more pronounced at the higher temperatures. Further heating in purified H$_2$ at 750°C for 1/2 h led to a drastic decrease in the number of particles. The average size of the particles also decreased. The decrease in the number of particles and also an overall decrease in their size cannot possibly be attributed to sintering by a global ripening mechanism. A part of the crystallites is lost via evaporation, or more likely has diffused onto or into the substrate, or has spread as a thin undetectable film. Figures 10a, 10b, 11a and 11b show some of the events. A large number of small and large particles disappeared completely without significant growth of the remaining particles (J). In fact, in some locations, the large particles decreased in size significantly while much smaller particles adjacent to them were affected only marginally or even remained unchanged (I). Larger particles were detected in regions where there were smaller particles before, suggesting growth by collision and coalescence of smaller particles (K). Small particles could be seen in regions adjacent to places where larger particles had disappeared (D), indicating decrease in size and subsequent migration of particles or vice versa. It is unlikely that the new small particles have grown in size from undetectable particles via atom capture because considerably larger particles could be seen to decrease in size or disappear nearby, which is not compatible with the growth of smaller particles from capture of single atoms. It is likely that these particles are migrating remnants of dissolving larger particles. Short distance migration of small particles (~1.5 nm) without coalescence (A) and a few instances of migration with subsequent coalescence (B) could be observed following heating for an additional 1/2 h at 750°C (Figs. 11b, 11c). However, the decrease in the number of particles was very small. Subsequent heating at 1/2 – and 1–h intervals resulted in migration and coalescence of nearby particles (B) (Figs. 11c, 11d). A few small particles also disappeared (particles marked with arrows). It is likely that these small particles collided with nearby larger particles and merged with them, since other small particles of about the same size or larger migrated (A) and a few nearby small particles grew (region 1 in Figs. 11c, 11d). Beyond 4 h, when longer heating intervals were employed, particle coarsening continued up to a total of 31 h of heating in H$_2$ (Figs. 11e–11g). In addition to coalescence of nearby particles, the decrease in size and/or disappearance of a number of small particles were also observed.

Figure 12 shows another sample heated at 750°C. The micrographs show the events observed on heating in H$_2$ at 750°C, subsequent to one cycle of heating in H$_2$ and O$_2$ at the same temperature. The micrographs show migration of particles (A), particles that were initially apart but later contacted each other, indicating migration and collision (B), dumbbell—shaped particles (E), separation into two of contacting particles (L), and generation of two particles from one (M). Also, a number of instances of overlapping particles (N) could be seen.

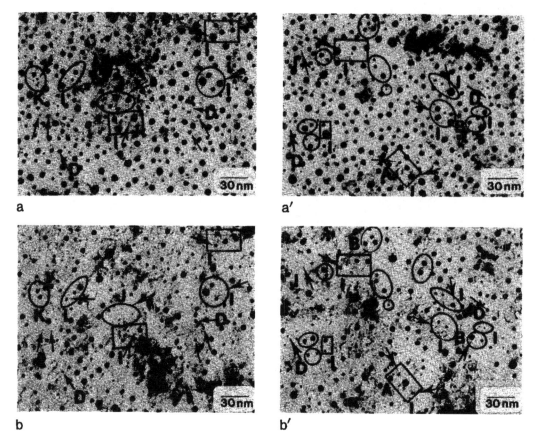

FIG. 10. Micrographs showing the sequence of changes in a Pt/Al$_2$O$_3$, sample of 1.5 nm initial film thickness on heating in purified hydrogen at 750°C. (a, a') Initial (1 h H$_2$, 750°C); (b, b') 1/2 h H$_2$, 750°C.

DISCUSSION

The results reported here have provided further evidence for crystallite migration and coalescence and their contribution to sintering of Pt/Al$_2$O$_3$ catalysts in H$_2$ and on alternate heating in H$_2$ and O$_2$. However, as reported before in the case of Pd/Al$_2$O$_3$ (36) and Fe/Al$_2$O$_3$ (39), ripening, especially of the localized kind, also often takes place concurrently. In fact, a variety of events takes place during sintering of supported metal catalysts and a few of them observed here in the case of Pt/Al$_2$O$_3$ are summarized schematically in Table 1. Even though short—distance migration and coalescence of nearby particles (particles that are one to a few diameters apart) is probably the major mechanism of particle growth, especially in the initial stages, a decrease in size of smaller particles because of localized ripening has also been observed. A number of small particles of about 1.5 to 2 nm in diameter or smaller disappeared. Since, as mentioned in the previous section, a number of particles of similar size remained unaffected or migrated and since larger particles have also been observed to migrate on the same micrograph, there appears to be no compelling reason to attribute the disappearance of such small particles to ripening only. It is possible and quite probable that these particles migrate relatively fast, especially because of their small sizes, and coalesce with the larger particles. Some small particles that remain stationary and unchanged

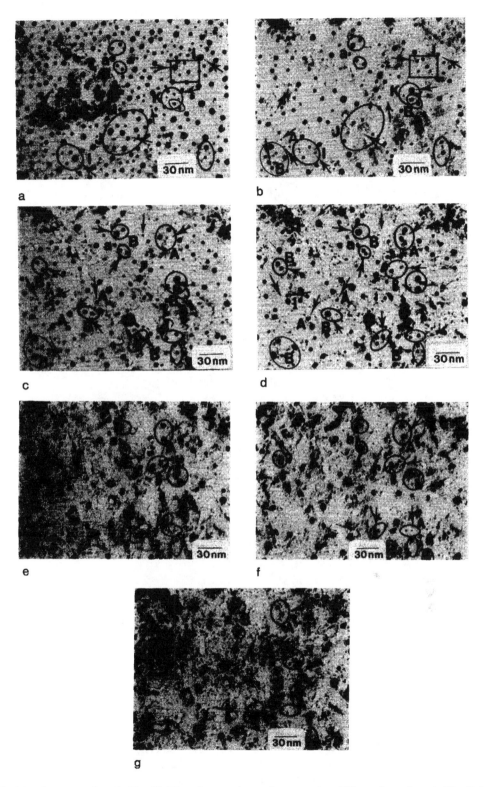

FIG. 11. Same sample as in Fig. 10. The micrographs are from a region different from those in Fig. 11. The following durations of heating are cumulative (a) Initial (1 h H_2, 750°C); (b) –1/2 h H_2; (c) 1 h H_2; (d) 3 h H_2; (e) 6 h H_2; (f) 9 h H_2; (g) 13 h H_2.

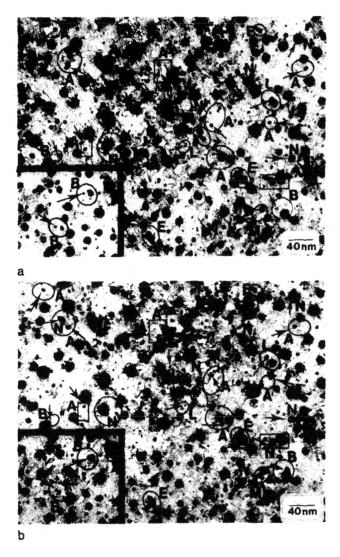

FIG. 12. Sequence of changes in a Pt/Al$_2$O$_3$ sample of 1.5 nm initial film thickness heated in purified H$_2$ and O$_2$ at 750°C. (a) Initial (15 h H$_2$, 750°C + 5 h O$_2$, 750°C + 17 h H$_2$, 750°C + 6 h H$_2$, 750°C); (b) 9 h H$_2$, 750°C.

might have been trapped in the valleys of the heterogeneous substrate surface, as also noted previously in the case of Pd/Al$_2$O$_3$ (36).

It is worth noting that in some cases, two particles approach each other to a kind of distance of minimum approach and growth subsequently occurs via transfer of atoms or molecules from one particle to the other through a very narrow bridge established between the particles (G in Figs, 2a, 2b and in Figs. 8a–8h). Such a growth by ripening between two adjacent particles via a whiskerlike bridge contact has also been observed with Fe/Al$_2$O$_3$ (39). It is also possible that the particles coalesce when pulled together because of the contraction of the connecting bridge. Arai *et al.* (37) have also postulated that the growth of Pt particles occurs via formation of a bridge between two adjacent particles and coalescence by an abrupt movement that brings the particles together, but only above 600°C and only with particles larger than 10 nm. It appears that when the particles are close to each other, growth via ripening is more of a localized nature, aided by long—range interparticle interactions (which apparently generate a localized curvature that facilitates the removal of atoms or molecules from one particle to the other). In fact, in addition, the micrographs suggest that the migration of crystallites is not diffusional (random) but directional, guided by long—range interparticle interactions. The particles seem to feel the presence of other particles nearby and, in general, migrate toward each other. Of course, migration of particles away from each other has also been observed, though only occasionally. It should be noted that there are also other kinds of interactions affecting the migration of the particles, such as the interactions between particles and support and the surface roughness of the support. Direction—selective migration of particles has been suggested before (36). When the interparticle force is strong, it is possible that even if a particle is trapped in a valley or is too large to migrate, it may still rotate and rearrange to orient its more highly curved protrusions along the periphery (and in extreme cases even form long whiskerlike protrusions) in the direction of a nearby particle to emit atoms or molecules toward it (G in Figs. 8a–8h). Such events, though isolated, have been observed with Fe/Al_2O_3 previously (39). Of course, in the case of supported iron particles the interparticle interaction forces may be stronger because of the ferromagnetic or superparamagnetic nature of the particles.

Finally, there are at least two possible reasons for failure to detect crystallite migration even in systems where it occurs. Since the particles in a fresh sample are, in general, small (less than 1 or 2 nm), they migrate relatively fast and even 1 or 2 h of observation (as is often used) may be too long to identify such a mechanism. Shorter heating intervals, especially in the initial stages, may be helpful. Even in the case of 1 or 2 h of heating, particle migration and coalescence may be detected or inferred if considerable care is taken to follow the same particles in various regions covering a large area of the sample. Second, if the observation is not continued for longer than 1or 2 h of heating, evidently it is not possible to detect the slower migration of the now relatively larger particles and/or other phenomena that may occur. Also in this case the observation made above regarding careful scanning of a large area might be useful.

In the case of model Pt/Al_2O_3 catalysts, as with other traditional supported metal catalysts, sintering is pronounced at higher temperatures, higher loadings, and at any temperature in the initial few (4–6) hours of heating a fresh catalyst. Sintering in as– received ultrapure hydrogen (which, however, contains traces of moisture and O_2) is enhanced relative to that in the same hydrogen further purified to eliminate/reduce the trace moisture and O_2. In a hydrogen atmosphere and below about 700°C, sintering seems to occur primarily by short—distance migration and coalescence of nearby particles as well as by localized ripening between a few neighboring particles. Significant migration of larger particles occurs around 500°C when H_2–treated samples are heated in O_2 or vice versa. At $T < 500$°C sintering is very slow in H_2 and on alternate heating in H_2 and O_2 and even slower in O_2.

CONCLUSION

Transmission electron microscopic evidence for the role of crystallite migration and coalescence among other mechanisms in the sintering of model Pt/Al_2O_3 catalysts is provided. Results of model Pt/Al_2O_3 catalysts heated in H_2 and alternately in H_2 and O_2 at temperatures in the range 500–750°C indicate that a large number of phenomena such as coalescence of nearby particles, crystallite migration (of both small, 1.5 nm, and large, about 8 nm, particles), migration followed by coalescence, decrease in size and disappearance of small particles near larger particles, disappearance of small and larger particles, decrease in size of large particles near unaffected smaller particles, decrease in size and subsequent migration of particles or vice versa, collision and inhibited coalescence of particles, and collision—coalescence—separation of particles occur. It appears that the particles on the support feel the presence of nearby particles probably via the interaction forces between them and migrate or emit atoms toward them, and consequently the migration is directional and the ripening is local.

REFERENCES

1. Mills, G. A., Weller, S., and Cornelius, E. B., *Proc. 2nd Int. Congr. CataL, Paris,* p. **2221** (1960).
2. Adler, S. F., and Keavney, J. J., *J. Phys. Chem.* **64**, 208 (1960).
3. Maat, H. J., and Moscou, L., *Proc. 3rd Int. Congr. Catal., Amsterdam* **2**, 1277 (1965).
4. Dalla Betta, R. A., McCune, R. C., and Sprys, J. W., *Ind. Eng. Chem. Prod. Res. Dev.* **15**, 169 (1976).
5. Johnson, F. L., and Keith, C. D., *J. Phys. Chem.* **67**, 200 (1963).
6. Kraft, M., and Spindler, H., *4th Int. Congr. Catal., Moscow,* p. 1252 (1968).
7. Kearby, K. K., Thorn, J. P., and Hinlicky, J. A., U.S. Patent 3,134,732 (1964).
8. Fiedorow, R. M. J., and Wanke, S. E., *J. Catal.* **43**, 34 (1976).
9. Ruckenstein, E., and Malhotra, M. L., *J. Catal.* **41**, 303 (1976).
10. Gollob, R., and Dadyburjor, D. B., *J. Catal.* **68**, 473 (1981).
11. Ruckenstein, E., and Dadyburjor, D. B., *Rev. Chem. Eng.* **1**(3), 251 (1983).
12. Ruckenstein, E., and Pulvermacher, B., *AIChE J.* **19**(2), 356 (1973).
13. Ruckenstein, E., and Pulvermacher, B., *J. Catal.* **29**, 224 (1973).
14. Wynblatt, P., and Gjostein, N. A., *Prog. Solid State Chem.* **9**, 21 (1975).
15. Flynn, P. C., and Wanke, S. E., *J. Catal.* **34**, 390 (1974).
16. Ruckenstein, E., and Dadyburjor, D. B., *J. Catal.* **48**, 73 (1977).
17. Ruckenstein, E., and Dadyburjor, D. B., *Thin Solid Films* **55**, 89 (1978).
18. Flynn, P. C., and Wanke, S. E., *J. Catal.* **37**, 432 (1975).
19. Baker, R. T. K., Thomas, C., and Thomas, R. B., *J. Catal.* **38**, 510 (1975).
20. Hermann, R., Adler, S. F., Goldstein, M. S., and Debaum, R. M., *J. Phys. Chem.* **65**, 2189 (1961).
21. Hughes, T. R., Houston, R. J., and Sieg, R. P., *Ind. Eng. Chem. Process Des. Dev.* **1**, 96 (1962).
22. Gruber, H. L., *J. Phys. Chem.* **66**, 48 (1962).
23. Somoijai, G., *Anal. Chem.* **1**, 101 (1968).
24. Chu, Y. F., and Ruckenstein, E., *Surf. Sci.* **67**, 517 (1977).
25. Chu, Y. F., and Ruckenstein, E., *J. Catal.* **55**, 281 (1978).
26. Chen, M., and Schmidt, L. D., *J. Catal.* **55**, 348 (1978).
27. McVicker, G. B., Garten, R. L., and Baker, R.T. K., *J. Catal.* **54**, 129 (1978).
28. Wang, T., and Schmidt, L. D., *J. Catal.* **66**, 301 (1980).
29. Fiedorow, R. M. J., Chahar, B. S., and Wanke, S. E., *J. Catal.* **51**, 193 (1978); Wanke, S. E., *in* "Sintering and Heterogeneous Catalysis" (G. C. Kuczynski, A. E. Miller, and G. A. Sargent, Eds.), *Mat. Sei. Res.,* Vol. 16, p. 223. Plenum, New York, 1984.
30. Baker, R. T. K., Prestridge, E. B., and Garten, R. L., *J. Catal.* **56**, 390 (1979); Baker, R. T. K., *Catal. Rev. Sei. Eng.* **19**(2), 161 (1979).
31. Yao, H. C., Sieg, M., and Plummer, H. K., Jr., *J. Catal.* **59**, 365 (1979).
32. Richardson, J. T., and Crump, J. G., *J. Catal.* **417** (1979).
33. Glassl, H., Kramer, R., and Hayek, K., *J. Catal.* **64**, 303 (1980).
34. Kuo, H. K., Ganesan, P., and DeAngelis, R. J., *J. Catal.* **66**, 171 (1980).
35. Straguzzi, G. 1., Aduriz, H. R., and Gigola, C. E., *J. Catal.* **66**, 171 (1980).
36. Chen, J. J., and Ruckenstein, E., *J. Catal.* **69**, 254 (1984).
37. Arai, M., Ishikawa, T., Nakayama, T., and Nishiyama, Y., *J. Colloid Interface Sei.* **97**, 254 (1984).
38. Kim, K. T., and Ihm, S. K., *J. Catal.* **96**, 12 (1985).
39. Sushumna, I., and Ruckenstein, E., *J. Catal.* **94**, 239 (1985) (Section 3.4 of this volume).
40. Ruckenstein, E., and Lee, S. H., *J. Catal.* **86**, 457 (1984).
41. Smith, D. J., White, D., Baird, T., and Fryer, J. R., *J. Catal.* **81**, 107 (1983).
42. Moss, R. L., *in* "Catalysis" (A Specialist Periodical Report) (C. Kemball and D. A. Dowden, Eds.), Vol. 4, p. 31. R. Soc. Chem., London, 1980.
43. Baird, T., *in* "Catalysis" (A Specialist Periodical Report) G. C. Bond and G. Webb, Eds.), Vol. 5, p. 172. R. Soc. Chem., London, 1981.
44. Burch, R. *in* "Catalysis" (A Specialist Periodical Report) (G. C. Bond and G. Webb, Eds.), Vol. 7, p. 149. R. Soc. Chem., London, 1983.
45. Dautzenberg, F. M., and Wolters, H. B. M., *J. Catal.* **51**, 26 (1978).
46. Menon, P. G., and Froment, G. F., *J. Catal.* **51**, 26 (1978).
47. Tauster, S. J., Fung, S. C., and Garten, R. L., *J. Amer. Chem. Soc.* **100**, 170 (1978).
48. References in "Metal–Support and Metal–Additive Effects in Catalysis" (B. Imelik, C. Nac– cache, G. Coudourier, H. Praliaud, P. Meriau– deau, P. Gallezot, G. A. Martin and J. Vedrine, Eds.). Elsevier, Amsterdam, 1982.

49. Bond, G. C., and Burch, R., *in* "Catalysis" (A Specialist Periodical Report) (G. C. Bond and G. Webb, Eds.), Vol. 6, p. 27. R. Soc. Chem., London, 1983.
50. Kunimori, K., Ikeda, Y., Soma, M., and Uchijima, T., *J. Catal.* **79**, 185 (1983).
51. Bassett, G. A., Proc. Eur. Reg. Congr. Electron Microsc., Delft, **1**, 270 (1960).
52. Skofronick, J. G., and Phillips, W. B., *J. Appl. Phys.* **38**, 4791 (1967).
53. Heinemann, K., and Poppa, H., *Thin Solid Films* **33**, 237 (1976).
54. Granquist, C. G., and Buhrman, R. A., *J. Catal.* **42**, 477 (1976).
55. Granquist, C. G., and Buhrman, R. A., *J. Catal.* **46**, 238 (1977).
56. Glassl, H., Kramer, R., and Hayek, K., *J. Catal.* **63**, 167 (1980).
57. Sushumna, I., and Ruckenstein, E., *in* "Hydrogen Effects in Catalysis" (Z. Paal and P. G. Menon, Eds.). Dekker, New York, 1988.
58. Sushumna, I., and Ruckenstein, E., *J. Catal.* **108**, 77 (1987) (Section 3.2 of this volume).

3.4 Role of Physical and Chemical Interactions in the Behavior of Supported Metal Catalysts

Iron on Alumina—A Case Study[*]

I. Sushumna and Eli Ruckenstein[†]

Department of Chemical Engineering, State University
of New York, Buffalo, New York 14260
Corresponding Author
[†]To whom correspondence should be addressed.

Received November 1, 1984; revised January 22, 1985

INTRODUCTION

Currently, metal catalysts are being prepared mostly in a highly dispersed, supported form replacing the old fused or unsupported powder types. While the commonly used refractory oxide supports, silica and alumina, increase the metal dispersion, they are not inert especially toward the non-noble metals (*1*). The chemical reactivity between the active metal and the oxide support in turn affects the particle size distribution and consequently the activity and selectivity of the catalyst. Iron, the most electropositive of the group VIII elements and consequently a very reactive metal, is employed extensively as a catalyst in such important industrial processes as ammonia synthesis, the recently rejuvenated Fischer–Tropsch synthesis, and other hydrogenation and dehydrogenation reactions (*2*). Traditionally, iron is deposited onto the oxide supports in the form of ferric salts, prior to calcination and reduction at high temperatures (*3, 12*). When present as Fe^{3+} ions, iron exhibits strong reactivity with the support, especially with alumina because of the chemical similarity between Fe^{3+} and Al^{3+} ions (*4, 5*). Consequently, at low loadings on alumina, iron is not reduced easily to the zero-valent state, because a large fraction of the iron ions interact strongly with the substrate. The strength of the interactions between the active-metal ion and the support and consequently the reducibility and the particle size of the supported iron catalyst depend upon the chemical precursor used, the catalyst pretreatment (particularly the temperature of decomposition), and the metal loading. For example, Garten and Ollis (*6*) have shown, for samples prepared by impregnation, that when the metal loading on alumina is low (0.05%), the ferric ions can be reduced only to the ferrous state, even when the reduction with hydrogen occurs at temperatures as high as 700°C. On the other hand, Brenner and Hucul (*7*) have observed that the decomposition at low temperatures (< 150°C) of zero-valent iron complexes, such as the iron carbonyls, adsorbed on alumina, yields highly dispersed iron catalysts. In this case the initial dispersion can be more than an order of magnitude higher than that of the catalyst prepared by the more traditional method of impregnation with ferric ions, but, because of the weaker interactions with the support, the zero

---

[*] *Journal of Catalysis* 94, 239–288 (1985). Republished with permission.

Role of Physical and Chemical Interactions in the Behavior of Supported Metal Catalysts   **185**

valent particles may sinter more easily. Raupp and Delgass (8) have shown that mild pretreatment of the impregnated catalysts by slow vacuum drying, prior to reduction, produces significantly smaller particles of iron in the case of Fe/SiO_2. The nonreducibility and/or the smaller particle sizes arising as a result of the interactions between metal and support can have important consequences on the activity and selectivity of the supported iron catalysts in the Fischer-Tropsch reaction (8–11).

The literature abounds in Mössbauer spectroscopic investigations of supported iron catalysts, especially regarding their role in the Fischer–Tropsch reaction (11–14). Mössbauer spectroscopy is a powerful tool to characterize catalysts involving ferromagnetic metals. Nonetheless, when a strong interaction between metal and support exists, the interpretation of the spectra becomes difficult regarding both the particle size distribution and the exact chemical compounds formed (15). In addition, information regarding the microstructure and the shape of the particles cannot be obtained (16). A complementary technique such as electron microscopy can provide both the size distribution and the shape of the particles and, via electron diffraction, can also identify the chemical compounds formed. Electron microscopy and electron diffraction have been used in the present paper to investigate the behavior of a model Fe/Al_2O_3 catalyst and results are presented regarding the physicochemical changes accompanying their heating at different temperatures in a reducing (H_2) atmosphere, in an oxidizing (O_2) atmosphere, and during an alternate heating in these atmospheres. The present investigation emphasizes the role of the surface phenomena and of the chemical interactions in the complex processes observed with model Fe/Al_2O_3 catalysts. Accordingly, the electron micrographs, which provide physical information on shape and size, are supplemented with chemical information on the compounds formed via electron diffraction.

EXPERIMENTAL

Preparation of model alumina supports. Electron-transparent and nonporous films of amorphous Al_2O_3 were prepared by anodically building up the oxide on a clean, thin, high-purity aluminum foil (99.999%, Alfa Products Inc.) and stripping the oxide film off by dissolving the unoxidized aluminum in a dilute mercuric chloride solution (21). The oxide films were washed in distilled water and picked up on gold electron-microscope grids. They were subsequently heated in air at 800°C for 72 h to transform the amorphous alumina into the γ or η form and also to ensure that no further changes would occur during its subsequent use.

Deposition of the active metal. Iron films of different thicknesses were deposited onto the alumina substrates by evaporating the corresponding amounts of 99.998% pure iron wire (Alfa Products Inc.) from a tungsten basket in an Edwards vacuum evaporator under a vacuum of better than 2×10^{-6} Torr. The thickness of the metal film deposited was estimated from the amount of metal evaporated and the distance between the source and the target. The substrate was kept at room temperature during deposition.

Heat treatment. Samples were heated in a quartz boat inside a 3.8-cm-diameter and 120-cm-long quartz tube. A predetermined gas flowed through the quartz tube during heating and its flow rate of 150 cm^3/min was maintained constant during the heat treatment. The gases used in the experiments were all ultrahigh purity (UHP) grade, supplied by Linde Division, Union Carbide Corporation. Hydrogen, 99.999% pure, contained less than 1 ppm O_2 and less than 2 ppm moisture. Helium, also 99.999% pure, contained less than 3 ppm moisture. For some experiments, the hydrogen was further purified by passing it through a Deoxo unit (Englehard Industries) followed by a silica gel column and then through a 5A molecular sieve bed immersed in liquid nitrogen. Helium was also purified by passing it through the liquid-nitrogen trap.

RESULTS AND DISCUSSION

After introducing the boat containing the sample into the reactor, each heat treatment followed the sequence: flushing the reactor with helium for at least 5 min; heating the specimen from room temperature to the desired value in a helium atmosphere, switching over the gas stream to hydrogen or oxygen as desired and heating for the predetermined length of time, and cooling down the specimen slowly to room temperature in a helium atmosphere before it is withdrawn for observation. Following each heat treatment, the same regions of the sample were photographed using a JEOL 100U transmission electron microscope. During each observation, the entire specimen area was scanned to check if the behavior was uniform throughout. Electron diffraction patterns were obtained for all the specimens after each treatment.

RESULTS AND DISCUSSION

Several samples have been investigated at each of the following loadings corresponding to film thicknesses of 5, 6, 7.5, 10, and 12.5 Å. Except for the cases in which the behavior was different, the results are reported for either 6- or 7.5-Å loadings. A large number of events have been observed at each of the temperatures investigated regarding the sintering and the wetting behavior as well as the changes in the shape of crystallites following heating in hydrogen, oxygen, and alternate heating in these two atmospheres. Table 1 summarizes the various events observed. Tables 2 to 5 provide the events observed under various conditions, the chemical compounds identified, and the corresponding micrograph numbers. The torus and core-and-ring structures, to which we refer frequently later in the text, are represented in Figs. 1a and b.

A. BEHAVIOR WITH AS-SUPPLIED HYDROGEN

A.1. Heating in Hydrogen

It is seen from Table 2 that distinct crystallites do not form easily at 300°C, whereas at 400°C they form after heating for only 1 h. At both temperatures, a compound whose d values lie between those of γ-Fe_2O_3(Fe_3O_4) and the higher aluminate, $Al_2Fe_2O_6$, is detected, in addition to other compounds. (γ-Fe_2O_3 and Fe_3O_4 have the same lattice constant and therefore cannot be easily differentiated by electron diffraction). It is to be noted that the major d values of $Al_2Fe_2O_6$ are greater than those of γ-Fe_2O_3(Fe_3O_4). Therefore, an increase in the major d values from those of γ-Fe_2O_3 toward those of $Al_2Fe_2O_6$ would indicate the formation of solid solutions of iron oxide in alumina ultimately leading to $Al_2Fe_2O_6$. Thus an increase in the d values toward those of $Al_2Fe_2O_6$ indicates the formation of Fe_2O_3 and its subsequent dissolution in Al_2O_3; conversely, a decrease in the d values of one such solid solution toward those of γ-Fe_2O_3(Fe_3O_4) indicates a partial precipitation of Fe_2O_3 from the solid solution and its eventual reduction to the metal and/or a nonstoichiometric lower oxide. In comparison to the former case, the latter can be considered a relatively reduced (less oxidized) state. The major d values of Al_2O_3 and of the lower aluminate, $Fe_2Al_2O_4$ are in the order $FeO > FeAl_2O_4 > Al_2O_3$ and therefore, the formation of solid solutions between FeO and Al_2O_3, ultimately leading to $FeAl_2O_4$, can also be inferred from the increase or decrease in the d values. One may note that FeO is not stable below 560°C as a separate phase. The stoichiometric aluminates, $FeAl_2O_4$ and $Al_2Fe_2O_6$, have been detected in some cases. When the d values have not corresponded to the stoichiometric aluminates or oxides, solid solution formation has been inferred for reasons discussed above.

At 400°C, the total number of crystallites remained almost constant for heating times greater than 1 h. A few nearby pairs of particles coalesced following their contact because of extension (A in Figs. 2f–i). The crystallites, however, underwent alternate changes in shape from a torus enclosing a cavity to one in which an annular ring separates the torus from a core in the cavity (Figs. 2e–i). The torus shape was found to be associated with a more oxidized state and the core-and-ring structure with a relatively less oxidized state as discussed in detail in Ref. (22).

TABLE 1
Types of Events Observed

Type	Event and figure number
A	Coalescence of nearby particles. Appearance of dumbbell-shape particles (2f–2i; 3a–c; 4a, 4b; 8a–c: 13b–d)
B	Migration with subsequent coalescence (3a–c: 7b, c; 13a–d) (or, coalescence followed by migration)
C	Migration without coalescence (4′c, d; 6d, e; 7b, c; 11a, b; 13b, c)
D. a	Decrease and disappearance of a few smaller particles (Localized ripening between a few particles. Direct ripening) (3a, b; 4′a–c)
D. b	Ripening between particles of almost the same size. Probably due to local curvature difference (3a, b; 4′a–c)
D. c	Transfer of molecules via a narrow, whisker-like bridge or a neck formed between two particles (3a, b; 4′a–c)
E	Decrease and disappearance of a large number of small particles. (Global ripening, Ostwald ripening; break-up to unresolved crystallites and their migration; extension to undetectable islands.) (4a–d; 12a–c)
F	Considerable sintering with redistribution of particles over some areas of the specimen (7b, c)

Schematic of the events
Before After

(*Continued*)

187

TABLE 1 (Continued)
Types of Events Observed

Type	Event and figure number	Schematic of the events (Before / After)
G	Decrease in size or disappearance of a particle with no obvious growth of nearby particles (3a, b; 4a, b; 6d, e)	
H	Disappearance of large particles without decreasing in size first; A large particle nearby a smaller particle decreases in size and/or disappears or coalesces with the latter. The smaller particle either increases in size or remains unaffected (3a–c; 13a–d)	
I	Disappearance of small as well as large particles (3a, b; 4a, b; 13a, b)	
J	Growth of some small particles (3b, c)	
K	Increase in contrast (darkening) of all particles. (Growth of all particles small and large); Light and thin particles grow bigger and darker. (6b, c; 7a, b)	
L	Appearance of new particles. (6e, f; 12a–d; 13a–c)	
M	Splitting of particles without complete separation of the units. (2g–i; 13b–d)	
N	Splitting of particles with separation and migration. (6d–f; 13a, b)	
O	Extension of small particles; recontraction of the extended particles. (2b–e; 13c, d)	
P	Appearance of periphery marks around particles. (7c; 9b, c)	

(Continued)

TABLE 1 (*Continued*)
Types of Events Observed

Schematic of the events
Before After

Type	Event and figure number	
Q	Appearance of a film extending out from around the particles. (12a, b)	
R	Substrate grain boundaries appear less distinct on heating in oxygen. A film appears to cover the substrate. The grain boundaries appear sharp again on heating in hydrogen subsequently. (12a, c, d; 15c, d, e)	
S. a	Considerable extension of the crystallite to result in a torus shape. (4e; 9a; 10b)	
S. b	Considerable extension and appearance of a torus shape with a small remnant particle in the cavity. (14b, d; 15a)	
T	Considerable extension to a kind of thick film with a remnant particle on top. (14b, d; 15d, e)	
U	Rupture of crystallites. The crystallite is composed of very small particles held close-together making it appear porous. (8c, d; 9d–f; 15d, e)	
V	Break-up of the crystallites with the subunits interconnected. The sub-units coalesce subsequently. (9a, b; 10b–d; 15a, b)	
W	Considerable break-up of the crystallites to very small pieces. (9g)	
X	Contraction of periphery marks to form small particles. (11c)	
Y	Faceting of particles. (11c)	

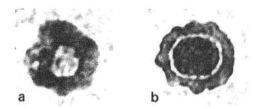

FIG. 1. Representation of (a) torus and (b) core-and-ring structures.

At 500°C, after heating in hydrogen for 6 h, both 6- and 12.5-Å specimens exhibited stability with respect to further growth (Tables $2_{6,500}$ and $2_{12.5,500}$).[2] However, whereas at the lower loading almost no metal was detected in the diffraction pattern even after a total of 27 h of heating in the as-supplied hydrogen, at the higher loading almost only metal was detected after only 6 h of heating in the same as-supplied hydrogen. This shows that only samples with high metal loadings can be reduced to the zero-valent state with hydrogen. As seen from the tables, the events observed and the compounds formed indicate that the chemical interactions between the crystallites and the substrate are very strong and result in the dissolution of the metal oxide in alumina and/or of alumina in the metal oxide. It appears that for the low loading used, it is almost impossible at 400 or 500°C to obtain the zero-valent metal with the as-supplied hydrogen. Even though there is some reducibility which causes the shape alternations (22), it is not sufficiently large to yield detectable α-Fe. The higher loading (12.5 Å) yielded, however, compact metallic particles of α-Fe which subsequently sintered considerably.

The results at 600°C for a loading of 6 Å are listed in Table $2_{6,600}$. Unlike at lower temperatures, the crystallites alternated in shape between a core-and-ring structure and a three-dimensional, compact structure, instead of a torus (Figs. 5a–d). At this temperature and loading, the compact structure is associated with $FeAl_2O_4$, Fe_2O_3, and only traces of α-Fe. In contrast, at 500°C and 12.5-Å loading, the compact particles were associated mostly with α-Fe.

Distinct crystallites formed more slowly at 700°C (Table $2_{6,700}$) than at 400 to 600°C (Table $2_{6,400} - 2_{6,600}$). Distinct crystallites could not be detected after an initial heating for 1 h (Fig. 6a). In addition, only rings of Al_2O_3 were present in the diffraction pattern. Possibly, very thin particles were present in an oxidized extended form on the substrate, undetectable by TEM and electron diffraction. The same observation is also valid for the 300°C treatment. At the lower temperature of 300°C, after about 2 h of heating the iron oxide molecules of the small (not well formed) clusters probably diffused into the substrate to form aluminate. At the higher temperature of 700°C, they are perhaps present initially as large, extended, thin islands of oxide, or even aluminate, too thin to be detected by electron diffraction. During the initial few hours of heating, the rate of aluminate formation is probably faster at 300°C than at 700°C because a larger fraction of molecules is in contact with alumina in the former than in the latter case. (In addition, since the clusters may not be well formed at 300°C, the molecules may be loosely held to one another.) The subsequent increase in size and contrast, with the number of crystallites remaining almost the same (Table $2_{6,700}$, Fig. 6c), occurs probably as a result of coalescence of the unresolved extended islands with the nearby resolved particles or by ripening. After a total of 18 h of heating, almost only $FeAl_2O_4$ is detected. The fact that only aluminate is detected indicates that, perhaps, it is present not only in the substrate, but on its surface and also in as well as on the particles.

---

[2] The first subscript refers to the initial film thickness in Å and the second to the temperature in °C.

TABLE 2
Heating in As-Supplied Hydrogen

Total heating time	Event observed	Compound identified
Table $2_{6,300}^{a}$ 1 h Up to 9 h	No distinct large particle formation No detectable change after heating in a few steps, except for changes in chemical composition. (2a)	γ-Al_2O_3, FeO · Al_2O_3 ($FeAl_2O_4$)and solid solution of Fe_2O_3 in Al_2O_3, approaching $Al_2Fe_2O_6$
Table $2_{6,400}$ 1 h 1 h	Scattered, light, small, and large particles with torus shape observed. (2a)[b]	FeO · Al_2O_3 and traces of Fe_3O_4(γ-Fe_2O_3)
	An increase in number of large and distinct crystallites. Some particles acquire the core-and-ring structure. (2b)	
2 h	Increase in size and darkening of small and large crystallites. Particles have a torus shape and are of uniform size. No change in the number of particles. (2c)	An increase in the intensity of $FeAl_2O_4$ rings and a decrease in the intensity of Fe_3O_4 rings
Up to 36 h	Considerable extension of some particles and their recontraction (O in 2c–e). Alternation in the shape of the crystallites (2e–i). Coalescence of nearby pairs of particles (A in 2f–i). Tendency to split and subsequent coalescence of these particles (M in 2g–i).	$FeAl_2O_4$ and a solid solution with d-values close to those of $Al_2Fe_2O_6$ when torus shape is observed. $FeAl_2O_4$ and a solid solution with d-values decreased from those corresponding to the case above and trace α-Fe when core-and-ring structure is observed
Table $2_{6,500}$ 1 h	Uniform size particles with torus shape (3a).	$FeAl_2O_4$, traces of $Al_2Fe_2O_6$
2.5 h	Increase in size of particles. Torus shape is maintained still. Very small cores are present in cavities of particles (1.b in 3c). Disappearance of small and large particles (1 in 3b, c). Disappearance of a large particle nearby a smaller particle (H in 3b, c) coalescence of nearby particles (A in 3a–c), dumbbell-shape particles (D. b in 3b, c), decrease in size and disappearance of some particles (D. b in 3b, c), migration and coalescence of particles (B in 3a–c), growth of some small particles (J in 3b, c) and extension and/or recontraction of particles are observed.	Increase in the intensity of $Al_2Fe_2O_6$
Up to 27 h	Marginal extension and contraction.	
Table $2_{12,5,500}$ 2 h	Particles with sharp, core-and-ring structure form. Small particles have a torus shape. (4a)	
3 h	All particles acquire torus shape. Some extension, and coalescence following contact. A few small particles decrease in size or disappear.	An increase in $Al_2Fe_2O_6$ and a decrease in $FeAl_2O_4$ Fe_3O_4(γ-Fe_2O_3), α-Fe
4 h	Particles contract. Particles acquire a core-and-ring structure. A number of small particles disappear (E in 4a, b). Coalescence of a few nearby pairs of particles (A in 4a, b).	

(Continued)

TABLE 2 (*Continued*)

Heating in As-Supplied Hydrogen

Total heating time	Event observed	Compound identified
5 h	Formation of compact particles. Decrease in size and/or disappearance of small particles, some of which coalesce with neighbors (A, E in 4b, c). Ripening between a few particles (D. a, D. b, D. c, in 4'a–c).	Mostly α-Fe
6 h	Disappearance of some more small particles.	Almost only α-Fe
Up to 24 h in a few steps	Compact, dark particles are formed. No other change. (4d)	Almost only α-Fe
46 h (22 in one step)	Particles are considerably extended and each particle encloses a cavity. The particles appear porous. Small fragments seen in the cavities of some particles (S.a in 4e).	Solid solution of Fe_2O_3 and Al_2O_3 with d-values slightly larger than those of Fe_3O_4 (γ-Fe_2O_3)
Table $2_{6,600}$ 2 h	Particles with core-and-ring structure form. The core is smaller and the annular gap is wider than at lower temperature. (5a, c)	More of $FeAl_2O4$ and less of Fe_3O_4(γ-Fe_2O_3)
4 h	The particles contract and tend to form compact particles. Very little change in the number of particles.	
Up to 41 h	Alternation in shape between a core-and-ring structure and a 3-dimensional compact structure. (5b, d)	More of Fe_3O_4 (γ-Fe_2O_3), traces of $Al_2Fe_2O_6$ and α-Fe
Table $2_{6,700}$ 1 h	Very few distinct crystallites are detected. (6a)	Al_2O_3
5 h	Light (low contrast) and mostly small particles are detected. (6b)	Faint rings of α-Fe, Al_2O_3
8 h	Increase in size and darkening of all particles. (6c)	$FeAl_2O_4$, faint rings of α-Fe and Fe_3O_4(γ-Fe_2O_3)
18 h	Growth of particles. Crystallites are dark and compact. Very little change in the number of resolved particles. (6d)	$FeAl_2O_4$
Up to 34 h	Migration (C), splitting (N), coalescence (A) and appearance of new particles (L) are detected. Traces of particle peripheries (P) and film around particles (Q) are also observed. (6e, f)	
Table $2_{6,800}$	(Sample previously heated at 600° C for 41 h). (7a)	
. 1 .	Very dark, compact particles are formed (K). No change in the number and position of the particles. (7b)	A compound with d-values slightly greater than those of Fe_3O_4 (γ-Fe_2O_3)
4 h	Considerable sintering and rearrangement of particles is observed (F). Migration of some large crystallites detected (C). Marks of particle peripheries seen (P). (7c)	Fe_3O_4 (γ-Fe_2O_3) and $Al_2Fe_2O_6$

[a] The first subscript refers to the loading in Å and the second to the temperature in °C.

[b] The numbers in parentheses refer to the figure numbers.

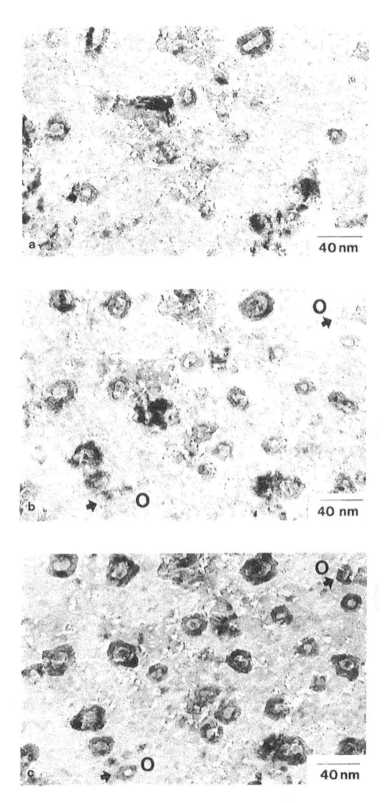

FIG. 2. Sequence of changes of a specimen heated at 400°C in as-supplied hydrogen. Initial loading, 6 Å. The micrographs show the same region after (a) 1/2 **h**, (b) 1 h, (c), 2 h. (*Continued*)

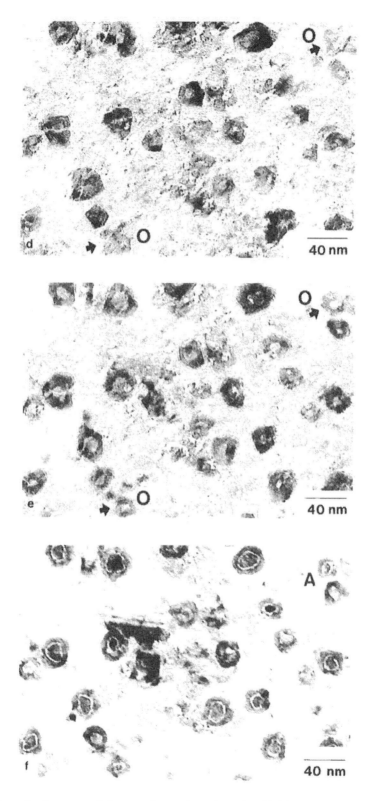

FIG. 2. (Continued) Sequence of changes of a specimen heated at 400°C in as-supplied hydrogen. Initial loading, 6 Å. The micrographs show the same region after (d) 3 h, (e) 6 h, (f) 12 h. (*Continued*)

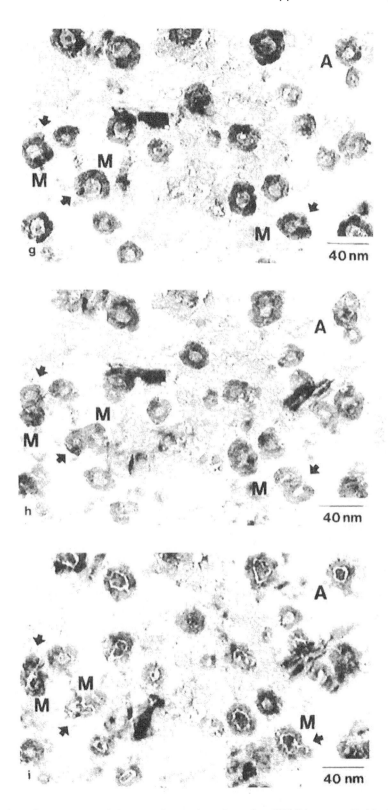

FIG. 2. (Continued) Sequence of changes of a specimen heated at 400°C in as-supplied hydrogen. Initial loading, 6 Å. The micrographs show the same region after (g) 21 h, (h) 33 h, (i) 36 h.

A.2. Alternate Heating in Oxygen and As-Supplied Hydrogen

Heating in oxygen caused, in general, extension of the crystallites over the substrate, as seen from Table 3. In addition, in the oxidized and extended state, the crystallites appear to be composed of a number of very small units held together (Figs. 8c, 9e, f). Except for the coalescence of a few nearby pairs of particles following their contact, as a result of extension on oxidation, the number of crystallites remained, in general, constant during alternate heating in oxygen and hydrogen. Heating in hydrogen following heating in oxygen caused the particles to contract and, at 400 and 500°C, to acquire the core-and-ring structure (Table $3_{6,400}$, $3_{6,500}$).

The 12.5-Å sample exhibited interesting behavior on alternate heating in oxygen and hydrogen at 500°C. As shown in Table $3_{12.5,500}$, the crystallites extended on oxidation and split into a few interconnected particles on subsequent heating in hydrogen (Figs. 9b; 10a–d). However, the subunits coalesced and contracted to form again compact particles (Fig. 9c) on continued heating. Marks of particle peripheries could be seen in the same figure, indicating extensive interaction with the substrate along the peripheries of the particles. The crystallites extend again on heating in oxygen and cover the area bounded by the periphery marks mentioned above. In the extended state, the crystallites appear to be very porous and are composed of a large number of smaller particles held together. Also, the substrate is covered with a very large number of small and thin crystallites seen clearly in a magnified micrograph (Fig. 9f). Subsequent heating in hydrogen causes the particles to contract and appear distinct, accentuating the fragmentation. The micrograph appears as if a kind of explosive shattering of the particles had occurred (Fig. 9g). In fact, the breakup occurs during heating in oxygen, as a result of the considerable extension which, in addition, probably occurs very rapidly causing the rupture. Mechanical fatigue might have also contributed to the fragmentation since the specimen had already been heated for about 200 h.

On heating in oxygen at 700°C (Table $3_{6,700}$). a thick film extends out from the periphery of the larger particles and also the grain boundaries of the substrate appear less distinct (Figs. 12b, c). The substrate might be covered by a film. In the same micrograph, a number of small particles have disappeared (E). It is possible that both the crystallites that disappeared as well as some unresolved ones

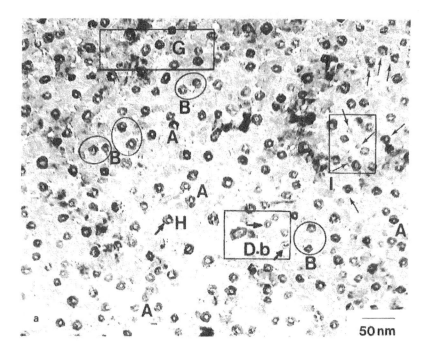

FIG. 3. Time sequence of the same region of the specimen heated previously at 300°C on subsequent heating in as-supplied hydrogen at 500°C. (a) 1 h. *(Continued)*

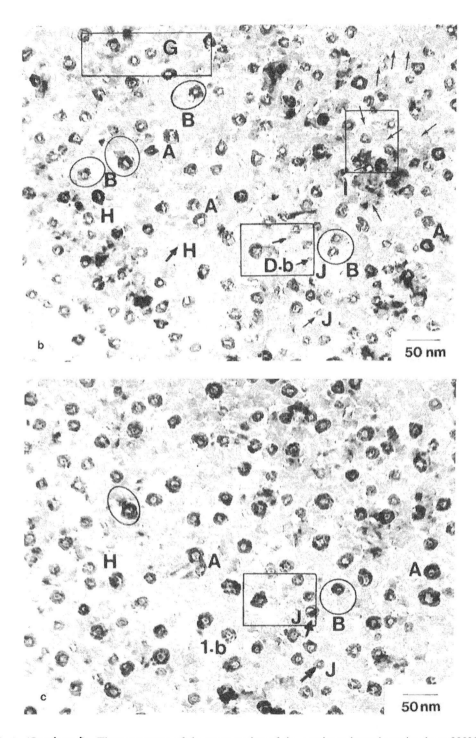

FIG. 3. (Continued) Time sequence of the same region of the specimen heated previously at 300°C on subsequent heating in as-supplied hydrogen at 500°C. (b) 2 h, (c) 5 h.

TABLE 3
Alternate Heating in Hydrogen and Oxygen

Heating atmosphere	Event observed	Compound identified
Table $3_{6,400}$ O_2	Torus shape particles with a tendency to fill in the cavity of the torus are observed.	$Fe_3O_4(\gamma\text{-}Fe_2O_3)$, $Al_2Fe_2O_6$
H_2	Particles acquire the core-and-ring structure. Coalescence of nearby particles and rupture of some particles are seen.	$FeAl_2O_4$, $\alpha\text{-}Fe$
Table $3_{6,500}$ O_2	Considerable filling in of the cavities of the torus. Some pairs of neighboring particles make contact following extension. (8c)	$Al_2Fe_2O_6$, $Fe_3O_4(\gamma\text{-}Fe_2O_3)$, trace $FeAl_2O_4$
H_2	Particles acquire a core-and-ring structure initially but on continued heating assume a torus shape. Particles in contact coalesce. (8a, b, d, e)	$Fe_3O_4(\gamma\text{-}Fe_2O_3)$, trace $FeAl_2O_4$
Table $3_{12.5.500}$ O_2	Extension. Tendency to fill in the cavity. Coalescence of nearby particles. Particles composed of loosely interconnected units. (9a, d, e, f; 10b)	d-Values approaching those of $Al_2Fe_2O_6$ and $FeAl_2O_4$
H_2	Splitting of particles into a few interconnected sub-units, a few of which coalesced subsequently (V). Periphery marks seen around particles (P). Fragmentation of particles (W). (9b, c, g; 10a, c, d)	$\alpha\text{-}Fe$, traces of Fe_3O_4 (-Fe_2O_3)
Table $3_{6,600}$ O_2	(Sample was previously heated in hydrogen at 800°C for 4 h, Fig. 7c). Migration of some particles (C). A film appears to extend out from underneath and around the particles. (11a, b)	$Fe_3O_4(\gamma\text{-}Fe_2O_3)$, Al_2FeO_6, traces of $FeAl_2O_4$
H_2	Contraction and formation of more circular particles. Periphery marks, observed before, contract to form small particles (X). (11c)	Mostly $Fe_3O_4(\gamma\text{-}Fe_2O_3)$, traces of $FeAl_2O_4$, traces of $\alpha\text{-}Fe$
Table $3_{6,700}$ O_2	A thick film extends from the periphery of the particles. Considerable extension or disappearance of small particles (E). Substrate grain boundaries appear less sharp (R). Very little extension of large particles. (12b, c)	Mostly $FeAl_2O_4$ and some Fe_3O_4 ($\gamma\text{-}Fe_2O_3$)
H_2	Appearance of a few, new, small particles (L). (12d)	$FeAl_2O_4$

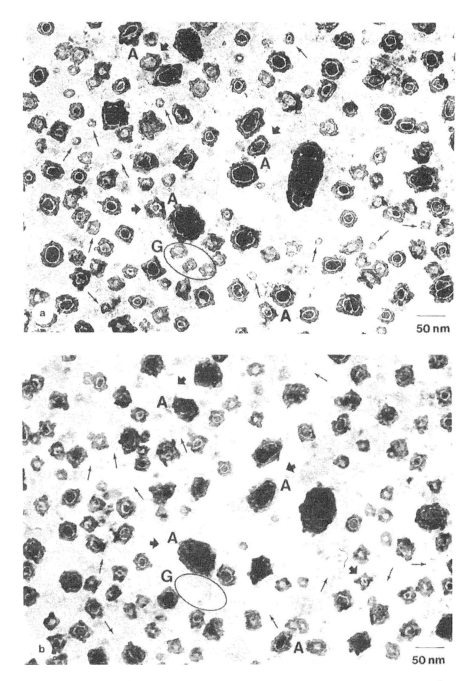

FIG. 4. Sequence of changes at 500°C in a specimen with higher loading. Initial loading, 12.5 Å. Two adjacent regions (Figs. 4 and 4′) are shown after (a) 2 h, (b) 4 h. *(Continued)*

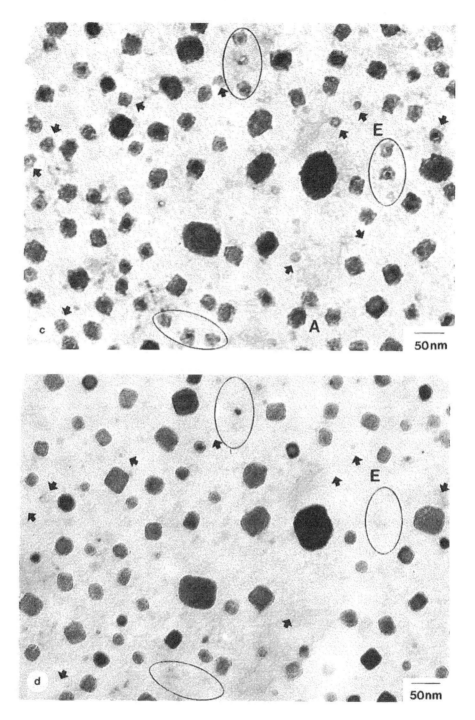

FIG. 4. (Continued) Sequence of changes at 500°C in a specimen with higher loading. Initial loading, 12.5 Å. Two adjacent regions (Figs. 4 and 4′) are shown after (c) 5 h, (d) 24 h.   (*Continued*)

Role of Physical and Chemical Interactions in the Behavior of Supported Metal Catalysts 201

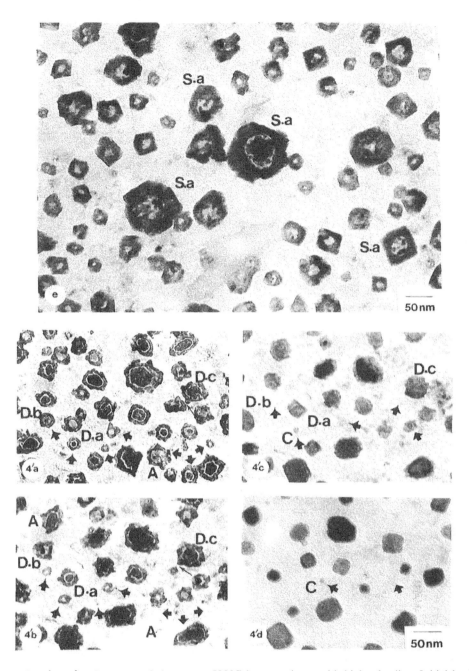

FIG. 4. (Continued) Sequence of changes at 500°C in a specimen with higher loading. Initial loading, 12.5 Å. Two adjacent regions (Figs. 4 and 4′) are shown after (e) 46 h.

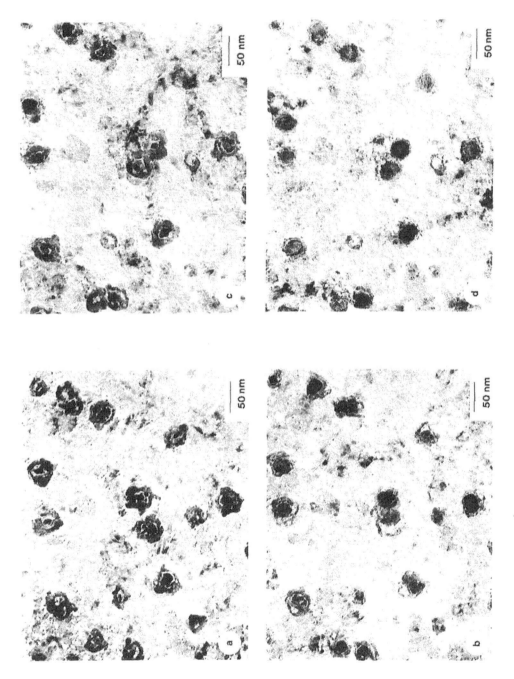

FIG. 5. Sequence of changes on healing a 6-Å initial loading specimen in as-supplied hydrogen at 600°C. (a) 2 h. (b) 6 h. (c) 14 h. (d) 30 h.

Role of Physical and Chemical Interactions in the Behavior of Supported Metal Catalysts 203

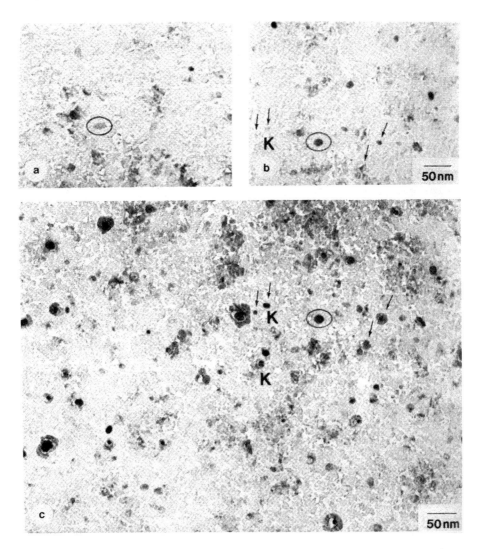

FIG. 6. Sequence of changes on heating in as-supplied hydrogen at 700°C. Initial loading, 6 Å. The same region is shown after (a) 1 h, (b) 5 h, (c) 8 h. *(Continued)*

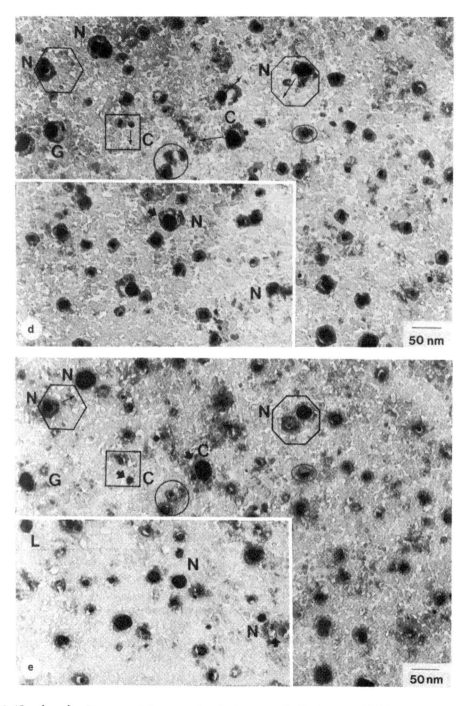

FIG. 6. (Continued) Sequence of changes on heating in as-supplied hydrogen at 700°C. Initial loading, 6 Å. The same region is shown after (d) 18 h, (e) additional 3 h at 500°C (21 h total). *(Continued)*

Role of Physical and Chemical Interactions in the Behavior of Supported Metal Catalysts 205

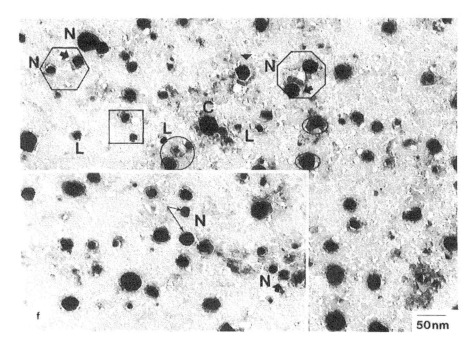

FIG. 6. (Continued) Sequence of changes on heating in as-supplied hydrogen at 700°C. Initial loading, 6 Å. The same region is shown after (f) 34 h, 700°C.

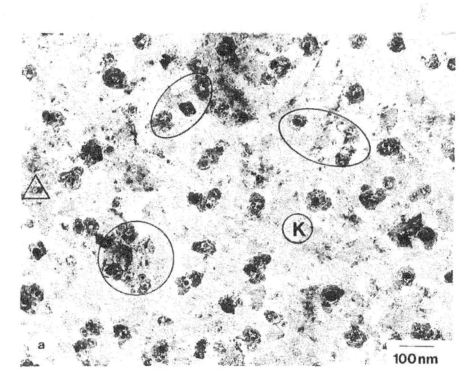

FIG. 7. Sequence of changes on heating the specimen of Fig. 5 at 800°C. The micrographs correspond to the same region: (a) 41 h, 600°C (included for reference). *(Continued)*

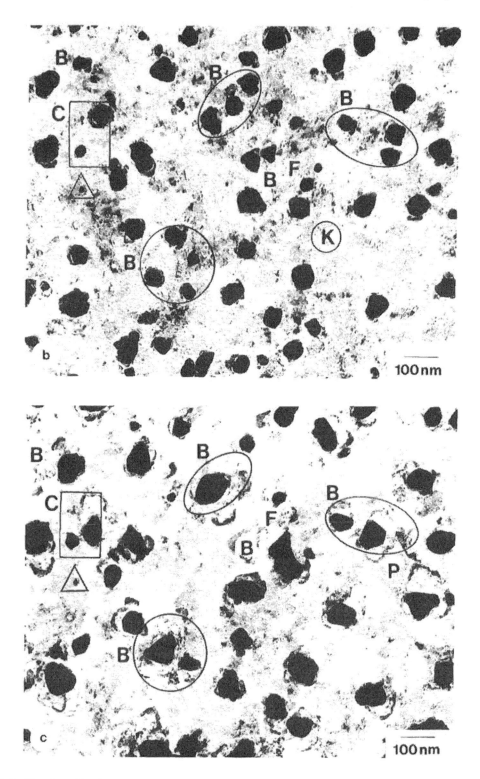

FIG. 7. (Continued) Sequence of changes on heating the specimen of Fig. 5 at 800°C. The micrographs correspond to the same region: (b) 2 h 30 min, 800°C; (c) 4 h, 800°C.

Role of Physical and Chemical Interactions in the Behavior of Supported Metal Catalysts 207

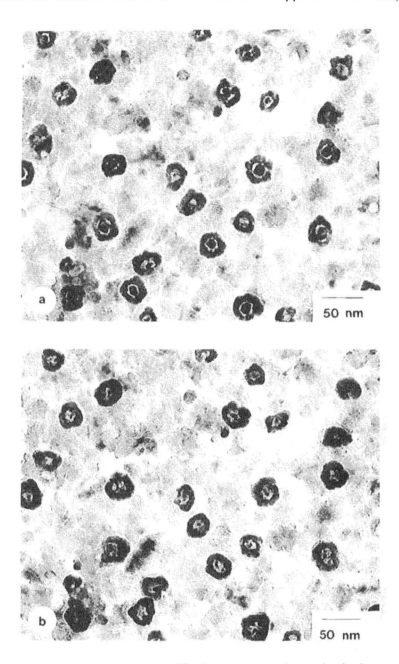

FIG. 8. Sequence of changes in the specimen of Fig. 3 on subsequent alternate heating in oxygen and hydrogen at 500°C. The region shown here is from the same specimen but of a different region than that of Fig. 3. The substrate is thicker in this region probably as a result of tearing and folding upon itself, initially. The particles are larger and the changes are sharper. The micrographs included show the changes subsequent to an oxidation step, (a) 5 h, H_2; (b) 10 h, H_2. *(Continued)*

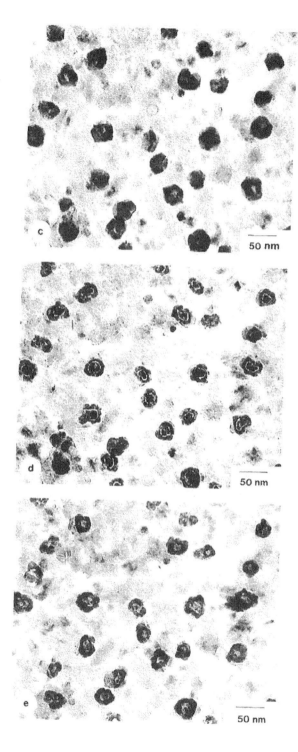

FIG. 8. (Continued) Sequence of changes in the specimen of Fig. 3 on subsequent alternate heating in oxygen and hydrogen at 500°C. The region shown here is from the same specimen but of a different region than that of Fig. 3. The substrate is thicker in this region probably as a result of tearing and folding upon itself, initially. The particles are larger and the changes are sharper. The micrographs included show the changes subsequent to an oxidation step, (c) 10 h, O_2; (d) 10 min, H_2; (e) 16 h, H_2. The behavior is similar at 400°C.

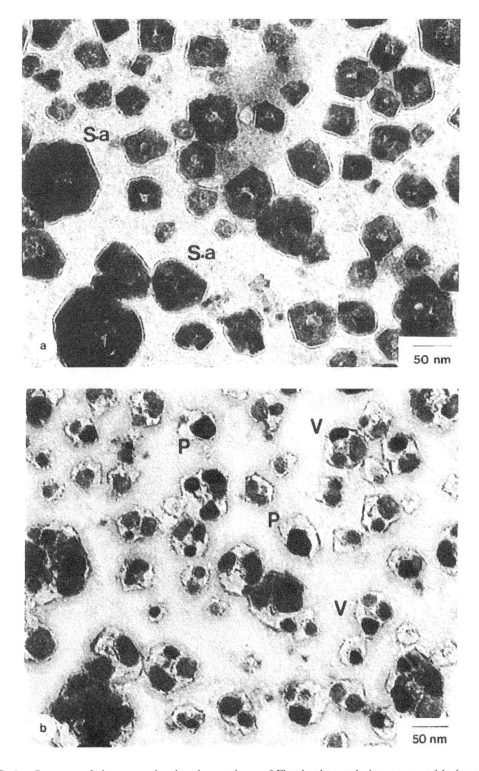

FIG. 9. Sequence of changes on heating the specimen of Fig. 4e alternately in oxygen and hydrogen at 500°C. (a) 2 h, O_2; (b) 2 h, H_2. (*Continued*)

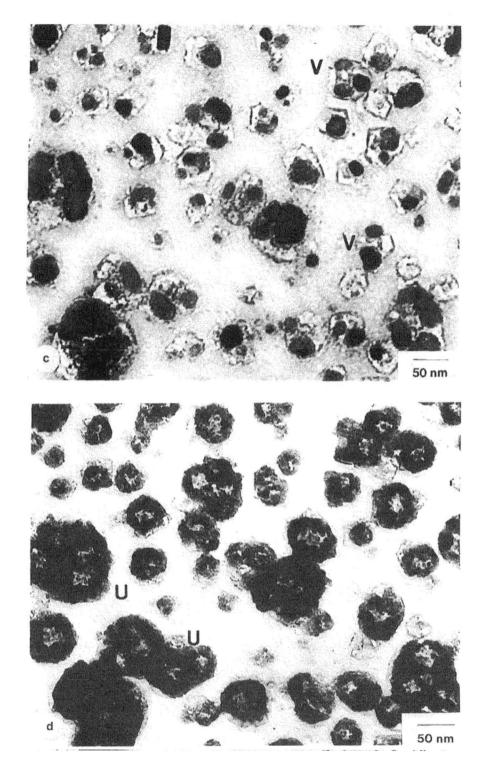

FIG. 9. (Continued) Sequence of changes on heating the specimen of Fig. 4e alternately in oxygen and hydrogen at 500°C. (c) 66 h, H$_2$; (d) 1 h, O$_2$. *(Continued)*

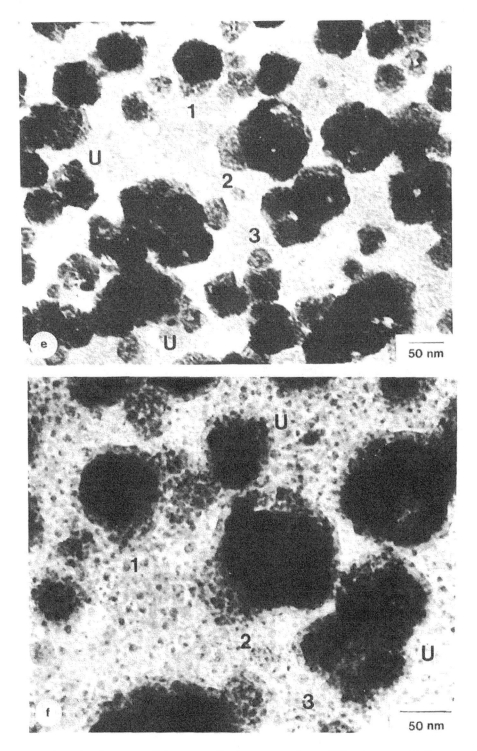

FIG. 9. (Continued) Sequence of changes on heating the specimen of Fig. 4e alternately in oxygen and hydrogen at 500°C. (e) 40 h, O_2, an adjacent region; (f) magnified micrograph of (e). *(Continued)*

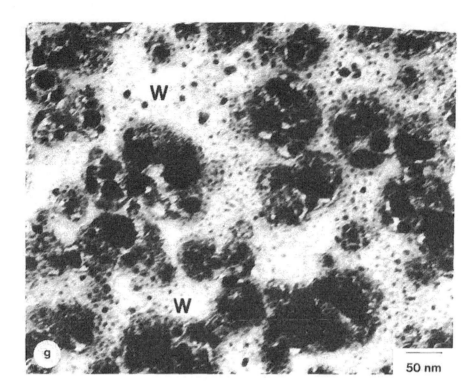

FIG. 9. (Continued) Sequence of changes on heating the specimen of Fig. 4e alternately in oxygen and hydrogen at 500°C. (g) 45 min, H_2.

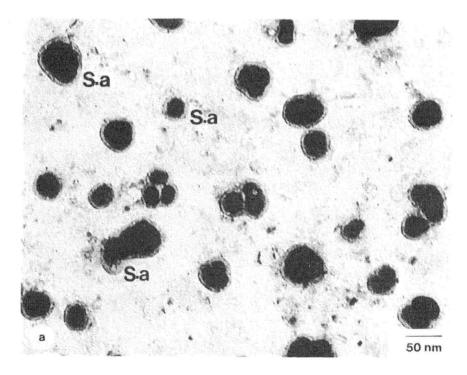

FIG. 10. As Fig. 9 but a different region of the same specimen better indicating the crystallite splitting. (a) 2 h, H_2. *(Continued)*

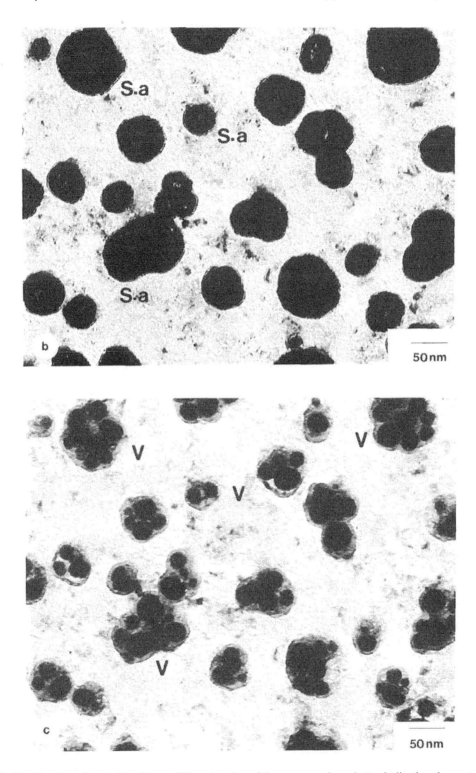

FIG. 10. (Continued) As Fig. 9 but a different region of the same specimen better indicating the crystallite splitting. (b) 1 h, O_2; (c) 1 h, H_2. *(Continued)*

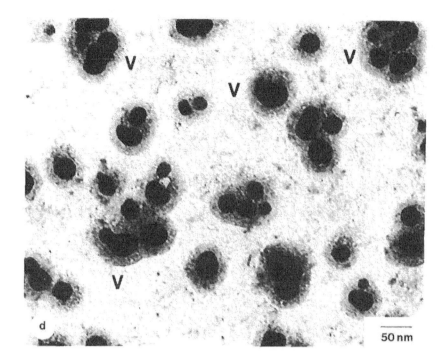

FIG. 10. (Continued) As Fig. 9 but a different region of the same specimen better indicating the crystallite splitting. (d) 2 h, H_2.

are present considerably extended on the surface covering the substrate as a thin film which co-exists with the three-dimensional crystallites. The thin film which appears to cover the substrate might be an oxide, or more likely, an aluminate, since almost only $FeAl_2O_4$ is detected in the diffraction pattern.

B. Behavior in Purified Hydrogen

B.1. Heating in Hydrogen

The behavior of the specimens when heated in purified hydrogen was different in a few respects from that observed with as-supplied hydrogen (Sect. A.1). Unlike the previous case, the initially formed crystallites are typically larger and have a bimodal distribution. Some of these initial particles are so thin that the substrate grain boundaries can be seen through them (a–c in Fig. 13a). The particles, especially the large ones, seem to prefer predominantly rectangular shapes to the usual circular ones. The events observed on heating a 7.5-Å initial film in purified hydrogen at progressively higher temperatures are listed in Table 4. Metallic iron is very clearly detected even at 400°C. At 500°C and higher temperatures almost only iron is detected, with trace quantities of oxide. In contrast, with the as-supplied hydrogen the metal could not be detected even at 700°C. In addition, at 500°C and even at low loadings, relatively compact particles formed, unlike heating in the as-supplied hydrogen which led to torus-shape particles. These results point out that the difficulty in reducing the iron catalysts to the metallic state is clearly a result of the chemical interactions with the support which are caused to a considerable extent by the oxygen and/or moisture present in the reducing chamber. The micrographs (Figs. 13a, b) show that a number of particles of about 100 Å and also a few particles considerably larger (about 350 Å) disappear (H) and, instead, a number of small crystallites of about 30 Å or less appear in all the regions, including locations where no particles were present before (L, N). The crystallites might have fragmented into smaller particles (including single atoms) which subsequently migrated on the substrate before colliding and coalescing with other particles. Of course, the migrating crystallites might have simultaneously lost some atoms too, thus decreasing further in size.

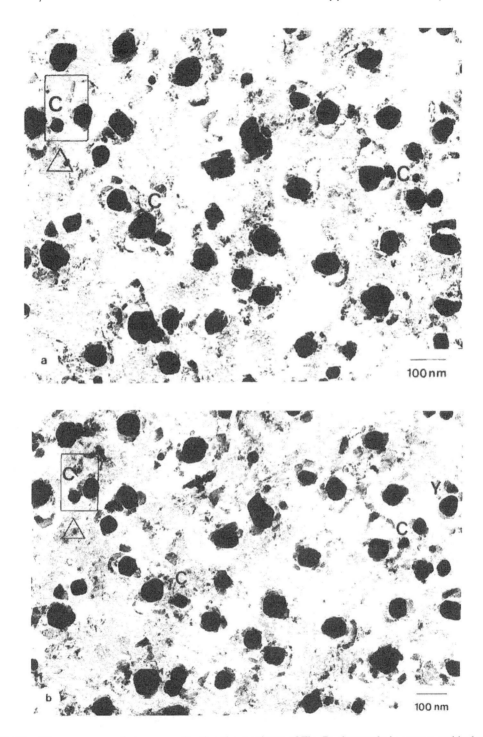

FIG. 11. Time sequence of changes on heating the specimen of Fig. 7c alternately in oxygen and hydrogen at 600°C. (a) 1/2 **h, O**$_2$; (b) 1 1/2 **h, O**$_2$; (*Continued*)

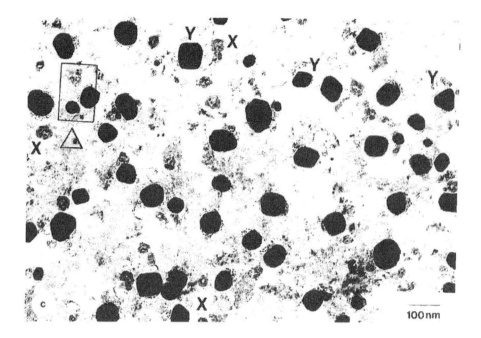

FIG. 11. (Continued) Time sequence of changes on heating the specimen of Fig. 7c alternately in oxygen and hydrogen at 600°C. (c) 1 h, H_2

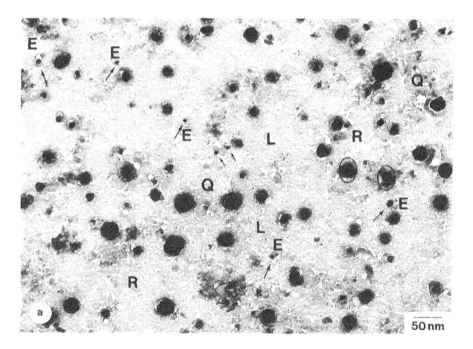

FIG. 12. Sequence of changes on healing the specimen of Fig. 6f alternately in oxygen and hydrogen at 700°C. (a) 33 h, H_2. *(Continued)*

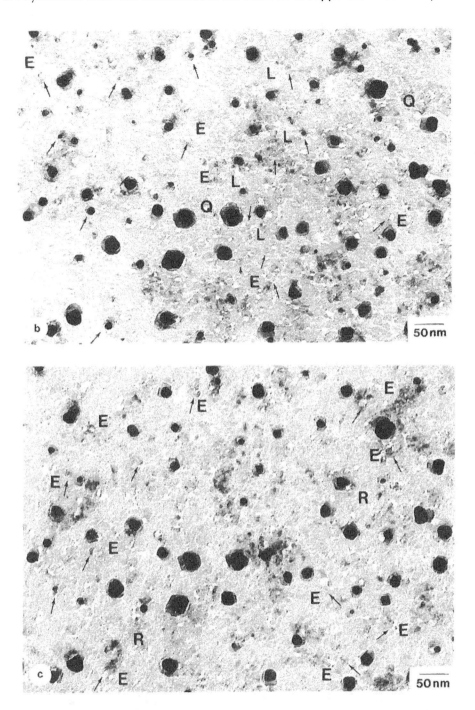

FIG. 12. (Continued) Sequence of changes on healing the specimen of Fig. 6f alternately in oxygen and hydrogen at 700°C. (b) 1 h. O_2; (c) 13 h, O_2. (Continued)

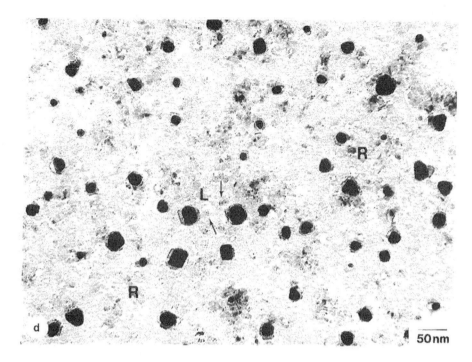

FIG. 12. (Continued) Sequence of changes on healing the specimen of Fig. 6f alternately in oxygen and hydrogen at 700°C. (d) 1 h, H$_2$.

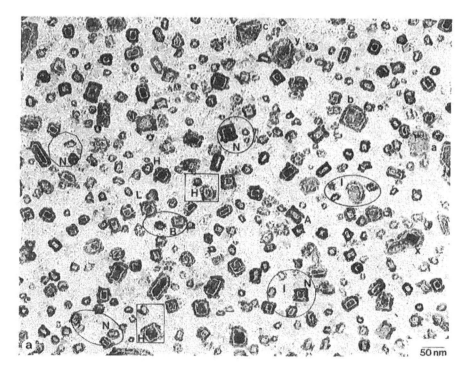

FIG. 13. Sequence of changes on healing a model Fe/Al$_2$O$_3$ specimen in purified hydrogen at progressively higher temperatures. Initial loading, 7.5 Å. The same region is shown after (a) 4 h, 400°C. *(Continued)*

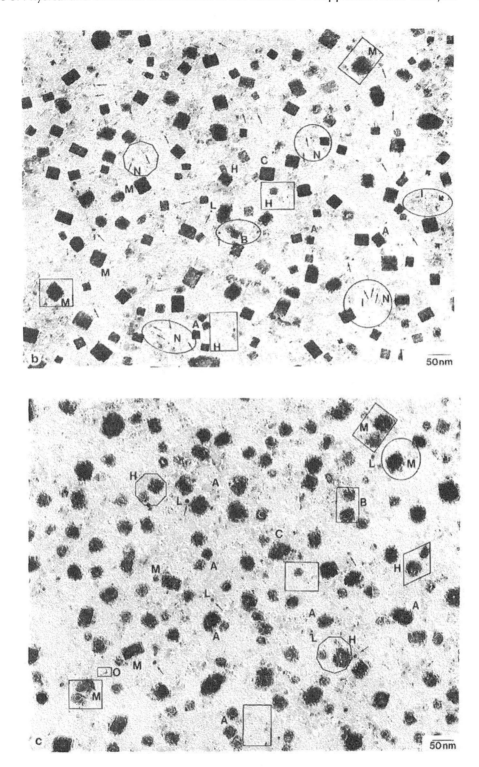

FIG. 13. (Continued) Sequence of changes on healing a model Fe/Al$_2$O$_3$ specimen in purified hydrogen at progressively higher temperatures. Initial loading, 7.5 Å. The same region is shown after (b) 2 h, 500°C; (c) 4 h, 600°C. *(Continued)*

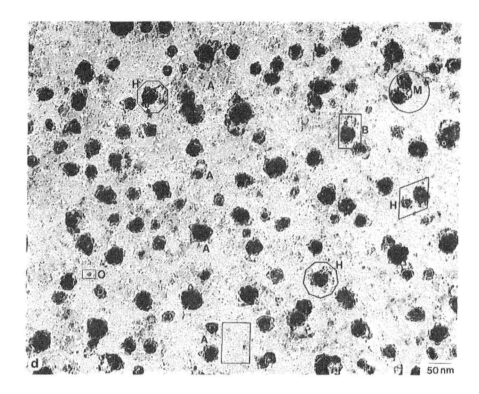

FIG. 13. (Continued) Sequence of changes on healing a model Fe/Al$_2$O$_3$ specimen in purified hydrogen at progressively higher temperatures. Initial loading, 7.5 Å. The same region is shown after (d) 2 h, 700°C.

TABLE 4
Heating in Purified Hydrogen

Total heating time[a]	Event observed	Compound identified
400°C, 4 h	Particles with very sharp core-and-ring structure and a broad size distribution are formed. Cores in some big particles have fringes indicating twinning (x, y in Fig. 13a). Smaller crystallites have a torus shape.	Fe$_3$O$_4$ (γ-Fe$_2$O$_3$) α-Fe
500°C, 2 h	Dark, compact particles are formed. Extremely narrow annular rings are still detected. Considerable sintering has occurred. A number of small and large particles disappeared (I). Also, a number of smaller crystallites appeared in other places (L). Coalescence of nearby particles (A) and disappearance of a large particle nearby an unaffected smaller particle (H) are detected (13a, b).	α-Fe
600°C, 4 h	Particles contract, Circular and restructured particles are formed. (13c)	α-Fe, traces of Fe$_3$O$_4$ (α-Fe$_2$O$_3$)
700°C, 2 h	Disappearance of some small particles, appearance of new ones (L), splitting (M), coalescence (A) and migration (B, C) are observed. (13d)	α-Fe

[a] 7.5 Å, progressive heating.

Role of Physical and Chemical Interactions in the Behavior of Supported Metal Catalysts 221

TABLE 5
Alternate Heating in Oxygen and Purified Hydrogen

Heating atmosphere	Event observed	Compound identified
Table $5_{6,500}$ O_2	Particle extension. Particles have a torus shape and a film appears to fill in the cavity. (14b)	d-values approaching those of $FeAl_2O_4$ and $Al_2Fe_2O_6$
H_2	Particle contraction. Faint annular ring is observed initially and compact particles are formed on continued heating.	$FeAl_2O_4$ decreases in amount. α-Fe and traces of Fe_3O_4 (γ-Fe_2O_3)
Table $5_{6,600}$ O_2	Crystallites extend and assume a torus shape, enclosing a large cavity and having a small residual particle in the cavity. (14d)	Fe, O_4 (γ-Fe_2O_3), $FeAl_2O_4$
H_2	Contraction and compact particle formation. Core-and-ring structure is not observed. (14a, c)	
Table $5_{7.5,500}$ O_2	(Heated before in hydrogen at 700°C, Fig. 13d). Considerable extension. Large cavity formation in the crystallite (S.b.). A small residual particle is present in the cavity of each crystallite. No change on prolonged heating. (15a)	$FeAl_2O_4$, Fe_3O_4 (γ-Fe_2O_3)
H_2	Splitting of some crystallites into a few interlinked particles, some having a core-and-ring structure (V). Sintering of the sub-units on further heating. (15b)	$FeAl_2O_4$, α-Fe, traces Fe_3O_4 (γ-Fe_2O_3)
Table $5_{7.5,700}$ H_2	Contraction and formation of more compact particles. Marks of particle peripheries (P). (15c)	$FeAl_2O_4$, Fe_3O_4 (γ-Fe_2O_3)
O_2	Considerable extension. No cavities in particles. A small residual particle is present on the extended parent crystallite (T). The substrate appears covered by a film (R). (15d)	$FeAl_2O_4$, $Al_2Fe_2O_6$,
H_2	Contraction and formation of compact particles. No core-and-ring structure. (15e)	$FeAl_2O_4$

B.2. Alternate Heating in Oxygen and Hydrogen

The extension following oxidation is more pronounced when purified hydrogen is used in the reduction step instead of the as-supplied hydrogen (Sect. A.2). In addition, a few differences in the behavior of the two cases have been observed, as listed in Tables 3 and 5. The events observed in Table $5_{7.5,700}$ are of particular interest. After 1 h of oxidation at 500°C, each crystallite extended considerably acquiring a torus shape (Fig. 15a). However, a small residual particle remained randomly located within the cavity. The leading edge of the torus has an undulating form and the torus appears to be composed of interlinked subunits. Heating this specimen subsequently in hydrogen at 500°C for only 1 h caused splitting, especially of the larger crystallites, into a few interconnected subunits, some of which have a core-and-ring structure (Fig. 15b). Subsequently, the subunits sintered to form compact particles (Fig. 15c). Heating in oxygen at 700°C caused pronounced wetting and extension of the crystallites. However, unlike the previous cases, cavities were not formed (Fig. 15d). Instead, each crystallite extended as a uniformly thick film with a residual particle on top. The substrate grain boundaries also appear less distinct, probably because of a film covering the substrate, as observed also in Section A.2 on oxidation at 700°C. Subsequent heating in hydrogen at 700°C caused each particle to contract to form a single compact particle, instead of splitting into a few as observed at 500°C.

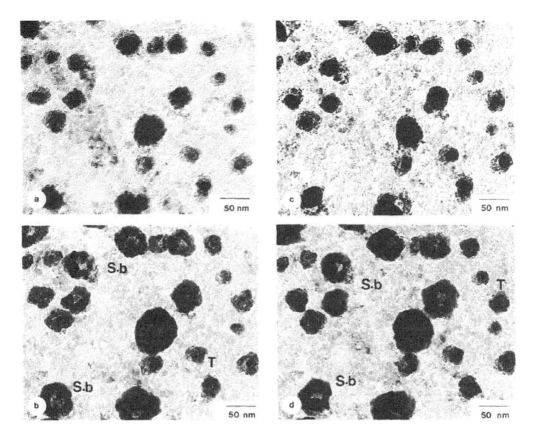

FIG. 14. Sequence of changes on heating a specimen alternately in oxygen and purified hydrogen. (a) 6 h, H_2, 600°C; (b) 1 h, O_2; (c) 5 h, H2, 600°C; (d) 1 h, O2, 600°C.

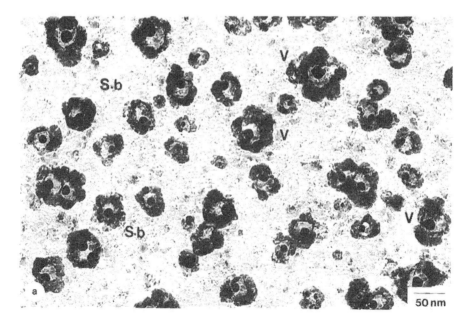

FIG. 15. Sequence of changes on heating the specimen of Fig. 13d alternately in oxygen and hydrogen. (a) 1 h, O_2, 500°C. *(Continued)*

Role of Physical and Chemical Interactions in the Behavior of Supported Metal Catalysts

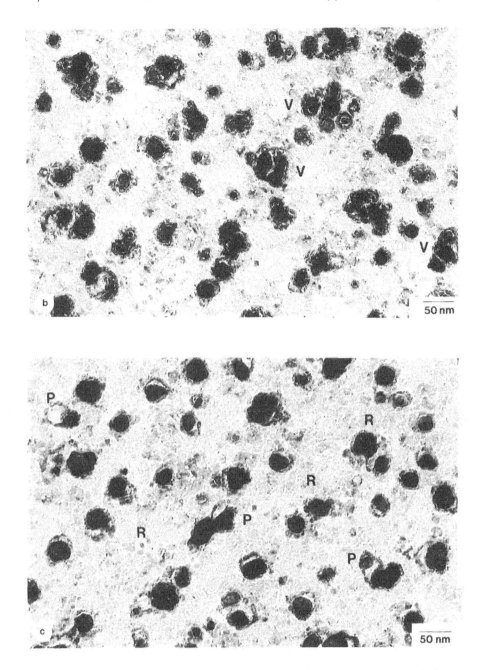

FIG. 15. (Continued) Sequence of changes on heating the specimen of Fig. 13d alternately in oxygen and hydrogen. (b) 30 min, H_2 500°C (reduction for a longer duration followed by oxidation at the same temperature was carried out between (b) and (c)); 2 h, H_2, 700°C; *(Continued)*

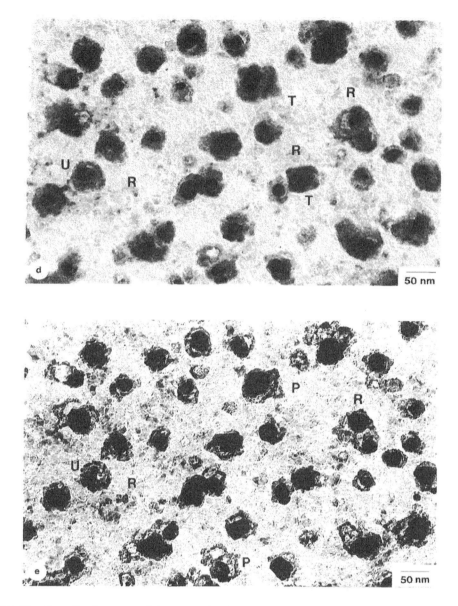

FIG. 15. (Continued) Sequence of changes on heating the specimen of Fig. 13d alternately in oxygen and hydrogen. (d) 12 h, O_2, 700°C (there is very little difference between 1 and 12 h oxidation); (e) 1 h, H_2, 700°C.

C. Effect of Moisture and Residual Oxygen Present in the As-Supplied Hydrogen

The effect of moisture on the crystallites and their composition was investigated by first purifying the as-supplied hydrogen and then bubbling it through distilled water. The effect of moisture appears to be similar to that of pure flowing oxygen. At 500°C, the crystallites extended and formed torus-shape particles. Solid solutions leading to $FeAl2O4$ were detected in the diffraction pattern. The effect of residual oxygen in the absence of moisture was examined by passing the as-supplied hydrogen through a desiccant. In contrast to the residual oxygen, which causes shape changes and some extension, the moisture causes much greater extension and easier aluminate formation. In other words, moisture considerably enhances the effects caused by the residual oxygen.

Role of Physical and Chemical Interactions in the Behavior of Supported Metal Catalysts **225**

DISCUSSION

The results presented in the previous section bring evidence for a variety of phenomena such as sintering, wetting and extension, shape changes, splitting, etc., which occur during heating of iron-on-alumina model catalysts. In what follows, some of these phenomena are discussed in more detail.

SINTERING BEHAVIOR

The process of sintering can occur by several mechanisms (17), such as migration of crystallites and their coalescence (23), emission of single atoms by the small crystallites and their capture by the large ones (24–27), and a combination of the two (27). Two possibilities have been suggested for sintering by single-atom emission and capture. In one of them, a large number of crystallites are involved and the small crystallites lose atoms to a surface phase of single atoms dispersed over the substrate, while the large ones capture atoms from this phase. This process, known as Ostwald ripening, occurs when the substrate surface phase of single atoms is supersaturated with respect to the large crystallites and undersaturated with respect to the small ones.

In the other case, called direct ripening (28), atoms released by a small particle move directly to a neighboring large crystallite, even though the surface phase of single atoms is, on the average, undersaturated with respect to all the crystallites present. In contrast to Ostwald ripening, which is global, the latter is local and can be considered as a fluctuation from the average behavior, determined by the local surface pressures generated by the particles involved.

As seen from Table 1, migration and coalescence of crystallites (Figs. 3a–c, 7b, c, 13a–d), disappearance of a large number of small particles (Ostwald ripening, Figs. 4a–d, 12a–c) and decrease and disappearance of a few particles near larger particles (direct ripening, Figs. 3a, b, 4'a–c) could all be detected in the present experiments. Interestingly, large crystallites appear to have migrated over large distances. For example, crystallites greater than 90 Å in size have migrated at 700°C (Figs. 6a–d). Also, crystallites of about 480 Å migrated over distances of about 600 Å at 800°C (Figs. 7b, c). Even though the number of particles whose migration could be detected is small, the few instances of migration of large particles over large distances, the redistribution of particles sometimes observed (Figs. 6d–f, 7b, c), the appearance of dumbbell-shape particles (Fig. 3c), and the appearance of small particles in regions unoccupied before (Figs. 6e, f; 12a–d; 13a–c) indicate that migration, more likely of the smaller crystallites, plays a role in the sintering phenomenon. Similarly, a large number of small as well as large crystallites disappear at temperatures as low as 400 or 500°C. In addition to the ripening mechanism, the disappearance of these small particles may be caused by their considerable extension to an undetectably thin film which subsequently coalesces with the nearby particles, or by their migration following fragmentation. In fact, in some cases, a surface film (instead of a surface phase of single atoms) appears to coexist with the three-dimensional crystallites. As pointed out in a previous section (Sect. B.l), the disappearance of the small particles from some regions and the appearance of other, new particles in regions far removed from the previous indicate that mechanisms other than just migration and coalescence (which is expected to occur over only shorter distances), or ripening (since contrary to what is predicted by ripening, large particles have vanished while smaller ones remained (H in Figs. 3b, c, 15a, b)) probably occur. Other mechanisms, such as fragmentation followed by migration of the smaller fragments (N in Figs. 6d–f), emission of atoms from the migrating crystallites, considerable extension to an undetectable film which splits and contracts to form new particles, etc., might have also occurred. The last process occurs, either because of alternate oxidation and reduction in the as-supplied hydrogen, and/or because of the heterogeneities of the substrate which interacts strongly with the particles in some regions and weakly in other regions.

Wetting Behavior

It is important to recognize that the wetting characteristics of the crystallites on the support affect the sintering and redispersion behavior, the thermostability of the size distribution, and the shape of the crystallites. The role of wetting in the redispersion of supported metal crystallites and in the changes of their shape has been emphasized in Refs. (*19, 20, 23, 29*).

Wetting Angle

The ability of a crystallite to wet a support and its extent are determined by the equilibrium of the three interfacial tensions, substrate-gas (σ_{sg}), crystal-lite–gas (σ_{cg}), and crystallite-substrate (σ_{cs}), as expressed by Young's equation

$$\sigma_{sg} - \sigma_{cs} = \sigma_{cg} \cos\theta, \tag{1}$$

where θ is the equilibrium wetting or contact angle. When, under some conditions such as oxidizing,

$$\sigma_{sg} - \sigma_{cs} > \sigma_{cg} \tag{1a}$$

is satisfied, no wetting angle can exist, since $\cos\theta > 1$, and the crystallite spreads over the substrate. On the other hand, when

$$\sigma_{sg} - \sigma_{cs} < \sigma_{cg} \tag{1b}$$

is satisfied, the crystallite forms an equilibrium contact angle, θ with the substrate, whose value lies in the range $0° < \theta < 180°$, depending upon both the difference ($\sigma_{sg} - \sigma_{cs}$) as well as σ_{cg}. Under vacuum, in inert, or reducing atmospheres, the metals which are commonly used as supported catalysts have, in general, a high surface tension σ_{cg}, as well as a high interfacial tension, σ_{cs}, with silica and alumina supports (*30, 31*). These lead to contact angles greater than 90° and the crystallites do not wet well the support. However, in an oxidizing environment both σ_{cg} and σ_{cs} are smaller, since these values are smaller for the oxide compared to those of the corresponding metal, and the contact angle can decrease. Accordingly, the crystallite will extend and wet the support belter.

Interaction Energy and Wetting

When two phases are brought into contact to form an interface, and there are no molecular interactions between them, the interfacial tension between the two should be expressed by the sum

$$\sigma_{cs} = \sigma_c + \sigma_s. \tag{2}$$

However, there are always attractive molecular interactions between two phases brought into contact and because of this, there is a corresponding decrease in the interfacial tension. These interactions are strong when they are of a chemical nature, for instance, the formation of compounds, or weak when they are of a physical nature, such as dispersion or polar interactions. The interfacial tension between crystallite and substrate is therefore given by

$$\sigma_{cs} = \sigma_c + \sigma_s - \left(U_{int} - U_{str}\right)$$

$$= \sigma_c + \sigma_s - U_{cs}, \tag{3}$$

where, U_{int} is the interaction energy per unit area of crystallite–substrate interface between the atoms or molecules of the crystallite and those of the substrate and U_{str} is the strain energy per unit area arising as a result of the mismatch of the two lattices.

When a chemical interaction takes place at the crystallite–substrate interface (leading to the formation of a compound), as in the case of iron on alumina in an oxidizing environment, U_{cs} will become large, thereby appreciably decreasing σ_{cs}. In fact, if the chemical bonding at the interface is very strong, U_{cs} will become so large that σ_{cs} can decrease to zero or (under nonequilibrium conditions) even below zero. Such a large decrease in the dynamic interfacial tension, σ_{cs}, and also a concomitant decrease in the value of σ_c (in an oxidizing environment the metal forms an oxide, and σ_c of the oxide is smaller than σ_c of the metal) lead to a rapid spreading of the crystallite over the substrate, since the driving force for spreading $(\sigma_{sg} - \sigma_{cs} - \sigma_{cg}\cos\theta)$ is considerably increased.[3] Initially, the reaction occurs only at the interface and then it proceeds in the bulk of either the substrate or the crystallite, more likely the former. Bulk reaction occurs because it decreases the free energy of the system and it continues until the rate of dissolution of the oxide molecules into the substrate is considerably limited by the diffusional resistance. Immediately following the reaction at the interface, the crystallite is probably present in a highly extended and unstable configuration because of the large driving force for spreading. Any thinning produced by perturbations (such as diffusion into the substrate, thermal fluctuations, etc.) will tend to grow, leading finally to the torus which, under the conditions of the experiment, might be the thermodynamically stable configuration. The rapid extension following the initial surface reaction is soon arrested since, subsequently, U_{cs} is relatively reduced. Indeed, after some of the oxide has already dissolved into the substrate, the subsequent surface interaction energy, U_{cs}, between the aluminate and the oxide is no longer as large as that between alumina and oxide, since, unlike the latter, the former does not lead to a chemical compound at the interface. Consequently, the driving force for spreading $\{\sigma_{sg} - (\sigma_s + \sigma_c - U_{cs}) - \sigma_{cg}\cos\theta\}$ becomes smaller. To summarize, σ_{cs} has a large decrease initially, even possibly becoming negative, and then increases. Correspondingly, the initial large driving force for spreading also decreases subsequently. Therefore, the considerably extended crystallite formed by the initial rapid extension subsequently reaches a new equilibrium wetting angle with the new layer of substrate (reaction product) at the (crystallite-substrate) interface. The time evolution of U_{cs} and σ_{cs} are shown schematically in Fig. 16.

If a small cavity had initially nucleated in the extended crystallite because of the rapid extension and the dissolution of a part into the substrate, the subsequent tendency to form a larger wetting angle, for reasons just described above, would make both the inner and the outer leading edges of

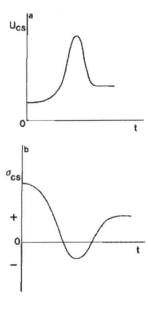

FIG. 16. Schematic of the variation with time of (a) the interaction energy, U_{cs}, and (b) the crystallite-substrate interfacial tension, σ_{cs}, when there is a chemical reaction between the crystallite and the substrate.

---

[3] Here θ is the instantaneous nonequilibrium wetting angle.

the cavity-containing crystallite to withdraw. This causes the expansion of the cavity and leads to the observed torus shape. It should be noted, however, that depending upon the substrate material (aluminate or alumina) ahead of each of the two leading edges being the same or different, the final equilibrium angles at the outer and inner edges of the torus may be the same or different. In addition, as a result of the rapid extension, the crystallite may undergo fragmentation and the torus may appear to be composed of interconnected smaller units (Figs. 9d, 9f). On the other hand, if the interactions at the leading edge with the substrate are very strong so as to bind the crystallite to the substrate and if for some reasons the initial cavity does not form in the extended crystallite, then the subsequent tendency to contract following reduction would generate opposing forces, since the crystallite tends to contract while its periphery is constrained from following through. This generates the annular gap and gives rise to the core-and-ring structure (32). This tendency is enhanced by temperatures which are sufficiently low so as not to break the binding at the crystallite periphery. If the temperature is high, however, the binding at the periphery can be broken and the particle will contract as a single unit to form a compact structure. Compact particles and marks of their previous particle peripheries on the substrate were indeed observed at the high temperatures of 700 and 800°C (Figs. 6d; 7c; 11a, b) (and also at 500°C but at higher loading or when pure hydrogen was used for reduction, Figs. 9c, 15c). From the above discussion it results that depending upon the kinetics of the different processes (such as dissolution, reaction, wetting etc.) either a torus shape or a core-and-ring structure may form initially. Subsequent alternation between these shapes, observed with Fe/Al_2O_3 heated in as-supplied hydrogen, is a result of alternate oxidation and relative reduction and is discussed in Refs. (22, 32). The events observed in the present experiments can be explained on the basis of the above strong chemical interaction and wetting considerations as follows.

Iron does not wet alumina in reducing or inert atmospheres when there is no reaction or mass transfer at the interface, its wetting angle near the melting point being 141° (31, 33). However, in either partially or fully oxidizing environments, such as hydrogen contaminated with oxygen and/or moisture (even only in traces) or flowing oxygen, the present results indicate that the crystallites are extended. Electron diffraction indicates that in both the oxidizing environments, in addition to some oxide (γ-Fe_2O_3 (Fe_3O_4)), an interaction product, probably a solid solution of iron oxide in alumina and/or vice versa, is formed. The reactivity of iron, in general, with oxygen is very high. In the case of supported iron (iron supported on alumina), oxygen, even present only in traces, is preferentially adsorbed at the interface (34), decreasing both σ_{cg} and σ_{cs}. This decreases θ and the crystallite extends on the alumina surface. An additional decrease in σ_{cs} occurs as a result of the interaction of the iron ions with the alumina support. Alumina, especially the γ form, possesses a defect structure and a de-hydroxylated surface (35) and is therefore very reactive. At low loadings, the iron ions move into the vacant sites of γ-Al_2O_3. The solubility of Fe^{3+} ions in γ-Al_2O_3 can be appreciable because of the chemical similarity of the cations and also the availability of defects in γ-Al_2O_3. As a result of this metal oxide reaction with alumina, a strong chemical bonding results between the crystallite and the support, and U_{cs} is tremendously increased. Such strong interactions lead to the observed extended forms and the torus or core-and-ring shapes as described in the previous paragraphs. However, although they prevent subsequent sintering, such strong interactions with the support cause difficulties in reducing the iron ions to the zero-valent metal. The reaction products, either the solid solutions of iron oxide with alumina or the stoichiometric aluminates, $FeAl_2O_4$ and $Al_2Fe_2O_6$, are very difficult to be reduced (as seen in Tables 2–5) compared to the pure oxide of iron. In fact, iron oxide itself requires only 300°C for reduction to the metallic state (12). The literature (6, 16) also indicates that ferric ions supported on Al_2O_3 or silica in low concentrations (about 0.1 wt%) can be only partially reduced by hydrogen to the ferrous state, even at 700°C. Direct evidence for the formation, at low loadings, of the difficult-to-reduce aluminates or solid solutions in the presence of even traces of oxygen and/or moisture has been indicated by electron diffraction in the present experiments (Tables 2–4). Their presence could explain the difficulty encountered in the literature to reduce Fe^{3+} ions below the Fe^{2+} state at temperatures as high as 700°C, if the reducing hydrogen contained even traces of oxygen or moisture. Other factors in the catalyst preparation procedure, such as the precursor used, pH, etc., may facilitate the formation of such compounds. On the other hand, if the loading is large, initially large particles form, which have a relatively smaller fraction of their atoms in contact with the substrate at the interface. The interaction compounds (aluminates or solid solutions) formed between the two will be in relatively small amounts compared to the oxide. Therefore they are not easily detected by electron diffraction and they do not pose difficulties in reducing the oxide to the metal, even at the relatively low temperature of 500°C. Hence, in such cases the oxide particles are reduced easily (α-Fe is detected by electron diffraction) and the reduced crystallites will have relatively weak interactions with the substrate. As a result, considerable sintering occurs and dense, compact particles are formed (Fig. 4d). In fact, when further purified hydrogen is used during the reduction step, formation of compact particles and reduction to α-Fe occur even for relatively low loadings even at the low temperatures of 500°C (Table 13). This occurs because of decreased interaction between crystallite and support in the absence of oxygen and/or moisture.

A thick film extends from around the particles on heating in oxygen at 700°C and only aluminate is detected in the diffraction pattern. A thick film of aluminate may cover the alumina substrate if $\sigma_{\text{aluminate-gas}} + \sigma_{\text{aluminate-alumina}} < \sigma_{\text{alumina-gas}}$. Similarly, a thick film of aluminate may cover the oxide particles if $\sigma_{\text{aluminate-gas}} + \sigma_{\text{aluminate-iron oxide}} < \sigma_{\text{iron oxide-gas}}$.

As seen in Figs. 14 and 15 the degree of extension of the crystallites, when they are heated in flowing oxygen, is higher at 700°C than at 600 and 500°C. The fact that the temperature coefficients of surface and interfacial tensions are negative for most materials (σ_{cg}, σ_{cs}, and σ_{sg} decrease with increasing temperatures) and the possibility of an enhanced reaction between the gas, the crystallite, and the substrate at higher temperatures may explain this observation. (Of course, this implies sufficient decreases with temperature of σ_{cs} and/or σ_{cg} compared to that of σ_{sg}.) In fact, it is possible that heating the crystallites in oxygen at temperatures higher than 700°C will cause even more extension, and eventually complete spreading. This can happen if the inequality (1a) is satisfied but will be observed only if the time needed for complete spreading is shorter than the heating time.

SMSI

The SMSI (strong metal-support interactions) referred to frequently in the recent literature (36) is in reality (as already noted by Ruckenstein and Pulvermacher (23)) a reflection of the quantity U_{cs}, which in turn is related via Eq. (3) to the interfacial tension σ_{cs}. Large values of U_{cs}, hence strong interactions (23) lead to small values of σ_{cs} (even to negative values of σ_{cs} under nonequilibrium conditions). As a result, a variety of surface phenomena associated with wetting and spreading, of the kind discussed in Ref. (23) and in this paper, occur.

Splitting Behavior

Splitting of crystallites has been observed under different conditions and in different forms (Tables 1–3, 5). When heated in as-supplied hydrogen at 400°C, some crystallites were observed to split into two and to merge again subsequently (M in Figs. 2g–i). The crystallites had a torus shape both before as well as after the splitting. Such splitting may be a result of the extension and the instability of the extended toroidal shape to local variations in curvature along the periphery. Following heating in oxygen, in the majority of cases, the crystallites appear to be composed of a number of smaller subunits held together (Figs. 9d, 9f). Such a breakup is more likely a result of the spreading generated stresses and rupture (20, 28). Palladium crystallites on alumina were also observed to become porous and develop cavities when heated in oxygen (18–20). It was shown that if the crystallites were small or if the temperature was high, the crystallites extended irregularly, exhibiting a kind of tearing phenomenon.

The extension of the crystallites following oxidation generates stresses in the crystallite and leads to the growth of existing cracks if these stresses exceed the critical stress for crack propagation given by Griffith's expression

$$\tau_{\text{c}} = \left(\frac{2E\gamma}{\pi r} \right)^{1/2},$$

where γ the surface tension, r is a characteristic length of the crack, and E is Young's modulus of elasticity. Further oxidation, especially at the tip of the cracks, facilitates the crack propagation and the crystallite is ruptured and appears to be made up of disjointed smaller units. However, because iron oxide has a tendency to wet alumina, the individual units of the crystallites may be very near to one another. Splitting is, therefore, not observed clearly in oxidizing atmospheres, but the crystallites appear porous. Subsequent heating in hydrogen tends to contract the particle and the splitting becomes evident as each individual unit tends to contract to a higher equilibrium wetting angle. Of course, on continued heating in hydrogen the individual crystallites coalesce and sinter together.

Redispersion of the crystallites may also occur as a result of rupture and contraction of an extended surface film. As seen in Figs. 12c and 15d a kind of thick film either of iron oxide or of aluminate may spread on alumina from the periphery of the crystallites. In an oxygen atmosphere there may even be a multilayer surface film on the surface of alumina, coexisting with the threedimensional crystallites. The film may not be continuous, but ruptured. When heated subsequently in hydrogen, the islands will contract to a higher equilibrium angle. During this process of contraction a part of the film may coalesce with the adjacent particles and a part may form new, smaller crystallites, effectively causing redispersion.

The fragmentation of the crystallites to a large number of smaller particles observed after repeated alternate heating in hydrogen and oxygen for a total of about 200 h (Figs. 9f and g) is again a result of the spreading generated stresses and rupture. If a chemical interaction occurs between the crystallites and the substrate, repeated alternations may modify the support and affect the wetting behavior of the crystallites as a result of the repeated diffusion of the iron oxide molecules in and out of the substrate. In addition, the fragmentation may be influenced by the mechanical fatigue that sets in the crystallites due to repeated extensions and contractions.

CONCLUSION

Several phenomena have been observed to occur during heating model iron/alumina catalysts in hydrogen containing trace amounts of oxygen and/or moisture, in additionally purified hydrogen, and in oxygen. Particle growth by migration and coalescence, by emission and capture of atoms involving only a few neighboring crystallites (direct ripening) as well as a large number of the crystallites (Ostwald ripening) have been observed. These are, however, only a part of the complex surface phenomena that occur. Coalescence of nearby particles, extension and disappearance of particles, appearance of new particles in regions unoccupied previously, and splitting of particles also occur. In addition, ppm quantities of oxygen and/or moisture in hydrogen are shown to affect the reducibility of the iron ions and also to cause stability with respect to growth of the crystallites. Electron diffraction indicated considerable chemical interaction of the crystallites with the support and indicated the formation of solid solutions of iron oxide with alumina or of the stoichiometric aluminates $FeO \cdot Al_2O_3$ and $Fe_2O_3 \cdot Al_2O_3$ As a result of these chemical interactions, the crystallites have been observed to undergo changes in shape alternately from a torus to a core-and-ring structure when heated, especially at low temperatures of 400 or 500°C, in ultrahigh pure hydrogen containing trace quantities of oxygen and/or moisture. Such alternations were not observed when hydrogen was further purified to minimize oxygen and moisture contents. The additionally purified hydrogen caused better reducibility to the zero-valent metal. On heating in as-supplied hydrogen, at the low loadings of 6 or 7.5 Å, very little Fe^0 forms even at 700°C, whereas almost complete reduction to metallic iron is achieved even at 500°C when the loading is high, about 12.5 Å. In contrast, with additionally purified hydrogen, metallic iron could be detected even at 400°C for a loading of only 7.5 Å. On heating in oxygen, the crystallites were observed to extend considerably, the extension being enhanced with increasing temperatures. Toroidal shapes with a small remnant particle in the cavity were observed on heating in oxygen at 500 and 600°C. The extent of wetting increased at 700°C. In addition, it was inferred from the electron micrographs that on heating in

Role of Physical and Chemical Interactions in the Behavior of Supported Metal Catalysts   **231**

oxygen at 700°C, a film whose thickness should be greater than monomolecular, coexists with three-dimensional crystallites. After prolonged alternate heating in hydrogen and oxygen, the crystallites were observed to split into a number of smaller particles, probably as a result of mechanical fatigue. The results are discussed in terms of the interactions between metal, gas, and support and the consequent wetting characteristics of the metal-support system. It is suggested that such strong chemical interactions may lead to a considerable decrease in the interfacial tension between crystallite and substrate and, therefore, to a tendency for the crystallites to spread out. It is stressed that the strong metal-support interaction, the sintering and redispersion phenomena, and the changes in the shape of the crystallites are all interlinked parts of a complex surface phenomenon that cannot be easily categorized as belonging to a simple "either or" phenomenon.

REFERENCES

1. Anderson, J. R., "Structure of Metallic Catalysts." Academic Press, New York, 1975.
2. Satterfield, C. N., "Heterogeneous Catalysis in Practice." McGraw–Hill, New York, 1980.
3. Ameise, J. A. Butt, J. B., and Schwartz, L. H., *J. Phys. Chem.* **82**(5), 558 (1978).
4. Guczi, L., *Catal. Rev.-Sci. Eng.* **23**(3), 329, (1981).
5. Yoshioka, T., Koezuka, J., and Ikoma, H., *J. Catal.* **16**, 264 (1970).
6. Garten, R. L., and Ollis, D. F., *J. Catal.* **35**, 232 (1974).
7. Brenner, A., and Hucul, D. A., *Inorg. Chem.* **18**(10), 2836 (1979).
8. Raupp, G. B., and Delgass, W. N., *J. Catal.* **58**, 337 (1979).
9. Raupp, G. B., and Delgass, W. N., *J. Catal.* **58**, 348 (1979).
10. Vannice, M. A., *J. Catal.* **37**, 462 (1975).
11. Perrichon, V., Charcosset, H., Barrault, J., and Forquy, C., *Appl. Catal.* **7**, 21 (1983).
12. Topsøe, H., Dumesic, J. A., and Morup, S., *in* "Applications of Mössbauer Spectroscopy" (R. L. Cohen, Ed.), Vol. 2. Academic Press, New York, 1980.
13. Raupp, G. B., and Delgass, W. N., *J. Catal.* **58**, 361 (1979).
14. Ameise, J. A., Grynkevich, G., Butt, J. B., and Schwartz, L. H., *J. Phys. chem.* **85**, 2484 (1981).
15. Hobson, M. C., Jr., and Gager, H. M., *J. Catal.* **16**, 254 (1970).
16. Garten, R. L., *J. Catal.* **43**, 18 (1976).
17. Ruckenstein, E., and Dadyburjor, D. B., *Reviews in Chem. Eng.* **1**, 251 (1983).
18. Chen, J. J., and Ruckenstein, E., *J. Catal.* **69**, 254 (1981).
19. Ruckenstein, E., and Chen, J. J., *J. Colloid Interface Sci.* **86**(1), 1 (1982) (Section 3.5 of this volume).
20. Chen, J. J., and Ruckenstein, E., *J. Phys. Chem.* **85**, 1696 (1981).
21. Chu, Y. F., and Ruckenstein, E., *J. Catal.* **41**, 385 (1976).
22. Sushumna, I., and Ruckenstein, E., *J. Catal.* **90**, 241 (1984).
23. Ruckenstein, E., and Pulvermacher, B., *J. Catal.* **29**, 224 (1973).
24. Chakraverty, B. K., *J. Phys. Chem. Solids* **28**, 2401 (1967).
25. Wynblatt, P., and Gjostein, N. A., *Progr. Solid State Chem.* **9**, 21 (1975).
26. Flynn, P. C., and Wanke, S. E., *J. Catal.* **34**, 390 (1974).
27. Ruckenstein, E., and Dadyburjor, D. B., *J. Catal.* **33**, 233 (1977).
28. Ruckenstein, E., and Dadyburjor, D. B., *Thin Solid Films* **55**, 89 (1978).
29. Ruckenstein, E., and Chu, Y. F., *J. Catal.* **59**, 109 (1979) (Section 3.1 of this volume).
30. Overbury, S. H., Bertrand, P. A., and Somorjai, G. A., *Chem. Rev.* **75**(5), 547 (1975).
31. Beruto, D., Barco, L., and Passerone, A., *in* "Oxides and Oxide Films" (Ashok K. Vijh, Ed.), Vol. Dekker, New York, 1981.
32. Ruckenstein, E., and Sushumna, I., "Proceedings, 9th Iberoamerican Symposium on Catalysis," Vol. II, p. 1074, July (1984).
33. Naidich, Ju. V., *in* "Progress in Surface and Membrane Science" (D. A. Cadenhed and J. F. Danielli, Eds.), Vol. 14, p. 354. Academic Press, New York, 1981.
34. Halden, F. A., and Kingery, W. D., *J. Amer. Ceram. Soc.* **59**, 557 (1955).
35. Maciver, D. S., Tobin, H. H., and Barth, R. T., *J. Catal.* **2**, 485 (1963); **3**, 502 (1964).
36. Imelik, B., *etal.* (Eds.), "Studies in Surface Science and Catalysis," Vol. 11. Elsevier, Amsterdam, 1982.

3.5 Wetting Phenomena during Alternating Heating in O₂ and H₂ of Supported Metal Crystallites*

Eli Ruckenstein[†] and J. J. Chen
Department of Chemical Engineering, State University
of New York, Buffalo, New York 14260
Corresponding Author
[†] To whom correspondence should be addressed.
Received February 16, 1981; accepted June 1, 1981

INTRODUCTION

The behavior of Al_2O_3 supported Pd crystallites during heating in hydrogen or oxygen has been investigated and reported recently by the authors (1, 2). Severe sintering of the Pd crystallites was found to occur during heating in H_2 above 650°C. During heating in O_2, the formation of pits on the crystallites, the coalescence of these pits into cavities, the extension by spreading of the oxidized crystallites on the alumina surface and fragmentation induced by spreading have been observed.

The scope of the present paper is to examine the behavior of Pd crystallites (20 to 200 Å in initial size) supported on alumina, under heat treatment in alternating oxidizing and reducing atmospheres. These experiments have been carried out because the alternation between the two extreme conditions can better emphasize the phenomena which occur in each of them and the effect of the chemical atmosphere. Palladium crystallites supported on thin alumina films have been heated alternately in 1 atm O_2 at 750°C and 1 atm H_2 at 550 or 600°C, for various time intervals. A JEOL 100U transmission electron microscope was used to examine the same areas of the specimens after each treatment in order to follow the changes in size, shape, and position of each crystallite. A variety of events has been identified during the experiments. Most Pd crystallites expanded by spreading on the alumina surface during heating in O_2 and recontracted during heating in H_2. Some large crystallites grew in size during heating in H_2. During heating in either O_2 or H_2, some small crystallites disappeared. Pits formed on the crystallites during heating in O_2 and disappeared during heating in H_2.

While extensive literature is available concerning the sintering of supported metal crystallites and its mechanisms [some references are (3–17)], no evidence seems to be available on the spreading of the crystallites during heating in O_2 and their contraction after an additional heating in H_2, as well as on pit formation during heating in O_2.

---

* *Journal of Colloid and Interface Science.* Vol. 86, No. 1, p. 1, March 1982. Republished with permission.

EXPERIMENTAL PROCEDURES

Thin alumina films, about 300 Å thick, were used as supports for the palladium crystallites. They were prepared, as suggested in Ref. (18), by anodization of high-purity aluminum foil followed by amalgamation to separate the alumina film from the unoxidized portion of the aluminum foil. The films were then washed in a large amount of distilled water, picked up on gold microscope grids, and dried in air. Before the deposition of palladium, the nonporous and amorphous alumina films were calcined in air at 800°C for 72 hr to transform amorphous alumina into crystalline γ-alumina. The purpose of this prolonged heat treatment was to ensure that no phase transformation and no motions of the grains of the substrate would occur during further heating.

A thin Pd film about 10 Å thick was vacuum-evaporated onto the support from a 99.999% purity Pd wire wound on a tungsten filament. The Pd/Al_2O_3 specimens were heated inside a quartz tube from room temperature to 800°C in a 1-atm H_2 stream and cooled to room temperature in a He stream. This treatment generated Pd crystallites.

The heating of the specimens in H_2 or O_2 was carried out at 1 atm in a quartz tube that is surrounded by a tubular furnace. The specimens contained in a ceramic boat were displaced to the middle of the tube after the tube had been heated up to the desired temperature with He flowing through. Then, H_2 or O_2 was introduced into the tube to replace He. The specimens were cooled down in a He stream after the prespecified time duration was reached. The time span of cooling down was reduced by letting compressed laboratory air to flow through the annular region between furnace and tube. All the gases, H_2, O_2, and He were of high-purity grade and were purchased from Union Carbide Corporation. The gases were passed through a column of 5A molecular sieves to remove any trace of water before entering the tube. The flow rate of gas was about 30 ml/min.

After each step of heat treatment, the same regions of the specimens were examined by a transmission electron microscope, to follow the changes in size, shape, and position of each crystallite.

RESULTS

Figure 1 lists schematically various kinds of events observed on the electron micrographs taken after each stage of alternate heating of the Pd/Al_2O_3 specimens in 1 atm O_2 or H_2. During heating in O_2, most of the crystallites have expanded by spreading on the alumina surface and became irregular in shape. Many of them contain pits or cavities. Some small crystallites, below about 65 Å in diameter, either disappeared or decreased in size. During heating in H_2, most of the crystallites which have extended during the previous heating in O_2 contracted back to a more circular shape with a decrease in size. The pits formed during the previous heating in O_2 disappeared. Disappearance of crystallites and crystallite migration have also been observed. Some large crystallites, with diameters greater than 90 Å before the previous oxidation grew in size upon heating in H_2. Micrographs 2a to g show the time sequence of the same region of a Pd/Al_2O_3 specimen and provide the examples listed in Table I for various kinds of events. Besides the events listed, occasionally new crystallites appeared during heating in H_2, e.g., crystallite N of micrograph 2e.

Figures 3a to e give the size distribution of (about 1000) crystallites measured after each reduction step. Due to the irregular shapes of the oxidized crystallites, the size distribution of crystallites after the oxidation steps was not measured. The decrease in size of many small crystallites during run 2 and the growth in size of some large crystallites shift the minimum diameter to the smaller sizes and lead to a bimodal distribution (Fig. 3b). During later stages many small crystallites disappeared during heating in O_2 and some disappeared during heating in H_2, while the large ones grew gradually during heat treatment in both O_2 as well as H_2. The size distributions of Figs. 3c to e reflect this development.

Table II lists the experimental conditions for each heat treatment, the events observed after each treatment on the electron micrographs of all the regions examined, and two average crystallite sizes after each reduction stage. These two average sizes are: the number-average diameter defined by

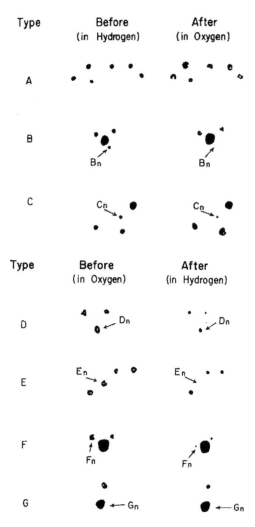

FIG. 1. Schematic illustration of various events observed during heating of Pd/γ-Al$_2$O$_3$ specimens in alternating O$_2$ and H$_2$ atmospheres. (Type A) During heating in oxygen, the crystallites change their shape from circular to irregular (with or without cavities) and extend by spreading on the alumina surface. (Type B) A small crystallite disappears while others change their shape during heating in oxygen. (Type C) A small crystallite decreases in size while others change their shape during heating in oxygen. (Type D) Previously oxidized crystallites contract back to more circular shapes and decrease in size during heating in hydrogen. (Type E) A previously oxidized crystallite disappears while others contract back to a more circular shape during heating in hydrogen. (Type F) A previously oxidized crystallite contracts and migrates to a new position during heating in hydrogen. (Type G) Previously large oxidized crystallites grow in size while the smaller ones contract during heating in hydrogen.

$$\bar{D} = \sum N_i \bar{D}_i / \sum N_i \quad [1]$$

and the surface-average diameter defined as

$$\bar{D}_s = \sum N_i \bar{D}_i^3 / \sum N_i \bar{D}_i^2. \quad [2]$$

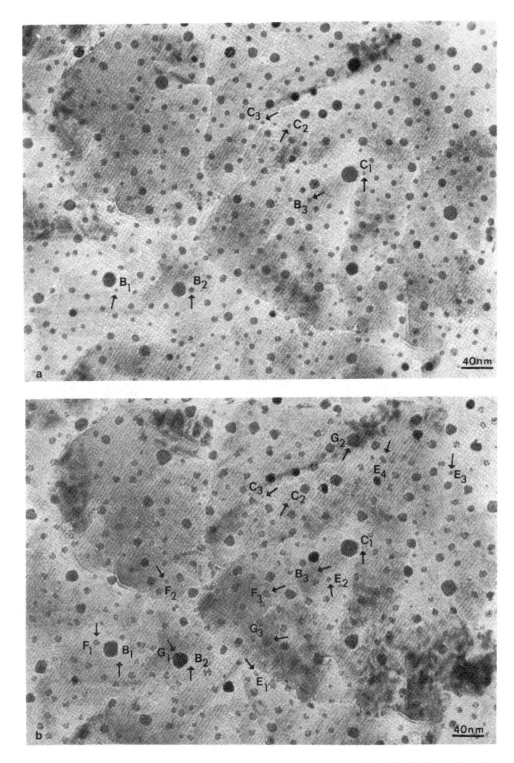

FIG. 2. Transmission electron micrographs showing the same region of a Pd/γ-Al$_2$O$_3$ specimen in time sequence (a) Fresh specimen; (b) after run 1. *(Continued)*

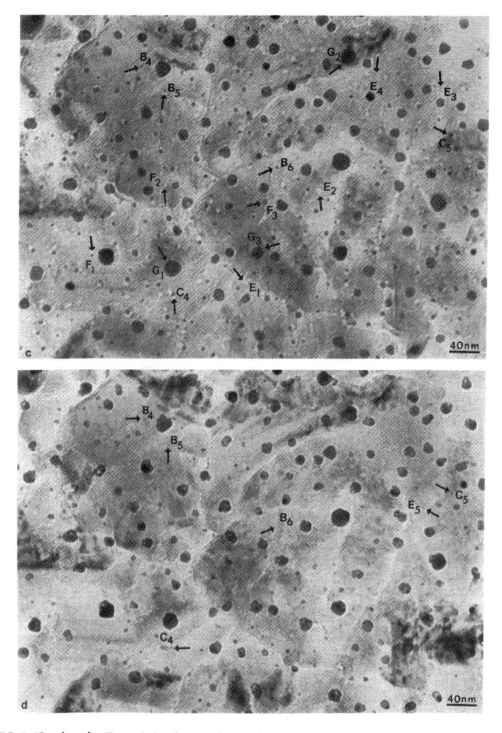

FIG. 2. (Continued) Transmission electron micrographs showing the same region of a Pd/γ-Al$_2$O$_3$ specimen in time sequence (c) after run 2; (d) after run 3. *(Continued)*

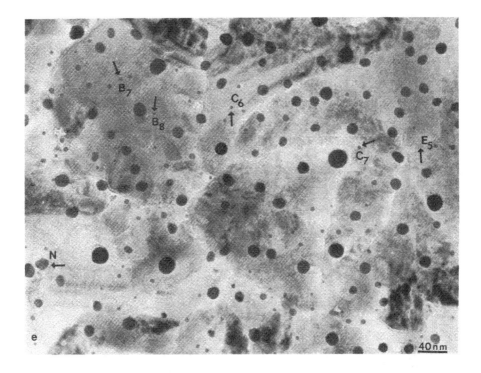

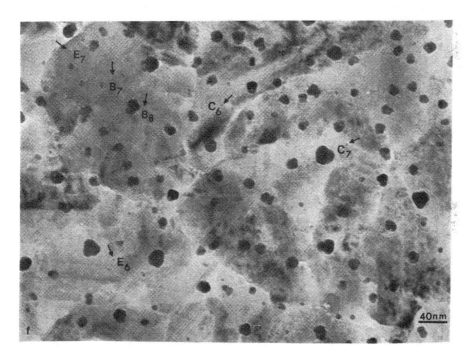

FIG. 2. (Continued) Transmission electron micrographs showing the same region of a Pd/γ-Al$_2$O$_3$ specimen in time sequence (e) after run 4; (f) after run 5. *(Continued)*

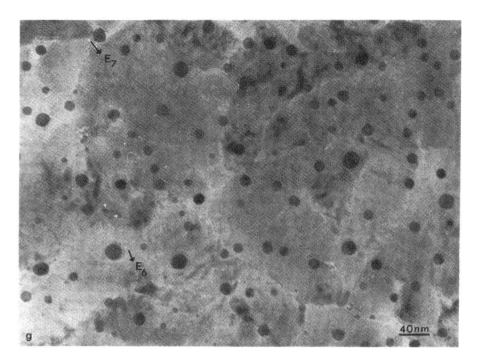

FIG. 2. (Continued) Transmission electron micrographs showing the same region of a Pd/γ-Al$_2$O$_3$ specimen in time sequence, (g) after run 6. Table II contains the characteristics of the runs.

Here N_i is the number of crystallites with diameters between $\bar{D}_i - \Delta D_i/2$ and $\bar{D}_i + \Delta D_i/2$, and $\bar{D}_i$ is the average diameter in each interval, ΔD_i.

DISCUSSION

Phenomena Observed during Heating in O$_2$

The phenomena observed during heating in O$_2$ include the formation of cavities on the crystallites, the extension by spreading of crystallites on the alumina surface, the disappearance of some small crystallites, and the decrease in size of some other small crystallites. The formation of pits as well as the spreading of the crystallites are associated with the oxidation of Pd to PdO. Indeed, these two phenomena have not been observed during heating in H$_2$ (1) and during heating in 1 atm O$_2$ at temperatures larger than 870°C, temperature at which the oxide becomes unstable.

The following mechanisms possibly explain pit (cavity) formation:

(a) The oxidation starts at some particular (active) sites on the heterogeneous crystallite surface. The lower surface tension of PdO molecules induces their migration toward the unoxidized areas of Pd since the metal has a higher surface tension than PdO. This process, driven by the surface tension gradient toward regions with higher surface tensions, is similar to the Marangoni effect in liquids (19). Migration of PdO molecules exposes fresh inner layers of Pd atoms to O$_2$. Thus the oxidation continues and a pit forms. The large cavities are probably generated by the coalescence of these pits. Since, as explained below, the crystallites extend, during heating in O$_2$, to a smaller equilibrium wetting angle, the stresses generated during this process may enhance the rate of growth of the pits.

Wetting Phenomena during Alternating Heating in O_2 and H_2

(b) The spreading generated stresses may be sufficiently large to lead to the propagation of small cracks existing on the surface of the oxidized crystallites and thus to generate pits. This process is enhanced by the oxidation of the freshly exposed atoms at the tip of the crack.

(c) A combination of mechanisms (a) and (b).

The extension by spreading of the oxidized crystallites can be explained in terms of wetting. Generally speaking, when a crystallite on the substrate surface is at thermodynamic equilibrium, the wetting angle θ is related to σ_{gs}, σ_{cg}, and σ_{cs}, the interfacial tensions between gas–substrate, crystallite–gas, and crystallite–substrate, via Young's equation[2]:

$$\sigma_{gs} = \sigma_{cs} + \sigma_{cg}\cos\theta \qquad [3]$$

As the crystallite is oxidized, σ_{cg} decreases substantially and σ_{cs} also decreases. Under these conditions, the wetting angle decreases and hence, the crystallite has the tendency to extend over the surface of the substrate. The irregular shape of the extended crystallites during the heating in O_2 is probably due to the nonuniform oxidation and heterogeneity of the substrate.

EMISSION OF PdO, OSTWALD RIPENING

Since some small crystallites decrease in size during heating in O_2, it is reasonable to conclude that emission of PdO to the free surface of the substrate occurs. This can generate a two-dimensional phase of PdO on the surface of the substrate (5, 21).

TABLE I

Examples of the Events of Fig. I

Type	Examples
A	Figs. 2a (before) and 2b (after)
	Figs. 2c (before) and 2d (after)
	Figs. 2e (before) and 2f (after)
B	B_1 to B_3 of Figs. 2a (before) and 2b (after)
	B_4 to B_6 of Figs. 2c (before) and 2d (after)
	B_7 and B_8 of Figs. 2e (before) and 2f (after)
C	C_1 to C_3 of Figs. 2a (before) and 2b (after)
	C_4 and C_5 of Figs. 2c (before) and 2d (after)
	C_6 and C_7 of Figs. 2e (before) and 2f (after)
D	Figs. 2b (before) and 2c (after)
	Figs. 2d (before) and 2e (after)
	Figs. 2f (before) and 2g (after)
E	E_1 to E_4 of Figs. 2b (before) and 2c (after)
	E_5 of Figs. 2d (before) and 2e (after)
	E_6 and E_7 of Figs. 2f (before) and 2g (after)
F	F_1 to F_3 of Figs. 2b (before) and 2c (after)
G	G_1 to G_3 of Figs. 2b (before) and 2c (after)

---

[2] For small crystallites the equilibrium problem is more complicated and has to be examined along the lines developed in Ref. (7). Because the present discussion is largely qualitative, Young's equation, which is valid for relatively large droplets, is used as the starting point.

The dissolution surface pressure of the crystallite, P_{sr}, is related to the radius r of the crystallite-substrate interface by a Kelvin-type equation

$$P_{sr} = P_{s\infty} \exp\left(\frac{\delta}{r}\right), \qquad [4]$$

where $P_{s\infty}$ is the dissolution surface pressure for a crystallite of infinite size, and

$$\delta = \sigma_l S_l / kT. \qquad [5]$$

Here, σ_l is the line tension (erg/cm) at the leading edge of the crystallite, S_l is the surface area per molecule at the crystallite-substrate interface, k is the Boltzmann constant, and T is the temperature in degrees Kelvin. With the simplified assumption of a two-dimensional ideal gas,

$$P_{s\infty} = n_{s\infty} kT, \qquad [6]$$

where $n_{s\circ}$ is the concentration of single PdO molecules in a two-dimensional phase in equilibrium with a crystallite of infinite size, expressed as molecules per unit area of substrate. Thus,

$$P_{sr} = n_{s\infty} kT \exp\left(\frac{\delta}{r}\right). \qquad [7]$$

Obviously, the dissolution surface pressure is larger when r is smaller. The two-dimensional phase of PdO molecules generates, near the leading edge of the crystallite, a surface pressure, P_s, which, assuming the two-dimensional phase to be an ideal gas, is given by

$$P_s = nkT. \qquad [8]$$

Here n is the concentration, at a given moment, of the molecules in the two-dimensional phase, near the leading edge of the crystallite.

Of course, if the equilibrium concentration of molecules at the leading edge of the crystallite, $n_{s\infty} \exp(\delta / r)$, is larger than n, or in other words, if. $P_{sr} > P_s$, single molecules would be emitted from the crystallite onto the substrate. The opposite happens when $P_{sr} < P_s$. Ostwald ripening occurs when $P_{sr} > P_{sb}$ for the small crystallites and $P_{sr} < P_{sb}$ for the large ones, where P_{sb} is the surface pressure in the bulk of the two-dimensional phase. In such cases, the small crystallites decrease in size, while the large ones increase Ostwald ripening (5)].

The disappearance of some small crystallites during heating in oxygen (event B, Fig. 1) can be explained as follows: A small crystallite has a much larger dissolution pressure than a large one does and thus its emission of molecules per unit length of leading edge and unit time is also higher. Somewhat larger crystallites will only decrease in size (event C, Fig. 1). Since the dissolution pressure is negligible for large crystallites, they may increase their radius by: (1) the spreading caused by the oxidation of Pd; (2) the coalescence with small migrating crystallites; and, if the surface pressure P_{sb} is sufficiently large, (3) by the capture of molecules of the two-dimensional phase of PdO molecules.

PHENOMENA OBSERVED DURING HEATING IN H_2

Upon heating in H_2, most of the crystallites which extended during heating in O_2 contracted back to more circular shapes because the oxide is reduced to metal. The sizes of the reduced crystallites are usually smaller than those of the oxidized ones (event D). This is because the equilibrium wetting

Wetting Phenomena during Alternating Heating in O_2 and H_2 241

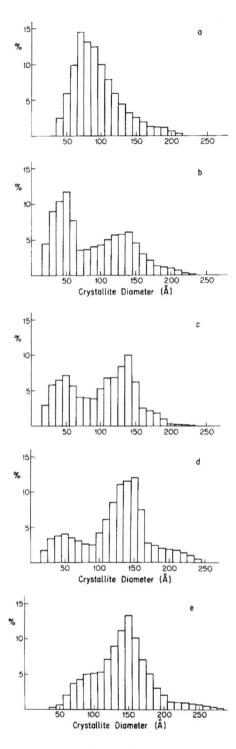

FIG. 3. Size distribution of Pd crystallites on Al_2O_3 substrate, (a) Fresh specimen; (b) after run 2; (c) after run 4; (d) after run 6; (e) after run 14.

TABLE II

Conditions of Heat Treatment, Average Crystallite Sizes, and Events Observed during Experiment

Run	Atmosphere (1 atm)	Temperature (°C)	Time interval I(hr)	Events observed	Number-average diameter (Å)	Surface-average diameter (Å)
Fresh specimen					94.30	119.42
1	O_2	750	0.5	A, B, C		
2	H_2	550	1.5	D, E, F, G	92.68	143.97
3	O_2	750	1.0	A, B. C		
4	H_2	550	3.0	D, E, F, G	105.58	145.58
5	O_2	750	2.0	A, B, C		
6	H_2	550	6.0	D, E, G	122.69	153.45
7	O_2	750	4.0	A, B		
8	H_2	550	12.0	D, E	130.75	156.89
9	O_2	750	4.0	A, B		
10	H_2	550	12.0	D, E	135.03	159.51
11	O_2	750	4.0	A, B		
12	H_2	550	12.0	D, G	137.24	161.02
13	O_2	750	10.0	A, B		
14	H_2	600	18.0	D	139.07	162.07

angle between Pd and Al_2O_3 is larger than that between PdO and Al_2O_3. The sizes of the reduced crystallites are even smaller than those of the corresponding crystallites before oxidation. This suggests that the crystallites might have lost some of their atoms during the oxidation step.

One of the two possible mechanisms for event E is that the emission of single PdO molecules from the crystallite during heating in O_2 is so substantial that the size of the crystallite resulting during heating in H_2 becomes smaller than the resolution of the electron microscope. An alternative mechanism is the migration of the crystallite and its capture by a neighboring large crystallite. The migration of resolved and unresolved particles and coalescence with the large one is a possible mechanism for the growth in size of a crystallite during heating in H_2.

The appearance of a new crystallite N of Fig. 2e could be due to the migration and coalescence of unresolved particles.

CONCLUSION

The behavior of small Pd crystallites is quite different from that of the larger ones during alternating heating in O_2 and H_2.

Many small crystallites disappear during heating in O_2 and some disappear during heating in H_2. While the disappearance of the small crystallites during heating in O_2 is probably associated with the emission of PdO by the particles to a two-dimensional phase, their disappearance during heating in H_2 is probably caused by migration and coalescence. Somewhat larger crystallites extend by spreading on the surface of the substrate during heating in O_2 and contract after an additional heating in H_2. Pits and cavities are generated in some of the crystallites during heating in O_2 and disappear after an additional heating in H_2. The large crystallites grow further during heating in H_2 probably because of the capture of the smaller, migrating, resolved, and unresolved particles. The extension as well as the contraction can be explained in terms of wetting. Pit formation is explained as either a surface tension gradient driven phenomenon enhanced by a spreading generated stress, or as a crack propagation phenomenon induced by the stress produced by spreading and enhanced by the oxidation of the freshly exposed atoms at the tip of the crack.

REFERENCES

1. Chen, J. J., and Ruckenstein, E., *J. Catal.* **9**, 254 (1981).
2. Chen, J. J., and Ruckenstein, E., *J. Phys. Chem.* **85**, 1606 (1981).
3. Skofronick, J. G., and Phyllips, W. B., *J. Appl.Phys.* **38**, 4791 (1967).
4. Phillips, W. B., Desloge, E. A., and Skofronick, J. G., *J. Appl. Phys.* **39**, 3210 (1968).
5. Chakraverty, B. K., *J. Phys. Chem. Solids* **28**, 2401 (1967).
6. Kern, R., Masson, A., and Metois, J. J., *Surface Sci.* **27**, 483 (1971).
7. Metois, J. J., Gauch, M., Masson. A., and Kern, R., *Surface Sci.* **30**, 43 (1972).
8. Ruckenstein, E., and Pulvermacher, B., *AIChE J.* **19**, 356 (1973).
9. Ruckenstein, E., and Pulvermacher, B., *j. Catal.* **29**, 224 (1973).
10. Flynn, P. C., and Wanke, S. E., *J. Catal.* **33**, 233 (1974).
11. Zanghi, J. C., Metois, J. J., and Kern, R., *Philos.Magn.* **29**, 1213 (1974).
12. Heinemann, K., and Poppa, H., *Thin Solid Films* **33**, 237 (1976).
13. Chu, Y. F., and Ruckenstein, E., *Surface Sci.* **67**, 517 (1977).
14. Metois, J. J., Heineman, K., and Poppa, H., *Philos. Mag.* **35**, 1413 (1977).
15. Baker, R. T. K., Thomas, C., and Thomas, R. B., *J. Catal.* **38**, 510(1975).
16. Ruckenstein, E., and Chu, Y. F., *Catal.* **59**, 109 (1979) (Section 3.1 of this volume).
17. Baker, R. T. K., Prestridge, E. B., and Garten, R. L., *J. Catal.* **59**, 293 (1979).
18. Ruckenstein, E.. and Malhotra, L., *J. Catal.* **41**, 303 (1976).
19. Adamson. A. W., "Physical Chemistry of Surfaces," 2nd ed. Interscience, New York (1967).
20. Ruckenstein, E., and Lee, P. S., *Surface Sci.* **52**, 298 (1975) (Section 3.1 of Volume I).
21. Ruckenstein, E., *in* "Studies in Surface Science and Catalysis" (J. Bourdon, Ed.), Vol. 4, p. 57. Elsevier, Amsterdam (1980).

3.6 The Behavior of Model Ag/Al$_2$O$_3$ Catalysts in Various Chemical Environments*

Eli Ruckenstein and Sung H. Lee

Department of Chemical Engineering, State University
of New York at Buffalo, Buffalo, New York 14260

Received June 8, 1987; revised August 18, 1987

INTRODUCTION

Silver is a well-known catalyst for the partial oxidation of ethylene to ethylene oxide. The catalytic oxidation of ethylene over silver has been reviewed by several authors (1–5). Most of the studies have been carried out to investigate the mechanism and mode of adsorption of oxygen on the surface of the catalyst, since this affects the selectivity and activity of the catalyst. The loss of active metal surface area caused by sintering has been investigated much less. The sintering behavior of supported metal particles is affected by many factors, such as the support material, reaction atmosphere, metal—support interaction, etc. In general, when alumina or silica are the supports, heating in H$_2$ causes sintering, while heating in an oxidizing atmosphere, such as O$_2$ or steam, can bring about redispersion (6–8). The behavior of supported silver has been reported by a number of authors. Presland *et al.* (9, 10) investigated the effect of oxygen and hydrogen on a silver film supported on an amorphous silica substrate. When heated in a hydrogen atmosphere which was either dry or saturated with water vapor, the silver films remained unchanged. In contrast, in an oxidizing atmosphere, hillocks were observed to form as the first stage; on further heating silver islands were generated. Seyedmonir *et al.* (11) investigated the dispersion of silver crystallites heated in He, H$_2$, and O$_2$. Very little or no change in dispersion was observed for silver crystallites supported on TiO$_2$, η–Al$_2$O$_3$, or SiO$_2$, on heating in H$_2$ or He at 400°C. The dispersion of silver crystallites supported on η–Al$_2$O$_3$ decreased, however, by 20–30% after heat treatment in O$_2$ at 400°C for 15 h. For Ag/TiO$_2$, no change in metal dispersion was observed in O$_2$. Recently, Plummer *et al.* (12) reported that silver particles supported on alumina coalesced and formed large particles when heated in air at 200–300°C. Employing *in situ* transmission electron microscopy, Heinemann and Poppa (13) observed crystallite migration followed by coalescence as well as ripening for silver particles supported on graphite, in the temperature range of 25–450°C and for a gas pressure of 10^{-9} Torr. Evidence for crystallite migration and coalescence for Ag/graphite, during heat treatment in hydrogen at temperatures between 320 and 515°C, was also presented by Baker and Skiba (14). There is, however, very little information about the behavior of silver crystallites supported on alumina, even though this is the most widely used catalyst for the partial oxidation of ethylene.

The aim of this paper is to investigate, in some detail, the behavior of alumina-supported Ag crystallites. For this purpose, model alumina-supported silver catalysts were employed. The specimens were heated in various gaseous atmospheres, O$_2$, H$_2$, C$_2$H$_4$, and a mixture of C$_2$H$_4$ and O$_2$, at

---

* *Journal of Catalysis* 109, 100–119 (1988). Republished with permission.

244

The Behavior of Model Ag/Al$_2$O$_3$ Catalysts in Various Chemical Environments

temperatures in the range encountered in industrial processes. The behavior of the silver crystallites supported on alumina has been examined by following the changes in the same regions of the specimen after each heat treatment.

EXPERIMENTAL

PREPARATION OF THE SAMPLE

Electron-transparent, nonporous, amorphous alumina films of approximately 300 Å thickness were prepared by anodization of chemically polished, thin, high-purity aluminum foils (99.999%, Alfa Products Inc.). The oxide film was stripped off by dissolving the unoxidized aluminum in a mercuric chloride solution (15). The oxide films were then washed in distilled water, picked up on gold electron microscope grids, and allowed to dry. They were subsequently heated in air at 800°C for 72 h to transform the amorphous alumina films into polycrystalline γ-alumina. High-purity silver wire (99.9%, Alfa Products) was vacuum-evaporated onto alumina films supported on gold electron microscope grids, under a pressure of less than 10^{-7} Torr, in an Edwards 306 vacuum evaporator.

THE HEAT TREATMENT

The specimens were heated in a quartz tube located in a furnace. After the specimen was introduced into the quartz tube, the tube was flushed with helium for at least 30 min. The specimen was then heated in a helium atmosphere. As soon as the desired temperature was reached, helium was replaced by the desired gas. After being heated at the desired temperature for the predetermined duration of time, the tube was cooled to room temperature in a helium atmosphere. The specimens were heated in the following gas streams: O$_2$, H$_2$, C$_2$H$_4$, and a mixture of C$_2$H$_4$ and O$_2$. The volume ratio in the mixture was 1:2. Most of the heat treatments were carried out at 250 or 300°C, temperatures in the range employed in industry. Some specimens, however, were heated at 400°C in O$_2$ or H$_2$. The flow rate of the gases was maintained at about 150 cm^3/min. The gases used in the experiments were purchased from Linde Division, Union Carbide Corporation. Hydrogen and helium were 99.999%, and oxygen was 99.99% pure. The purity of ethylene was 99.5%. Hydrogen was further purified by being passed through a Deoxo (Engelhard Industries) unit and a molecular sieve bed immersed in liquid nitrogen. Helium was also purified by being passed through a molecular sieve bed immersed in liquid nitrogen. After each heat treatment, the specimen was examined in a JEOL 100U transmission electron microscope.

RESULTS

THE BEHAVIOR OF THE SILVER CRYSTALLITES ON ALTERNATE HEATING IN H$_2$ AND O$_2$

Figure 1 is a micrograph of a specimen after deposition of a silver film of 10 Å thickness. Detectable small crystallites with a large number density were formed soon after deposition even at room temperature. Similar micrographs were also obtained for loadings of specimens with 7.5 and 5 Å.

When the specimen of Fig. 1 was heated in H$_2$ at 300°C for 1 h, large crystallites formed (Fig. 2a). The electron diffraction pattern indicated that the crystallites were present as Ag. After an additional 1 h of heating in H$_2$ at 300°C, many small particles disappeared (A in Figs. 2a, 2b). Some particles (B) migrated over the substrate. In other regions (C), several neighboring particles disappeared and new larger particles appeared in locations where no particles were observed before, most likely as a result of migration and coalescence. One may also note that some elongated particles changed their shape to almost a circular one (D in Figs. 2a, 2b). The elongated particles (D) of Fig. 2a were a result of collision and incomplete coalescence of two particles. Upon further heating of the specimen in H$_2$ at 300°C for 1 h more, events similar to the preceding ones occurred again (Fig. 2c). Some small

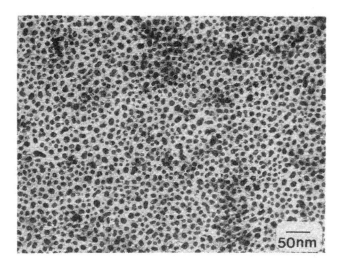

FIG. 1. Ag/Al$_2$O$_3$ specimen after deposition of a 10 Å–Ag film on Al$_2$O$_3$.

particles (A) disappeared, while other particles (B) migrated over the substrate. In several regions, considerable sintering occurred, resulting in a redistribution of the crystallites (C in Figs. 2b, 2c). In addition, a few particles decreased in size (E in Figs. 2b, 2c).

The specimen was then heated in oxygen. Heating for 0.5 h at 300°C brought about severe sintering (Fig. 2d), much more severe than in H$_2$. Neighboring particles coalesced to form larger particles (F in Figs. 2c, 2d), and many small crystallites, 100–150 Å in size, disappeared because of either migration and coalescence or direct ripening. In addition, a large number of extremely small crystallites, 50 Å or less in size, appeared all over the substrate. Only Ag was detected in the electron diffraction pattern. After heating in O$_2$ at 300°C for 0.5 h more, most of the large crystallites remained unaffected, a few crystallites disappeared (G in Figs. 2d, 2e), but the number of small crystallites increased. These small crystallites were either circular or wormlike. The electron diffraction pattern indicated that the particles remained as Ag throughout the heating in O$_2$.

The specimen was subsequently heated in H$_2$. After heating for 0.5 h in H$_2$ at 300°C (Fig. 2f), the small, wormlike crystallites remained unchanged, while most of the small, circular, crystallites disappeared (H in Figs. 2e, 2f). The large particles remained unaffected. Heating in H$_2$ for 1 h more at 300°C resulted in little change.

The specimen was subsequently heated at 400°C. Heating in H$_2$ at 400°C for 0.5 h resulted in considerable sintering with a redistribution of the crystallites (Fig. 2g). In spite of the severe sintering of the circular particles, the small, wormlike crystallites remained unchanged throughout heating (Ic in Figs. 2f, 2g). On further heating in H$_2$ at 400°C for 0.5 h more, sintering continued to occur, resulting in a further redistribution of the crystallites (Fig. 2h). The existing wormlike particles remained unaffected, while their number increased, and new, small, circular particles appeared (I in Figs. 2g, 2h). One may also note that some crystallites became faceted (J in Fig. 2h).

Subsequently, the specimen was heated in O$_2$ at 400°C for 0.5 h (Fig. 2i). Some particles disappeared, probably as a result of migration and coalescence with other particles. The faceted particles became circular, and new, small crystallites appeared. On further heating in O$_2$ at 400°C for 0.5 h more (Fig. 2j), many large particles disappeared, while new, small crystallites appeared. It is possible that a fraction of the particles which disappeared was permanently lost to the gas stream, while the rest remained on the substrate as small crystallites.

The specimen was again heated in H$_2$ at 300°C for 1 h. As shown in the micrograph (N in Fig. 2k), the circular, small crystallites increased in size, while much smaller crystallites disappeared.

The Behavior of Model Ag/Al$_2$O$_3$ Catalysts in Various Chemical Environments 247

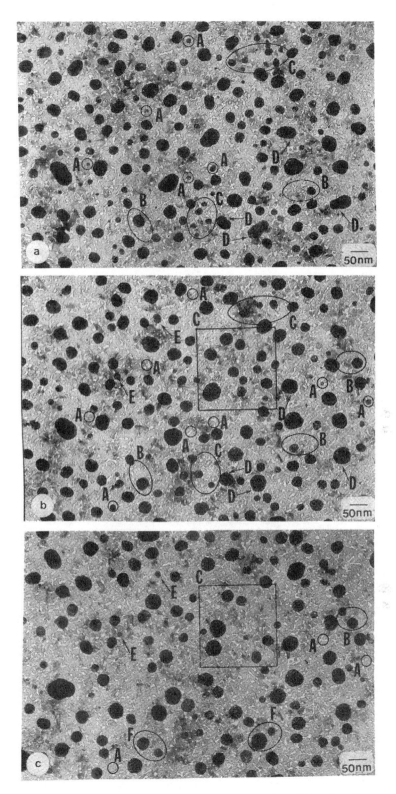

FIG. 2. Changes on heating the specimen of Fig. 1 alternately in O$_2$ and H$_2$. The micrographs are from a different region than that of Fig. 1. (a) H$_2$, 300°C, 1 h; (b) H$_2$, 300°C, 1 h; (c) H$_2$, 300°C, 1 h. *(Continued)*

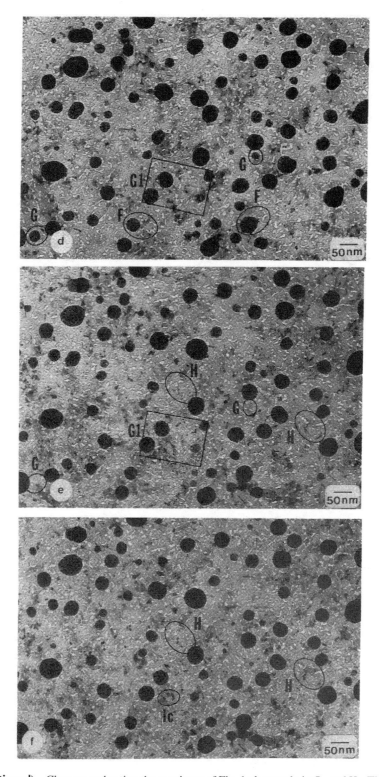

FIG. 2. (Continued) Changes on heating the specimen of Fig. 1 alternately in O_2 and H_2. The micrographs are from a different region than that of Fig. 1. (d) O_2, 300°C, 0.5 h; (e) O_2, 300°C, 0.5 h; (f) H_2, 300°C, 0.5 h,

(*Continued*)

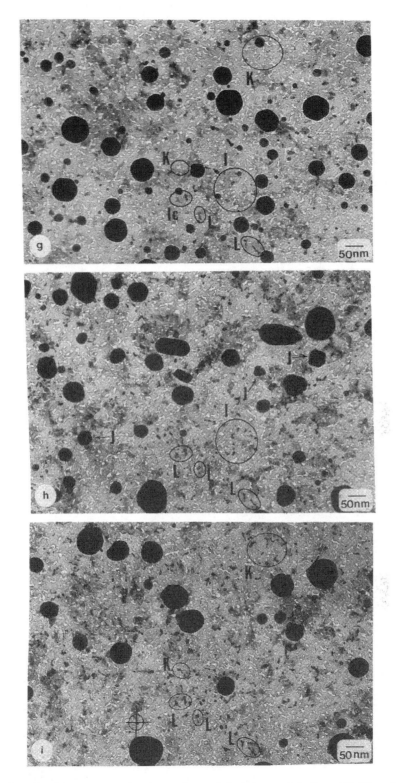

FIG. 2. (Continued) Changes on heating the specimen of Fig. 1 alternately in O_2 and H_2. The micrographs are from a different region than that of Fig. 1. (g) H_2, 1.5 h (1 h at 300°C and 0.5 h at 400°C); (h) H_2, 400°C, 0.5 h; (i) O_2, 400°C, 0.5 h. *(Continued)*

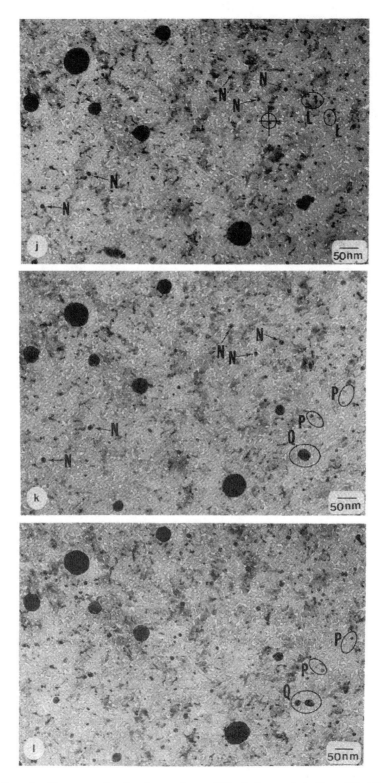

FIG. 2. (Continued) Changes on heating the specimen of Fig. 1 alternately in O_2 and H_2. The micrographs are from a different region than that of Fig. 1. (j) O_2, 400°C, 0.5 h; (k) H_2, 300°C, 1 h; (l) H_2, 300°C, 1 h.

However, the wormlike crystallites and most of the large particles remained unaffected. On further heating in H$_2$ at 300°C for 1 h more (Fig. 2l), some of the small crystallites of circular shape migrated over the substrate (P in Figs. 2k, 2l).

THE BEHAVIOR OF THE SILVER PARTICLES FOLLOWING OXIDATION OF ETHYLENE

To investigate the behavior of silver particles in the actual process of ethylene oxidation, specimens of 7 Å loading were heated in O$_2$, C$_2$H$_4$, and a mixture of O$_2$ and C$_2$H$_4$, respectively. Since ethylene oxidation is usually carried out in the range 200–300°C, the specimens were heated at 250°C.

Heating in O$_2$ at 250°C. Figure 3a represents the initial state after heating in H$_2$ at 300°C for 4 h. The specimen was subsequently heated in O$_2$ at 250°C. During 20 min heating in O$_2$, considerable sintering occurred, resulting in a redistribution of the particles (Fig. 3b). However, as observed earlier at 300 and 400°C, small crystallites also formed. Some of them had a wormlike shape and

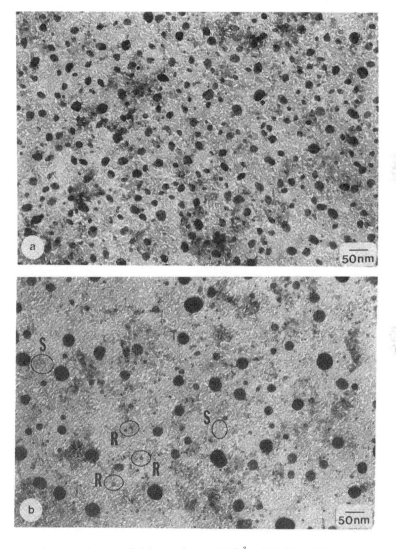

FIG. 3. Sequence of changes in an Ag/Al$_2$O$_3$ specimen with 7 Å initial film thickness heated in O$_2$ at 250°C. (a) H$_2$, 300°C, 4 h; (b) O$_2$, 20 min. *(Continued)*

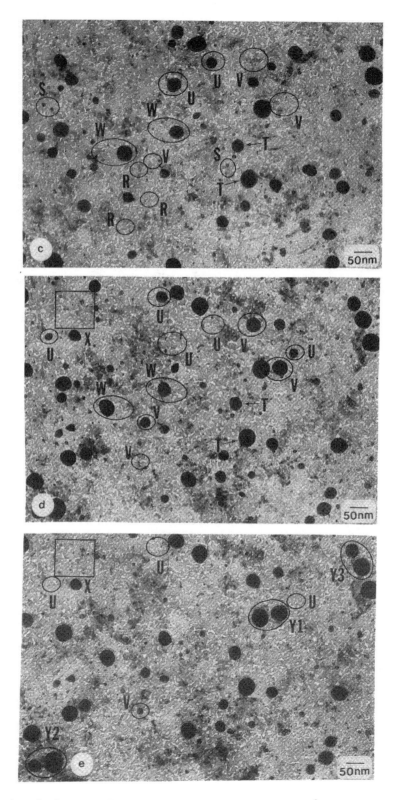

FIG. 3. (Continued) Sequence of changes in an Ag/Al$_2$O$_3$ specimen with 7 Å initial film thickness heated in O$_2$ at 250°C. (c) O$_2$, 20 min; (d) O$_2$, 20 min; (e) O$_2$ 30 min. *(Continued)*

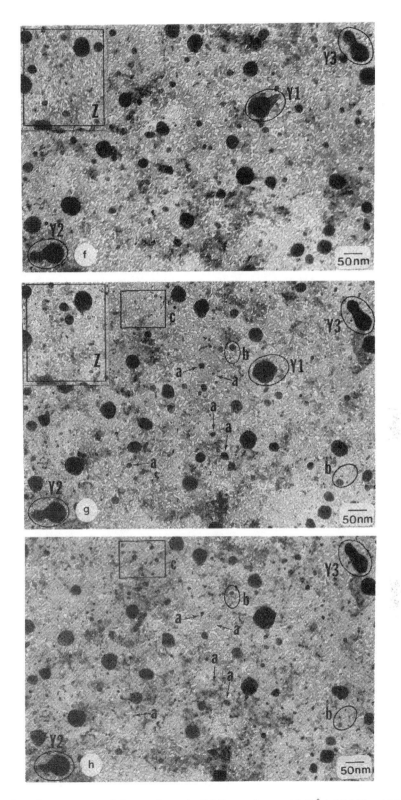

FIG. 3. (Continued) Sequence of changes in an Ag/Al$_2$O$_3$ specimen with 7 Å initial film thickness heated in O$_2$ at 250°C. (f) O$_2$, 30 min; (g) O$_2$, 1 h; (h) O$_2$, 1 h.

were present in the white regions of the substrate. Considerable sintering continued to occur on further heating in O_2 for 20 min more (Fig. 3c). Many small crystallites (R) disappeared, while new, small crystallites (S) appeared in regions where no crystallites were observed before. During heating for an additional 20 min in O_2 (Fig. 3d), many of the large particles (T) remained unaffected. However, some large particles (U) disappeared, probably because they migrated over the substrate and coalesced with other particles, while new crystallites of large size (V) appeared in regions where no crystallites were observed before (most likely via the coalescence of two or more migrating crystallites). Crystallites, marked W, migrated over the substrate. After further heating for an additional 0.5 h in O_2, the events observed were similar to those that occurred during the preceding heating in O_2 (Fig. 3e). In addition, a number of small crystallites appeared (X in Figs. 3d, 3e). Even though the large particles were relatively stable, a large number of silver atoms, undetectable in the electron microscope, were probably present on the substrate. Subsequently, small detectable crystallites probably formed via migration and coalescence of undetectable clusters. On heating for an additional 0.5 h in O_2 at 250°C (Fig. 3f), many small crystallites, 50 Å or less in size, disappeared while new crystallites approximately 150 Å in size appeared. In addition, large neighboring crystallites merged in some regions (Yl, Y2, and Y3). In region Yl, a large amount of the material of one particle is transferred to the other one. However, in regions Y2 and Y3, the two neighboring particles are connected through a bridge. The behavior of the smaller crystallites was similar to that in the preceding heating. After heating in O_2 for 1 h more (Fig. 3g), small particles of 50 Å or less appeared again. In region Yl, the transfer of material from one particle to the neighboring one through the bridge connecting them was nearly completed. For the particles in region Y2, which were connected through a bridge, there was no significant change (Fig. 3g). On subsequent heating for 1 h more in O_2, a large amount of material was, however, transferred into the other particle through the bridge (Fig. 3h). For the particles in region Y3, the bridge between the two particles increased in width on further heating (Figs. 3g, 3h). A similar behavior was observed before for Pt and Fe (*16–18*).

After a total of 4 h of heating in O_2 (Fig. 3h), the small crystallites increased in size, while 150–200 Å particles decreased or disappeared (a in Figs. 3g, 3h). In region b, a particle decreased in size, while several smaller crystallites appeared in the vicinity of the decreasing particle, probably as a result of the coalescence of the atoms and/or clusters which were lost by the decreasing particles.

Heating in C_2H_4 at 250°C. Figure 4a is a micrograph of a specimen heated in H_2 at 300°C for 4 h. Large crystallites of about 400 Å formed. In addition, very small crystallites of about 30 Å as well as wormlike crystallites appeared. The specimen was subsequently heated in C_2H_4 at 250°C. Heating for 20 min in C_2H_4 did not cause any significant change (Fig. 4b). Even the small particles remained unaffected. After 20 min more of heating in C_2H_4, the crystallites remained again unchanged. Upon heating for a total of 1 h 40 min in C_2H_4 (Fig. 4c), most of the crystallites changed their shape from circular to ellipsoidal without changing their locations. Some crystallites (d in Fig. 4c) had a straight edge over half of their periphery. In region e, two neighboring crystallites migrated over the substrate and formed a doublet. After heating for up to 3.5 h in C_2H_4, the crystallites in contact merged completely to form single particles (e in Fig. 4d). Two neighboring particles migrated and subsequently contacted each other (f in Figs. 4c, 4d), while nearby particles coalesced to form single particles (g in Figs. 4c, d). Most of the crystallites, however, remained almost unchanged. Their shape remained ellipsoidal but became somewhat more elongated. In addition, some particles rotated (h in Figs. 4c, 4d), and the particles which had a straight edge over a part of their peripheries acquired ellipsoidal shapes (d in Fig. 4d). The small crystallites did not sinter and remained in their original locations without any change (indicating that they were very stable during heating in C_2H_4), and no new small crystallites were generated. In contrast, during heating in

O_2 or H_2, the very small crystallites underwent various changes such as disappearance and coalescence, and new crystallites appeared as well. The specimen of Fig. 4d was heated further in O_2 at 250°C for 0.5 h. As shown in the micrograph (Fig. 4e), considerable sintering occurred. Most of the particles acquired circular shapes, and some small crystallites, of 80 Å or less in diameter, appeared in regions where no crystallites were observed before (i in Figs. 4d, 4e).

Heating in C_2H_4 + O_2 at 250°C. Figure 5a shows the initial state of the specimen after heating in H_2 at 300°C for 4 h. Subsequently, the specimen was heated in C_2H_4 + O_2 at 250°C. During heating in C_2H_4 + O_2 for 20 min, considerable sintering occurred and the crystallites acquired irregular shapes (Fig. 5b). Sintering was most probably caused by O_2, and the change in the crystallites' shape was brought about by C_2H_4. Very small crystallites also appeared. After further heating for 20 min more in C_2H_4 + O_2 (Fig. 5c), the behavior was very similar to that during the

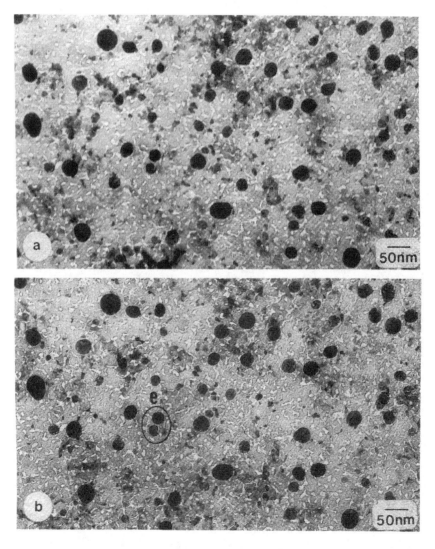

FIG. 4. Sequence of changes in an Ag/Al$_2$O$_3$ specimen with 7 Å initial film thickness heated in C_2H_4 at 250°C. (a) H_2, 300°C, 4 h; (b) C_2H_4, 20 min. *(Continued)*

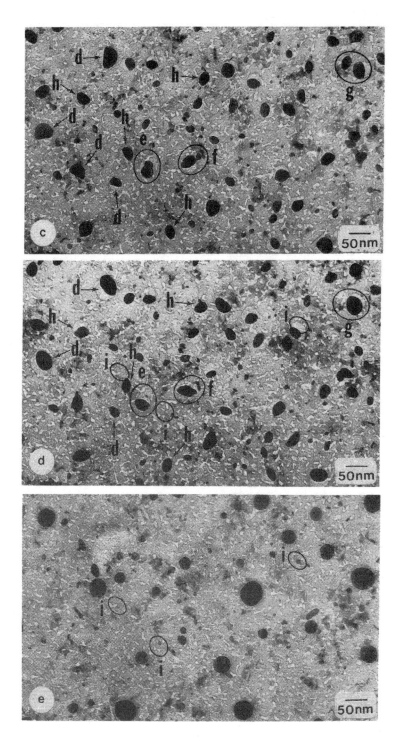

FIG. 4. (Continued) Sequence of changes in an Ag/Al$_2$O$_3$ specimen with 7 Å initial film thickness heated in C$_2$H$_4$ at 250°C. (c) C$_2$H$_4$, 80 min; (d) C$_2$H$_4$, 110 min; (e) O$_2$, 250°C, 30 min.

preceding heating. Upon heating for up to 1.5 h in $C_2H_4 + O_2$ (Fig. 5d), most of the large crystallites remained without changing their locations. Some particles about 200 Å in size migrated over the substrate (j in Figs. 5c, 5d), and in region k two neighboring particles coalesced and migrated over the substrate. In region m, new crystallites appeared without any evidence of migration. Also, the number of small crystallites decreased. During additional heating for a total of 2.5 h in $C_2H_4 + O_2$ (Fig. 5e), most of the large particles remained in their locations and events similar to those which occurred in regions j and m during the previous heat treatment continued to occur (j, m in Figs. 5d, 5e).

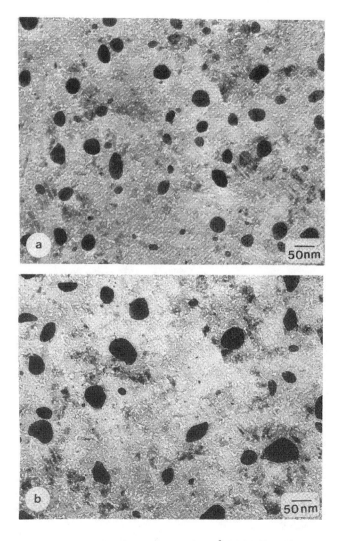

FIG. 5. Sequence of changes in an Ag/Al$_2$O$_3$ specimen with 7 Å initial film thickness heated in a mixture of C_2H_4 and O_2 at 250°C. (a) H_2, 300°C, 4 h; (b) C_2H_4 and O_2, 20 min. (*Continued*)

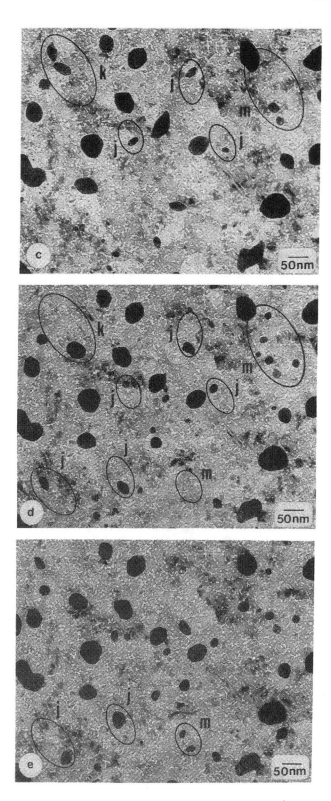

FIG. 5. (Continued) Sequence of changes in an Ag/Al$_2$O$_3$ specimen with 7 Å initial film thickness heated in a mixture of C$_2$H$_4$ and O$_2$ at 250°C. (c) C$_2$H$_4$ and O$_2$, 20 min; (d) C$_2$H$_4$ and O$_2$, 50 min; (e) C$_2$H$_4$ and O$_2$, 1 h.

DISCUSSION

Several mechanisms have been proposed for the sintering of supported metal catalysts. One of them, known as the crystallite migration model, considers sintering to occur by migration of metal crystallites over the surface of the support and their subsequent coalescence following collision (*19, 20*). In a second mechanism, the crystallites smaller than a critical size lose atoms or molecules to a two-dimensional phase of single atoms on the substrate, while those of a larger size gain atoms from the two-dimensional phase (*21–23*). This process, called Ostwald ripening, involves a two-dimensional surface phase of migrating atoms which is supersaturated with respect to the large crystallites and undersaturated with respect to the small ones. However, the process of emission of atoms by a small particle and their capture by a larger neighboring particle can also occur directly, without the involvement of a pool of single atoms covering a large surface area. In contrast to Ostwald ripening, which is global, the latter involves only a few neighboring particles and is called direct ripening (*24*).

As mentioned earlier, evidence for crystallite migration and coalescence as well as for ripening in graphite-supported silver catalysts has been reported by a few authors (*13, 14*). In Ref. (*24*) it was shown that the ripening identified in Ref. (*13*) was direct ripening. Let us now examine a number of events reported in this paper in some detail.

SINTERING BEHAVIOR

While, as expected, very small particles have migrated (p in Fig. 21), it is worth noting that even large crystallites 200–400 Å in diameter migrated over the substrate (B in Figs. 2a–2c; W in Figs. 3c, 3d; and j in Figs. 5c–5e). Some particles migrated and subsequently coalesced (e in Figs. 4b–4d). The "localized" disappearance of some small crystallites indicates that direct ripening could play a role in sintering (A in Figs. 2a–2c; N in Figs. 2j, 2k; and R in Figs. 3b, 3c). The migration of the smaller crystallites and their subsequent coalescence with larger ones, however, cannot be ruled out as a possible mechanism for the disappearance of the small crystallites. The decrease in size of a small crystallite near a larger crystallite provides clear evidence for direct ripening (E in Figs. 2b, 2c). In some regions (b in Figs. 3g, 3h), small particles appeared in the vicinity of large particles which decreased in size, suggesting the possibility that the small particles resulted from the atoms or clusters emitted by the larger particles. Note also that new small crystallites appeared not only in regions close to large particles but also all over the substrate (c in Figs. 3g, 3h). This may mean that the very small particles which were lost by the large particles to the substrate migrated long distances before colliding with other particles (undetectable by electron microscopy) to generate the "new" detectable particles.

An observation which deserves to be noted was the appearance of small crystallites. The micrographs showed that during heating in O_2 at 300°C for 0.5 h following heating in H_2 at the same temperature, a large number of small crystallites appeared, even though other particles sintered to form larger particles (Figs. 2c, 2d). After an initial severe sintering of the crystallites, most of the large particles were unaffected by further heating in O_2 (G1 in Figs. 2d, 2e). Similar behavior was also observed on another specimen heated in O_2 at 250°C (Figs. 3g, 3h). Figs. 2g and 2h show that heating in H_2 at 400°C resulted both in the appearance of very small crystallites (region I) and in the sintering of large particles to generate even larger particles.

The appearance of small crystallites was observed in O_2, H_2, and a mixture of C_2H_4 and O_2. Their number was the largest in O_2, intermediate in H_2, and the smallest in the mixture of C_2H_4 and O_2. Indeed, upon heating in C_2H_4 alone, small crystallites did not appear on the substrate. Harriott (25) reported that when the catalyst was exposed to the ethylene oxidation reaction, a large amount of silver was present as small particles which could not be detected by X-ray diffraction. The small crystallites of the present investigation are either circular or wormlike. While the circular crystallites sintered easily, the wormlike particles remained unchanged (H in Figs. 2e, 2f; L in Figs. 2g–2j).

It should be noted that the wormlike particles are located in the white regions of the micrographs and have the shape of the white regions (K in Figs. 2g, 2i). The white regions of the substrate may be depressions of some kind in the substrate. When the migrating atoms or clusters of atoms fell into these depressions, they could not escape from them easily. More atoms or clusters of atoms were captured until the depressions were filled with atoms. As a result, the captured particles acquired the shape of the white regions and remained unchanged upon further heating.

Regarding the appearance of small particles, let us now examine a few possible explanations. The presence of silver atoms or clusters of atoms on the substrate may be a consequence of the diffusion of oxygen into the silver particles. The dissolved oxygen could oxidize silver. However, when the temperature is higher than the decomposition temperature of the oxides, only transient, unstable oxide phases can form. From the unstable oxides, some oxygen can penetrate further into the subsurface of the particles, generating internal microstresses. The internal microstresses can be released by partial disintegration, leading to an increase of atoms or clusters of atoms over the substrate. Indeed, silver oxides are not stable at the temperatures employed in the present experiment, 250–400°C. Ag_2O and $AgIO_2$ are reported to decompose at 230 and 100°C, respectively (26). Evidence for the existence of the subsurface oxygen in silver has been reported (27–29). Backx *et al.* (*27*) investigated oxygen adsorption on the (111) surface of Ag and reported that below–100°C, oxygen adsorbs on the (111) surface as molecules; above–100°C, the oxygen molecules dissociate and only atomic oxygen could be detected. In addition, at temperatures above 150°C, the absorbed atomic oxygen diffuses into the subsurface region. From the change in the work function and from field emission pattern, Czanderna *et al.* (*28*) found that oxygen starts to diffuse into the bulk of silver at temperatures between 100 and 200°C. In addition, at the temperatures employed in the present experiments, 250–300°C, which are close to the Tammann temperature ($\sim0.5\ T_{melting}$ in K) for silver, 345°C, the lattice atoms have higher mobility and can detach from the particle more easily. The migration, collision, and coalescence of the species resulting from disintegration may be responsible for the appearance of the small crystallites.

The appearance of the extremely small crystallites during heating in H_2 might be a result of the coalescence of undetectable migrating atoms or clusters of atoms which were generated during previous heating in O_2. In addition, oxygen, which was adsorbed on the surface and subsurface of silver when the specimen was exposed to air, might have contributed to the presence of silver atoms or clusters of silver atoms on the substrate.

In a previous paper from this laboratory (7) concerning the behavior of Fe/Al_2O_3 in O_2 at high temperatures, evidence for the coexistence of a thin film with crystallites was presented. During subsequent heating in H_2, the thin film was reduced to the metal and became unstable, and a large number of small crystallites were generated via the rupture of the film. A detailed explanation of this behavior was given in Ref. (7) and was based on the observation that the interactions of the metal with the substrate are weaker than those of the oxide with the substrate. As a result, the metal atoms prefer the more favorable interactions within the particle to those within the film. The appearance of a large number of small particles in the present experiments may perhaps be explained along similar lines. Transient, unstable, silver oxides form, and undetectable thin films of silver oxide detach from the silver particles. Because the silver oxides are unstable at the temperatures employed, the resulting silver films rupture to form a large number of small crystallites. The appearance of small crystallites during heating in H_2 may be due to the rupture of the undetectable patches which were formed during heating in O_2.

More likely, the large number of detectable small particles may be a result of the migration and coalescence of undetectable particles present on the substrate from the beginning. The number of detectable particles is larger in oxygen than in hydrogen because sintering is more severe in the former case. A final explanation involves the following considerations: At the temperatures employed, the number of atoms of the two-dimensional phase in equilibrium with the crystallites is relatively large and dependent on the gaseous environment of the experiment. On cooling the sample in He to room temperature for TEM observations, the two-dimensional phase becomes supersaturated and

The Behavior of Model Ag/Al₂O₃ Catalysts in Various Chemical Environments

may condense, particularly on the large crystallites. Some atoms may condense on undetectable crystallites to form detectable ones, or even may nucleate new crystallites. However, since the diffusion coefficient at room temperature is small, it hinders the capture of atoms by the large particles as well as the formation of nuclei and their growth.

COMPARISON OF SINTERING IN VARIOUS ATMOSPHERES

The experimental results show that sintering of Ag particles supported on alumina occurs both in O_2 and H_2, being more severe in O_2. Similar observations have been also made by previous investigators, for both discrete particles and continuous films (9–11, 30).

When the specimen was heated in C_2H_4, the silver particles were very resistant to sintering. The particles changed their shape from circular to ellipsoidal and very few coalesced to form large particles; most of the particles, including the small ones, remained in their locations. Presland *et al.* (30) also reported that silver particles sintered slowly during heating in ethylene. The absence of migration and the shape change of the silver particles during heating in C_2H_4 are a result of deposition of carbon species on the surface and around the particles. Recently, the shape change of Fe, Co, and Ni particles supported on alumina and heated in atmospheres containing CH_4 and/or CO (31) was examined. The authors suggested that the shape change of the particles was due to the deposition of coke on the surface of the particles and its penetration inside. It was also reported (12) that when carbon was deposited on the specimen consisting of silver particles supported on alumina, the silver particles, because of their encapsulation, did not migrate over the substrate. McBreen and Moskowits (32), employing surface-enhanced Raman scattering, detected the presence of amorphous carbon when supported silver catalysts were heated in C_2H_4 or in a C_2H_4 and O_2 mixture up to 172°C. It is likely that in the present experiments also, amorphous carbon was deposited on the silver surface and caused the shape change as well as the encapsulation of the silver particles.

The behavior of silver particles during heating in a mixture of oxygen and ethylene at 250°C was a combination of the behaviors in O_2 and in C_2H_4. Sintering in the mixture of oxygen and ethylene was less severe than in oxygen but more severe than in ethylene, and the silver particles acquired irregular shapes.

Comparing the results of heating in various atmospheres, one may conclude that sintering of the silver particles supported on Al_2O_3 decreases in the order $O_2 > H_2 > C_2H_4$. The degree of sintering in the mixture of C_2H_4 and O_2 is between that in O_2 and C_2H_4.

CONCLUSION

The present results provide evidence for various phenomena that occur during the heat treatment of Ag/Al₂O₃ model catalysts. On heating in O_2 at 250 and 300°C, various events, such as migration of crystallites, followed by coalescence, disappearance of small and large particles, decrease in the size of the small particles, coalescence of nearby particles, connection of two neighboring particles by a bridge and material transfer through the bridge were observed to occur. In addition, a large number of small crystallites of two different shapes, circular and wormlike, appeared. The circular small particles were observed to grow, coalesce, or disappear, while the wormlike particles were located in valleys of the substrate and remained unchanged on further heating. A few possible explanations have been provided for the presence of a large number of small particles on the substrate.

The events observed during heating in H_2 were similar to those observed in O_2, but sintering was less severe. The presence of a large number of small particles is probably a result of the previous heating in O_2. During heating in C_2H_4 at 250°C, most of the silver crystallites, both small and large, remained in their locations and only changed their shape from circular to ellipsoidal. Carbon deposition on the surface of the particles seems to be responsible for both the immobility and shape change of the silver particles. In the mixture of C_2H_4 and O_2, sintering of the silver crystallites was less severe than in O_2 and the particles acquired irregular shapes because of the presence of C_2H_4.

REFERENCES

1. Margolis, L. Ya., *in* "Advances in Catalysis" (D. D. Eley, H. Pines, and P. B. Weisz, Eds.), Vol. 14, p. 429. Academic Press, New York, 1963.
2. Voge, H. H., and Adams, C. R., *in* "Advances in Catalysis" (D. D. Eley, H. Pines, and P. B. Weisz, Eds.), Vol. 17, p. 151. Academic Press, New York, 1967.
3. Dadyburjor, D. B., Jewur, S. S., and Ruckenstein, E., *in* "Catalysis Review" (H. Heinemann and J. J. Carberry, Eds.), Vol. 19, p. 293. Dekker, New York, 1979.
4. Carra, S., and Forzatti, P., *in* "Catalysis Review" (H. Heinemann and J. J. Carberry, Eds.), Vol. 15, p. 1. Dekker, New York, 1977.
5. Hucknall, D. J., "Selective Oxidation of Hydrocarbons," Chap. 2. Academic Press, New York, 1974.
6. Ruckenstein, E., and Lee, S. H., *J. Catal.* **86**, 457 (1984).
7. Ruckenstein, E., and Sushumna, I., *J. Catal.* **97**, 1 (1986).
8. Ruckenstein, E., and Hu, X. D., *J. Catal.* **100**, 1 (1986).
9. Presland, A. E. B., Price, G. L., and Trimm, D. L., *Surf. Sci.* **29**, 424 (1972).
10. Presland, A. E. B., Price, G. L., and Trimm, D. L., *Surf. Sci.* **29**, 435 (1972).
11. Seyedmonir, S. R., Strohmayer, D. E., Guskey, G. J., Geoffroy, G. L., and Vannice, M. A., *J. Catal.* **93**, 288 (1985).
12. Plummer, H. K., Jr., Watkins, W. L. H., and Gandhi, H. S., *Appl. Catal.* **29**, 261 (1987).
13. Heinemann, K., and Poppa, H., *Thin Solid Films* **33**, 237 (1976).
14. Baker, R. T. K., and Skiba, P., Jr., *Carbon* **15**, 233 (1977).
15. Chu, Y. F., and Ruckenstein, E., *J. Catal.* **41**, 385 (1976).
16. Sushumna, I., and Ruckenstein, E., *J. Catal.* **94**, 239 (1985) (Section 3.4 of this volume).
17. Sushumna, I., and Ruckenstein, E., *J. Catal.*, **109**, 433 (1988) (Section 3.3 of this volume).
18. Arai, M., Ishikawa, T., Nakayama, T., and Nishiyama, Y., *J. Colloid Interface Sci.* **97**, 254 (1984).
19. Ruckenstein, E., and Pulvermacher, B., *AIChEJ.* **19**, 356 (1973).
20. Ruckenstein, E., and Pulvermacher, B., *J. Catal.* **29**, 224 (1973).
21. Chakraverty, B. K., *J. Phys. Chem. Solids* **28**, 2401 (1967).
22. Wynblatt, P., and Gjostein, N. A., *Prog. Solid State Chem.* **9**, 21 (1975).
23. Ruckenstein, E., and Dadyburjor, D. B., *J. Catal.* **33**, 233 (1977).
24. Ruckenstein, E., and Dadyburjor, D. B., *Thin Solid Films* **55**, 89 (1978).
25. Harriott, P., *J. Catal.* **21**, 56 (1971).
26. "Handbook of Chemistry and Physics," 62nd ed. CRC Press, Boca Raton, Florida, 1981–1982.
27. Backx, C., De Groot, C. P. M., and Biloen, P., *Surf Sci.* **104**, 300 (1981).
28. Czandema, A. W., Frank, O., and Schmidt, W. A., *Surf. Si.* **38**, 129 (1973).
29. Kagawa, S., Iwamoto, M., and Morita, S., *J. Chem. Soc. Faraday Trans. 1* **78**, 143 (1982).
30. Presland, A. E. B., Price, G. L., and Trimm, D. L., *J. Catal.* **26**, 313 (1972).
31. Lee, S. H., and Ruckenstein, E., *J. Catal.* **107**, 23 (1987) (Section 3.10 of this volume).
32. McBreen, P. H. and Moskovits, M., *J. Catal.* **103**, 188 (1987).

3.7 Role of Wetting in Sintering and Redispersion of Supported Metal Crystallites*

Eli Ruckenstein

Faculty of Engineering and Applied Sciences, State University of New York at Buffalo, Buffalo, New York 14214, USA

Received 7 November 1978; manuscript received in final form 18 June 1979

1. INTRODUCTION

Supported metal crystallites are composed of a large number of small metal crystallites having sizes between 2 and a few tens of nm distributed over a substrate. The supported metal crystallites are of interest in crystal growth, in thin solid film technology and in catalysis. In industrial catalysts the metal crystallites are distributed over the internal surface area of a highly porous substrate (alumina or silica). The heating in vacuum or in a chemical atmosphere of the supported metal leads, in general, to the sintering of the crystallites. This sintering occurs either because the crystallites migrate over the surface of the substrate and coalesce when they collide [1–5] or/and because of Ostwald ripening [6–9], In the latter case single metal atoms are assumed to be the only mobile particles and the small crystallites lose atoms to the surface of the substrate while the large crystallites capture such atoms. Of course, for Ostwald ripening to occur, the surface concentration of single metal atoms on the substrate has to be higher than the surface concentrations in equilibrium with the larger crystallites but smaller than the surface concentrations in equilibrium with the smaller crystallites.

Aging observed during catalytic processes can be associated, at least in part, with the decay of the exposed surface area of the metal because of the sintering of the crystallites [9–12]. There are, however, circumstances under which redispersion of the crystallites can also occur [13–6]. This happens, for instance, when platinum crystallites supported on alumina are heated in air at about 500°C [16]. An interesting observation was made recently when Pt crystallites deposited on a thin film of alumina were heated alternatively in an oxidizing and in a reducing atmosphere [17]. Transmission electron microscopy has shown that several cycles of alternating heating in O_2 and H_2 at 750°C and 1 atm are needed before Pt crystallites redisperse during the oxidation step. Thereafter, redispersion or sintering is produced periodically by changing the chemical atmosphere from an oxidizing to a reducing one. Redispersion occurred during heating in oxygen and sintering during heating in hydrogen.

The scope of this paper is to explain sintering and redispersion in terms of the thermodynamics of wetting and to identify conditions under which sintering by migration of crystallites or sintering by Ostwald ripening occurs. It will be shown that the Ostwald ripening mechanism is associated with a particular type of phase separation. When such a phase separation cannot occur, migration and coalescence is the mechanism of sintering. (We ignore here the possibility of metal vaporation followed by vapor transport.)

---

* *Journal of Crystal Growth* 47 (1979) 666–670. Republished with permission.

2. THERMODYNAMICS OF WETTING OF A SUBSTRATE BY A THICK FILM

Let us consider a film transferred from a large reservoir to a substrate and assume that the film is thick. This means that the range of the interaction forces between one atom at the free surface of the film and the substrate is much smaller than the thickness of the film. The specific free energy of formation of the film on a uniform substrate is given, in this case, by the expression.

$$\sigma_\infty = \sigma_{mg} + \sigma_{ms} - \sigma_{gs}, \tag{1}$$

where σ_{mg} is the interfacial tension between film and gas, σ_{ms} is the interfacial tension between film and substrate, and σ_{gs} is the interfacial tension between gas and substrate. The subscript ∞ indicates that the film is thick. If $\sigma_\infty < 0$ the material composing the film wets the substrate. In the opposite case the material does not wet the substrate and therefore islands are generated. In order to decrease the free energy of the system these islands tend to coalesce into a single large island forming with the substrate an angle given by Young's equation

$$\sigma_{gs} - \sigma_{ms} = \sigma_{mg} \cos\theta.$$

In vacuum, the metals have high surface tensions. Platinum has, for instance, a surface tension of 2340 dyne/cm at 1310°C. The oxides have a surface tension several times smaller than the metals. Therefore the metals do not wet the oxides. If the metal composing the film is oxidized, its surface tension decreases and σ_∞ may become negative. Under these conditions the metal oxide spreads over the oxide substrate.

3. THERMODYNAMICS OF WETTING OF A SUBSTRATE BY A THIN FILM

In this case the range of the interaction forces between one atom at the free surface of the film and the substrate is larger than the thickness of the film and, hence, the film can no longer be treated as a bulk phase. The specific free energy of formation of the film, σ, is in this case a function of the thickness h of the film. To compute this dependence pairwise additivity is assumed and the Lennard-Jones potential is used for the interaction between two atoms. The interaction potential has the form

$$u(R) = -\epsilon \left[2(\Omega/R)^6 - (\Omega/R)^{12} \right], \tag{2}$$

where R is the distance between two atoms, $\epsilon = |u(\Omega)|$ and Ω is the coordinate of the minimum in the curve $u = u(R)$. Although the interaction potential is poorly represented by eq. (2) for metals and oxides its simplicity offers some insight into the problem. Performing the computations (see Appendix), one obtains

$$\sigma = \sigma_\infty + \frac{\pi}{6} \frac{\epsilon_{ms}\Omega_{ms}^6 n_m n_s - \epsilon_{mm}\Omega_{mm}^6 n_m^2}{h^2}$$

$$- \frac{\pi}{360} \frac{\epsilon_{ms}\Omega_{ms}^{12} n_m n_s - \epsilon_{mm}\Omega_{mm}^{12} n_m^2}{h^8} \tag{3}$$

Hence

$$\sigma = \sigma_\infty + \alpha/h^2 - \beta/h^8 \equiv \sigma_\infty + f(h). \tag{4}$$

Here, h is the thickness of the film, σ_∞ is the value of σ for $h \to \infty$ and n is the number of atoms per cm^3; the subscript m refers to the metal and the subscript s to the substrate. Eq. (3) implies the assumption that the specific entropy of formation of the film is equal to that of the thick film. Because of the continuum approach used in its derivation, eq. (3) holds only for film thicknesses sufficiently large compared to molecular spacing. It is worth noting that eq. (3) disregards the strain energy produced by the misfit between substrate and film.

For the dispersion interactions Dzyaloshinskii et al. [19] developed a theory based on macroscopic field equations which account for many body interactions. The result is the same concerning the dependence on h. The coefficient α is however expressed in terms of dielectric permeabilities.

The film is stable if $\sigma < 0$. Mechanical stability requires the additional condition $d\sigma/dh < 0$. Many situations are possible. Of particular interest is the case in which the thick film does not wet the substrate but a very thin film of thickness h_0 can wet the substrate. This means that $\sigma_\infty > 0$, but $\sigma(h_0) < 0$ and $(d\sigma/dh)_{h_0} < 0$. This may happen when α and β are negative and σ_∞ is positive but not too large. In fig. 1, σ is plotted versus h for the latter case. The curve has a minimum at $h = h_m$. If the loading of the substrate corresponds to a thickness h such that $h_c < h < h_m$, then both σ and $d\sigma/dh$ are negative and the film is stable. If, however, $h > h_m$ the mechanical stability condition is violated and the film is unstable. In this case a kind of phase separation occurs; one "phase" is a film of thickness h_m, while the other "phase" consists of a single crystallite containing the excess of atoms (molecules). Thus the free energy of the system decreases.

For σ_∞ positive, $\sigma(h)$ can become zero only if α is a negative quantity. Because for metals α is of the order of -10^{-13} erg, assuming that σ_∞ is of the order of 10^2 erg/cm^2, σ becomes zero for a thickness of a few atomic dimensions. Because the repulsion is short range we expect the existence of a negative minimum of σ at about a monolayer (or perhaps at less than a monolayer). Eq. (3) is not strictly valid from a quantitative point of view for these low thicknesses. It remains, however, valid from a qualitative point of view.

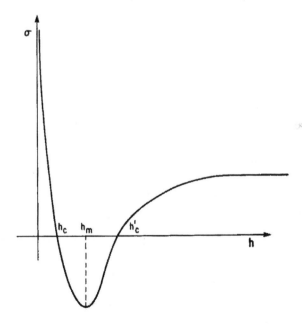

FIG. 1. Plot of σ versus h for the case in which the thick film does not wet the substrate but a very thin film can wet the substrate.

4. DISCUSSION

Eq. (3) shows that there are circumstances under which no wetting, partial wetting or total wetting occurs. If $\sigma(h) > 0$, no spreading will occur for any h while if $\sigma(h)$ and $d\sigma(h)/dh$ are negative spreading will always take place. However, if $\sigma < 0$ and $d\sigma/dh < 0$ in a given range of thicknesses $h_c < h < h_m$, then total spreading will occur in that range of thicknesses and partial spreading for thicknesses larger than h_m. Partial spreading is a consequence of that "phase separation" described in the previous section.

Let us now consider a large number of crystallites on a substrate. If σ_∞ is so large that $\sigma > 0$ for all values of h even when α is negative, no spreading (and hence no redispersion) can occur. Because of the tendency for a minimum free energy the crystallites will migrate over the surface of the substrate and coalesce to achieve the smallest possible area. Of course the migration of the larger crystallites is extremely slow and only sufficiently small crystallites migrate at an appreciable rate. Although from a thermodynamic point of view there is a tendency for the formation of a single crystallite, the kinetic process may be so slow that a large number of crystallites can survive for the entire life of the supported metal. Metals supported on oxides have in vacuum (and probably also in a reducing atmosphere) large values of σ_∞. In these circumstances $\sigma(h)$ cannot become negative for any value of h. In this case no redispersion can occur and migration and coalescence is the only kinetic mechanism of sintering compatible with thermodynamics. If, however, σ_∞, although positive, is not too large one can have for negative values of α, $\sigma(h) \leqslant 0$ for $h_c \leqslant h \leqslant h'_c \ll \infty$ and redispersion is possible under the form of a film. The evaluations made in the previous section suggests that the film is mono-molecular. This may happen in an oxygen atmosphere, because the surface tension of the metal oxide is much smaller than that of the metal. Starting from metal crystallites, the heating in oxygen generates metal oxide which spreads through the leading edge of the crystallites over the substrate. As long as the surface concentration in the film which is formed is lower than the saturation concentration corresponding to the largest crystallite present, all the crystallites will lose molecules to the substrate. However, if the surface concentration becomes larger, then only the crystallites having sizes smaller than a critical value will lose molecules to the surface of the substrate, while those with larger sizes will gain such molecules from the surface of the substrate (Ostwald ripening). The overall behavior is redispersion as long as the exposed surface area of the metal (or the oxidized metal) increases, or is sintering if the exposed area decays. The ultimate thermodynamic result is a single crystallite in contact with a thin film. However, for kinetic reasons, the time needed to achieve this final state can be longer than the life of the catalyst.

The first several cycles of heating of Pt in O_2 and H_2 in the experiments described in section 1 enhance, probably by generating some porosity in the crystallites, the oxidation of the crystallites during the heating in O_2 and produce some reconstruction of the substrate. Only after this "initiation" process, platinum oxide spreads, during the heating in oxygen, through the leading edge of the crystallites over the surface of the substrate probably as a mono-molecular (or submonomolecular) film, and redispersion occurs. When platinum oxide is reduced, during heating in H_2, to platinum, the two-dimensional fluid recondenses into metal crystallites, being either recaptured by the remaining crystallites or forming new ones. The recondensation happens because the metal does not wet the substrate. This explains why redispersion and sintering can be produced periodically by heating alternatively in O_2 and H_2.

The main conclusion is that a necessary condition for sintering to occur by Ostwald ripening is the change of the sign of the free energy of formation $\sigma(h)$ from positive to negative values when the thickness h decreases and becomes sufficiently small. If $\sigma(h)$ is positive for all values of h, then migration and coalescence is the only possible mechanism of sintering.

APPENDIX

Assuming pair-wise additivity, the interaction potential between one molecule of the film located at a distance z from a semi-infinite substrate is given by [18]

$$U_{12}(z) = \int n_1 u(R) 2\pi R^2 dR \sin\theta d\theta, \tag{A.1}$$

where n_1 is the number of molecules of substrate per unit volume and θ is the angle between the radius vector and the normal to the surface. Using the power law $u(R) = -\gamma_{12}R^{-m}$, one obtains

$$U_{12}(z) = -\frac{2\pi\gamma_{12}n_1}{(m-2)(m-3)} \frac{1}{z^{m-3}}. \tag{A.2}$$

The potential energy of interaction per unit area between the film and substrate is given by

$$\int_{\delta_{12}}^{h} n_2 U_{12}(z) dz \equiv \int_{\delta_{12}}^{\infty} n_2 U_{12}(z) dz - \int_{h}^{\infty} n_2 U_{12}(z) dz$$

$$= -\frac{2\pi\gamma_{12}n_1 n_2}{(m-2)(m-3)(m-4)} \left(\frac{1}{\delta_{12}^{m-4}} - \frac{1}{h^{m-4}} \right), \tag{A.3}$$

where n_2 is the number of molecules per unit volume of the film and δ_{12} is the distance between two molecules of species 1 and 2 in contact. The free energy of formation of the film per unit area is obtained by subtracting from expression (A.3) the potential energy of interaction per unit area between the film and a semi-infinite-space filled with the molecules of the film (instead with those of the substrate). Hence [18] one obtains

$$\sigma = const + \frac{2\pi}{(m-2)(m-3)(m-4)} \frac{\gamma_{12}n_1 n_2 - \gamma_{22}n_2^2}{h^{m-4}}$$

$$\equiv const + \frac{A_{12} - A_{22}}{12\pi h^{m-4}}, \tag{A.4}$$

where A_{ij} is the Hamaker constant.

Dzyaloshinskii et al. [19] developed a theory based on macroscopic field equations which accounts for many body interactions. The result obtained for the dispersion interactions is the same concerning the dependence on h. The coefficient α is however expressed in terms of dielectric permeabilities.

If the Lennard–Jones potential is used for $u(R)$, then eq. (3) is obtained.

REFERENCES

[1] A. Masson, J.J. Métois and R. Kern, Surface Sci. **27** (1971) 483.
[2] Y.F. Chu and E. Ruckenstein, Surface Sci. **67** (1977) 517.
[3] J.J. Métois, K. Heineman and H. Poppa, Phil. Mag. **35** (1977) 1413.
[4] Y.F. Chu and E. Ruckenstein, J. Catalysis **55** (1978) 281.
[5] E. Ruckenstein and B. Pulvermacher, J. Catalysis **29** (1973)224.
[6] B.K. Chakraverty, J. Phys. Chem. Solids **28** (1967) 2401.

[7] P. Wynblatt and N.A. Gjostein, Progr. Solid State Chem. **9** (1975) 21.

[8] E. Ruckenstein and D.B. Dadyburjor, J. Catalysis **48** (1977) 73.

[9] H.J. Maat and L. Moscou, in: Proc, 3rd Intern. Congr. on Catalysis, Vol. **11** (1965) p. 1277.

[10] H.L. Gruber, J. Phys. Chem. **66** (1962) 48.

[11] G.A. Mills, S. Weller and E.B. Cornelius, in: Proc. 2nd Congr. on Catalysis (1960) p. 2221.

[12] R.A. Herrmann, S.F. Adler, M.S. Goldstein and R.M. deBaun, J. Phys. Chem. **65** (1961) 2189.

[13] S.F. Adler and J.J. Keavney, J. Phys. Chem. **64** (1960) 208.

[14] F.L. Johnson and C.D. Keith, J. Phys. Chem. **67** (1963) 200.

[15] S.W. Weller and A.A. Montagna, J. Catalysis **20** (1971) 394.

[16] E. Ruckenstein and M.L. Malhotra, J. Catalysis **41** (1976) 303.

[17] E. Ruckenstein and Y.F. Chu, J. Catalysis **59** (1979) 309.

[18] J. Frenkel, Kinetic Theory of Liquids (Clarendon, Oxford, 1946).

[19] E.I. Dzyaloshinskii, E.M. Lifshitz and L.P. Pitaevskii, Advan. Phys. **10** (1961) 165.

3.8 Optimum Design of Zeolite/ Silica–Alumina Catalysts*

Sung H. Lee and Eli Ruckenstein

Department of Chemical Engineering, State University
of New York at Buffalo, Amherst, N.Y. 14260

(Received September 3, 1985; in final form January 9, 1986)

INTRODUCTION

The effectiveness of porous catalysts depends on the reaction rate constants and diffusion coefficients and is low when the latter coefficients are small. Ruckenstein [1] suggested that the effectiveness of a porous catalyst can be increased by using a diluted catalyst, which consists of small spherical particles of active porous catalyst imbedded in a more porous, inactive particle. Since the diffusion coefficient is greater in the more porous material but the volume available for reaction decreases with increasing dilution, the overall reaction rate can exhibit a maximum as a function of the volume fraction of the active component. Recently, Nandapurkar and Ruckenstein [2] extended the above considerations to the catalytic cracking of hydrocarbons by the zeolite/silica-alumina catalysts. In this case, some steps of the reaction occur in the more porous (silica-alumina) medium. They again concluded that the diluted catalyst pellets can be more active than the pellets of the same size consisting of only zeolite and that there exists an optimum volume fraction of diluent (silica–alumina) for which the rate of formation of the desired compound is maximum.

Dadyburjor [3] has carried out calculations for a catalyst with infinite flat plate geometry, which contains flat plates of active material sandwiched between plates of less active material. He concluded that for a first order irreversible reaction there exists an optimum distribution which maximizes the overall reaction rate. In a more recent paper, Dadyburjor [7] extended the calculations to a number of reaction schemes which approximate the hydrocarbon cracking over zeolite/silica-alumina catalysts. The calculations showed that for a parallel reaction scheme the selectivity exhibits a maximum when the zeolite is located at the outer surface, while for the reactions in series it exhibits a maximum for a dilute uniform surface distribution of zeolite. These calculations have been carried out for fixed volume fractions of the more active component. However, the rate of formation of the desired product as well as the selectivity can be affected by the volume fraction of the active component [1, 2]. It is, therefore, of interest to analyze the simultaneous effect on the rate of formation of the desired product and selectivity of both (i) the overall volume fraction and (ii) the different distributions of the more active component in the less active one. In addition, we will consider that the active particles and the pellets are spherical. For the sake of simplicity, two kinds of nonuniform distributions

---

* *Chem. Eng. Commun.* Vol. 46 pp. 043–064 (1986). Republished with permission.

are considered: the composite-surface and composite-center distributions. The composite-surface distribution has a uniform dilute region (composed of less porous material imbedded in the more porous material) around a core containing only the more porous material. The composite-center distribution has a uniform dilute region as a core surrounded by the pure, more porous material. These two types are sketched in Figure 1. The catalytic cracking of hydrocarbons over the zeolite/silica-alumina catalyst is examined as an example. This catalyst contains small zeolite particles imbedded within the more porous silica–alumina matrix and constitutes an interesting example of a diluted catalyst. In the commercial catalytic cracking process, the volume fraction of zeolite is in the range of 0.10 ~ 0.15 [8]. Therefore, the effect of the nonuniform distribution on the rate of formation has been calculated when the amount of zeolite is in that range.

Two simplified reaction schemes, $A \to C \to D$ and $\begin{smallmatrix} A \to B \to C \\ \searrow \swarrow \\ D \end{smallmatrix}$, are used to describe the reaction. In the first reaction scheme, which is similar to that employed by Nace et al. [4], component A represents the oil and component C, the gasoline, constitutes the desired product. Component D represents the gases which result from the secondary cracking of gasoline. In the zeolite/silica–alumina catalysts, the rate constant for the reaction $A \to C$ in the active zeolite is much greater than that in the silica–alumina matrix, while the rate constant for the reaction $C \to D$ is smaller in the zeolite than in the silica–alumina [5]. A composite-surface distribution is, in this case, preferable to a uniform distribution as well as to a composite-center distribution, because the greater concentration of zeolite in the surface region leads to a higher local rate of formation of C and a part of C diffuses outside without being transformed to D in the core region. In addition, there is an optimum dilution in the surface region as a result of the competition between reaction and diffusion. The second reaction scheme

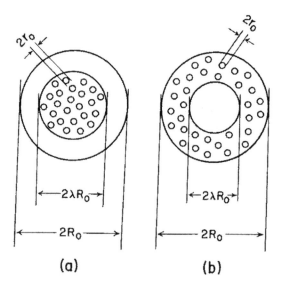

FIGURE 1 Sketches of the nonuniform distribution composite catalyst with spherical geometry: (a) Composite-center distribution, (b) Composite-surface distribution.

Optimum Design of Zeolite/Silica–Alumina Catalysts

$A \rightarrow B \rightarrow C$
$\quad \searrow \swarrow$
$\quad D$ contains an additional step, namely, the formation of an intermediate compound B.

The compounds C and D are the same as in the previous reaction scheme, but the compound D results from both C and B. This reaction scheme was also considered by Nandapurkar and Ruckenstein [2], for the cases in which the reaction $A \rightarrow B$ occurs in the silica–alumina matrix and the reaction

$B \rightarrow C$
$\quad \searrow \swarrow$
$\quad D$ takes place in the zeolite. Here, one considers that both reactions, $A \rightarrow B$ and $\begin{smallmatrix} B \rightarrow C \\ \searrow \swarrow \\ D \end{smallmatrix}$,

occur, to different extents, in the zeolite as well as in the silica–alumina matrix. If the rate constant of the reaction $A \rightarrow B$ is greater in the silica–alumina matrix than in the zeolite, then the composite-center distribution leads to a higher activity than both the uniform distribution and composite-surface distribution. This happens because the formation of component B in the outer region can lead to a greater amount of component C in the inner region.

ANALYSIS

Let us consider a spherical catalyst pellet containing small, spherical active particles imbedded in its matrix. Two length scales characterize the system, a large scale for the reaction and diffusion process in the silica–alumina matrix and a small scale for the reaction and diffusion process in the small zeolite particles. The continuum approach used in this analysis is satisfactory only if the large length scale is sufficiently large compared to the size of the small particles, (such as to contain a sufficiently large number of zeolite particles), but also sufficiently small compared to the size of the nonhomogeneous catalyst particle. For each of the two kinds of distributions, the reaction-diffusion equations (based upon the selected reaction schemes) as well as the appropriate boundary conditions for both the small and large spheres can be easily obtained. For convenience, they are summarized in Table I.

1 $\begin{smallmatrix} A \rightarrow B \rightarrow C \\ \searrow \swarrow \\ D \end{smallmatrix}$ COMPOSITE-CENTER DISTRIBUTION

In this reaction scheme B is the intermediate product and C is the desired product. First order reactions and isothermal conditions are assumed. K_A, K_{B1}, K_{B2} and K_C are the reaction rate constants. The rate constant for the consumption of B is given by:

$$K_B \equiv K_{B1} + K_{B2.}$$ (27)

As mentioned earlier, for this reaction, the composite-center distribution has a better performance than both the uniform distribution and composite-surface distribution.

The reaction-diffusion equations in the small spheres are provided by Eqs. (3) through (5) of Table I and the boundary conditions are given by Eqs. (8) and (9) of the same Table. The solutions of Eqs. (3) through (5) for the boundary conditions (8) and (9) are:

for component A,

$$p_A = p_{A0} \frac{r_0}{r} \frac{\sinh(\alpha r)}{\sinh(\psi_{Aa})},$$ (28a)

TABLE I
Reaction-Diffusion Equations and Boundary Conditions

		Composite-center distribution		Composite-surface distribution

| | Reaction Scheme | | | |

$$A \xrightarrow{K_A} B \xrightarrow{K_{B1}} C$$
$$K_{B2} \searrow \quad \swarrow K_C$$
$$D$$

(1)

$$A \xrightarrow{K_A} C \xrightarrow{K_C} D \qquad (2)$$

Small sphere — Reaction-diffusion equations

For A:

$$\frac{d}{dr}\left(4\pi r^2 D_{Aa}\frac{dp_A}{dr}\right) = 4\pi r^2 K_{Aa}p_A \qquad (3)$$

For A:

$$\frac{d}{dr}\left(4\pi r^2 D_{Aa}\frac{dp_A}{dr}\right) = 4\pi r^2 K_{Aa}p_A \qquad (6)$$

For B:

$$\frac{d}{dr}\left(4\pi r^2 D_{Ba}\frac{dp_B}{dr}\right)$$

$$= 4\pi r^2\left(-K_{Aa}p_A + K_{Ba}p_B\right) \qquad (4)$$

For C:

$$\frac{d}{dr}\left(4\pi r^2 D_{Ca}\frac{dp_C}{dr}\right)$$

$$= 4\pi r^2\left(-K_{Aa}p_A + K_{Ca}p_C\right) \qquad (7)$$

For C:

$$\frac{d}{dr}\left(4\pi r^2 D_{Ca}\frac{dp_C}{dr}\right)$$

$$= 4\pi r^2\left(-K_{B1a}p_B + K_{Ca}p_C\right) \qquad (5)$$

Boundary conditions

$$\frac{dp_i}{dr} = 0 \text{ at } r = 0 \qquad (8)$$

$$p_i = p_{i0} \text{ at } r = r_0 \qquad (9)$$

(*Continued*)

TABLE I (*Continued*)
Reaction-Diffusion Equations and Boundary Conditions

			Composite-center distribution		**Composite-surface distribution**

Large sphere   Reaction-diffusion equations in inner region

Composite-center distribution

For A:

$$\frac{d}{dR}\left(4\pi R^2 D_{Ae}\frac{dq_A}{dR}\right) = 4\pi R^2 \left\{\varepsilon K_{Ad}q_A + 4\pi r_0^2 nD_{Aa}\left(\frac{dp_A}{dr}\right)_{r=r_0}\right\} \tag{10}$$

For B:

$$\frac{d}{dR}\left(4\pi R^2 D_{Be}\frac{dq_B}{dR}\right) = 4\pi R^2 \left\{-\varepsilon K_{Ad}q_A + \varepsilon K_{Bd}q_B + 4\pi r_0^2 nD_{Ba}\left(\frac{dp_B}{dr}\right)_{r=r_0}\right\} \tag{11}$$

For C:

$$\frac{d}{dR}\left(4\pi R^2 D_{Ce}\frac{dq_C}{dR}\right) = 4\pi R^2 \left\{-\varepsilon K_{B1d}q_B + \varepsilon K_{Cd}q_c + 4\pi r_0^2 nD_{Ca}\left(\frac{dp_C}{dr}\right)_{r=r_0}\right\} \tag{12}$$

Composite-surface distribution

For A:

$$\frac{d}{dR}\left(4\pi R^2 D_{Ad}\frac{dq_A}{dR}\right) = 4\pi R^2 K_{Ad}q_A \tag{13}$$

For C:

$$\frac{d}{dR}\left(4\pi R^2 D_{Cd}\frac{dq_C}{dR}\right)$$

$$= 4\pi R^2 \left(-K_{Ad}q_A + K_{Cd}q_C\right) \tag{14}$$

(*Continued*)

TABLE I (*Continued*)

Reaction-Diffusion Equations and Boundary Conditions

		Composite-center distribution		**Composite-surface distribution**

Large sphere — Reaction-diffusion equations in outer region

For *A*:

$$\frac{d}{dR}\left(4\pi R^2 D_{Ad}\frac{dC_A}{dR}\right) = 4\pi R^2 K_{Ad}C_A \tag{15}$$

For *B*:

$$\frac{d}{dR}\left(4\pi R^2 D_{Bd}\frac{dC_B}{dR}\right)$$

$$= 4\pi R^2\left(-K_{Ad}C_A + K_{Bd}C_B\right) \tag{16}$$

For *C*:

$$\frac{d}{dR}\left(4\pi R^2 D_{Cd}\frac{dC_C}{dR}\right)$$

$$= 4\pi R^2\left(-K_{B1d}C_B + K_{Cd}C_c\right) \tag{17}$$

For *A*:

$$\frac{d}{dR}\left(4\pi R^2 D_{Ae}\frac{dC_A}{dR}\right) = 4\pi R^2$$

$$\times\left\{\varepsilon K_{Ad}C_A + 4\pi r_0^2 n D_{Aa}\left(\frac{dp_A}{dr}\right)_{r=r_0}\right\} \tag{18}$$

For *C*:

$$\frac{d}{dR}\left(4\pi R^2 D_{Ce}\frac{dC_C}{dR}\right) = 4\pi R^2$$

$$\times\left\{-\varepsilon K_{Ad}C_A + \varepsilon K_{Cd}C_C\right.$$

$$\left. + 4\pi r_0^2 n D_{Ca}\left(\frac{dp_C}{dr}\right)_{r=r_0}\right\} \tag{19}$$

(*Continued*)

TABLE I (*Continued*)
Reaction-Diffusion Equations and Boundary Conditions

		Composite-center distribution		Composite-surface distribution	
Boundary conditions	$\dfrac{dq_i}{dR} = 0$ at $R = 0$		(20)		
	$C_i = C_{i0}$ at $R = R_0$		(21)		
	$q_i = C_i$ at $R = \lambda R_c$		(22)		
	$D_{ie}\dfrac{dq_i}{dR} = D_{id}\dfrac{dC_i}{dR}$ at $R = \lambda R_0$		(23)	$D_{id}\dfrac{dq_i}{dR} = D_{ie}\dfrac{dC_i}{dR}$ at $R = \lambda R_0$	(24)

$i = A, B, C$

n is the number of the small particles per unit volume in the composite region, defined by.

$$n = 3(1-\varepsilon)/(4\pi r_0^3) \tag{25}$$

D_{ie} is the effective diffusivity of component i in the composite region, defined by:

$$D_{ie} = \varepsilon D_{id} + (1-\varepsilon)D_{ia} \tag{26}$$

where ψ_{Aa} is the Thiele Modulus of component A in the small particle, defined by:

$$\psi_{Aa} = r_0\sqrt{K_{Aa}/D_{Aa}}, \tag{28b}$$

and

$$\alpha = \sqrt{K_{Aa}/D_{Aa}}; \tag{28c}$$

for component B,

$$p_B = \frac{r_0}{r}\left\{\left(p_{A0}\frac{K_{Aa}}{D_{Ba}}\frac{1}{\alpha^2-\beta^2} + p_{B0}\right)\frac{\sinh(\beta r)}{\sinh(\psi_{Ba})} - p_{A0}\frac{K_{Aa}}{D_{Ba}}\frac{1}{\alpha^2-\beta^2}\frac{\sinh(\alpha r)}{\sinh(\psi_{Aa})}\right\}, \tag{29a}$$

where

$$\psi_{Ba} = r_0\sqrt{K_{Ba}/D_{Ba}} \tag{29b}$$

and

$$\beta = \sqrt{K_{Ba}/D_{Ba}}; \tag{29c}$$

and for component C,

$$p_C = \frac{r_0}{r}\left\{F\frac{\sinh(\alpha r)}{\sinh(\psi_{Aa})} - G\frac{\sinh(\beta r)}{\sinh(\psi_{Ba})} + H\frac{\sinh(\gamma r)}{\sinh(\psi_{Ca})}\right\}, \tag{30a}$$

where

$$\psi_{Ca} = r_0\sqrt{K_{Ca}/D_{Ca}}, \tag{30b}$$

$$\gamma = \sqrt{K_{Ca}/D_{Ca}}, \tag{30c}$$

$$F = \frac{K_{Aa}}{D_{Ba}}\frac{K_{B1a}}{D_{Ca}}\frac{p_{A0}}{(\alpha^2-\beta^2)(\alpha^2-\gamma^2)}, \tag{30d}$$

$$G = \frac{K_{Aa}}{D_{Ba}}\frac{K_{B1a}}{D_{Ca}}\frac{p_{A0}}{(\beta^2-\gamma^2)(\alpha^2-\beta^2)} + \frac{K_{B1a}}{D_{Ca}}\frac{p_{B0}}{\beta^2-\gamma^2} \tag{30e}$$

and

$$H = \frac{K_{Aa}}{D_{Ba}}\frac{K_{B1a}}{D_{Ca}}\frac{p_{A0}}{(\alpha^2-\gamma^2)(\beta^2-\gamma^2)} + \frac{K_{B1a}}{D_{Ca}}\frac{p_{B0}}{\beta^2-\gamma^2} + p_{C0}. \tag{30f}$$

Optimum Design of Zeolite/Silica–Alumina Catalysts

The reaction-diffusion equations in the inner, composite region are given by Eqs. (10) through (12), listed in Table I. The first term on the right hand side of Eq. (10) represents the rate of consumption of component A in the silica–alumina matrix, while the second term is the rate of consumption of component A in the small particles. Similarly, the first two terms on the right hand side of Eqs. (11) and (12) represent the production and consumption rates of each component in the silica–alumina matrix, whereas the last terms in Eqs. (11) and (12) are the production rates in the small particles. Equations (10) through (12) must be solved for the following conditions:

$$p_{A0} = q_A \text{ at } r = r_0,$$ (31a)

$$p_{B0} = q_B \text{ at } r = r_0$$ (31b)

and

$$p_{C0} = q_C \text{ at } r = r_0.$$ (31c)

These boundary conditions are reasonable provided that the number density of the small particles is sufficiently large.

Substituting Eq. (28) in Eq. (10) and taking into account the boundary condition (31a), Eq. (10) can be rewritten as:

$$\frac{d}{dR}\left(4\pi R^2 D_{Ae}\frac{dq_A}{dR}\right) = 4\pi R^2 K_{Ae,A}q_A,$$ (32a)

where

$$K_{Ae,A} = K_{Ad}\varepsilon + 4\pi r_0 n D_{Aa}\left\{\psi_{Aa}\coth(\psi_{Aa}) - 1\right\}.$$ (32b)

The effective diffusion coefficient D_{Ae} in the composite region increases with increasing ε and a linear dependence, given by Eq. (26) of Table I, is further used in the calculations.

Similarly, Eq. (11) can be rewritten in the form:

$$\frac{d}{dR}\left(4\pi R^2 D_{Be}\frac{dq_B}{dR}\right) = 4\pi R^2\left(-K_{Ae,B}q_A + K_{Be,B}q_B\right),$$ (33a)

where

$$K_{Ae,B} = K_{Ad}\varepsilon + 4\pi r_0 n K_{Aa}\left\{\psi_{Aa}\coth(\psi_{Aa}) - \psi_{Ba}\coth(\psi_{Ba})\right\}/\left(\alpha^2 - \beta^2\right)$$ (33b)

and

$$K_{Be,B} = K_{Bd}\varepsilon + 4\pi r_0 n D_{Ba}\left\{\psi_{Ba}\coth(\psi_{Ba}) - 1\right\}.$$ (33c)

Finally, Eq. (12) can be rearranged as:

$$\frac{d}{dR}\left(4\pi R^2 D_{Ce} \frac{dq_C}{dR}\right) = 4\pi R^2 \left(-K_{Ae,C}q_A - K_{Be,C}q_B + K_{Ce,C}q_C\right),\tag{34a}$$

where

$$K_{Ae,C} = -4\pi r_0 n K_{Aa}\left(K_{B1a}/D_{Ba}\right)$$
$$\times\left\{\frac{\psi_{Aa}\coth(\psi_{Aa})-1}{(\alpha^2-\beta^2)(\alpha^2-\gamma^2)} + \frac{\psi_{Ba}\coth(\psi_{Ba})-1}{(\beta^2-\alpha^2)(\beta^2-\gamma^2)} + \frac{\psi_{Ca}\coth(\psi_{Ca})-1}{(\gamma^2-\alpha^2)(\gamma^2-\beta^2)}\right\},\tag{34b}$$

$$K_{Be,C} = K_{B1d}\varepsilon + 4\pi r_0 n K_{B1a}\left\{\psi_{Ba}\coth(\psi_{Ba}) - \psi_{Ca}\coth(\psi_{Ca})\right\}/(\beta^2-\gamma^2)\tag{34c}$$

and

$$K_{Ce,C} = K_{Cd}\varepsilon + 4\pi r_0 n D_{Ca}\left\{\psi_{Ca}\coth(\psi_{Ca}) - 1\right\}.\tag{34d}$$

Further, the reaction-diffusion equations in the outer region (containing only silica–alumina) can be described by Eqs. (15) through (17) of Table I.

The boundary conditions at the center are given by Eq. (20) of Table I, while those at the outer surface of the pellet are given by Eq. (21). Equations (22) and (23) of Table I represent the matching conditions at the interface between the two regions, for the concentrations and fluxes, respectively.

The overall rate of formation of component C is given by:

$$\bar{r}_C = -4\pi R_0^2 D_{Cd}\left(\frac{dC_C}{dR}\right)_{R=R_0}\tag{35}$$

and the selectivity is defined as the ratio:

$$S = \frac{\bar{r}_C}{-\bar{r}_A},\tag{36}$$

where, $-\bar{r}_A$, the overall rate of consumption of component A, is given by:

$$-\bar{r}_A = 4\pi R_0^2 D_{Ad}\left(\frac{dC_A}{dR}\right)_{R=R_0}.\tag{37}$$

In order to calculate $\bar{r}_C$ and S, Eqs. (32) through (34) and (15) through (17) have been solved analytically and the results are substituted in Eqs. (35) and (36). However, the coefficients in these solutions have been calculated numerically, because of their complexity. Analytical results can be, however, obtained for the uniform distribution. Equations (10) through (12) apply throughout the entire catalyst pellet for the uniform distribution. The appropriate boundary conditions at the center are given by Eq. (20) of Table I, while the boundary conditions at the surface of the pellet are:

$$q_A = q_{A0} \quad \text{at} \quad R = R_0,\tag{38a}$$

$$q_B = q_{B0} \quad \text{at} \quad R = R_0\tag{38b}$$

Optimum Design of Zeolite/Silica–Alumina Catalysts

and

$$q_C = q_{C0} \quad \text{at} \quad R = R_0. \tag{38c}$$

The solutions of Eqs. (32) through (34) for the boundary conditions (20) and (38) yield the following analytical expressions for q_A, q_B and q_C:

$$q_A = q_{A0} \frac{R_0}{R} \frac{\sinh(\alpha_e R)}{\sinh(\psi_{Ae})}, \tag{39a}$$

where

$$\alpha_e = \sqrt{K_{Ae,A} / D_{Ae}} \tag{39b}$$

and

$$\psi_{Ae} = R_0 \sqrt{K_{Ae,A} / D_{Ae}}; \tag{39c}$$

$$q_B = \frac{R_0}{R} \left\{ \left(\frac{K_{Ae,B}}{D_{Be}} \frac{q_{A0}}{\alpha_e^2 - \beta_e^2} + q_{B0} \right) \frac{\sinh(\beta_e R)}{\sinh(\psi_{Be})} - \frac{K_{Ae,B}}{D_{Be}} \frac{q_{A0}}{\left(\alpha_e^2 - \beta_e^2\right)} \frac{\sinh(\alpha_e R)}{\sinh(\psi_{Ae})} \right\}, \tag{40a}$$

where

$$\beta_e = \sqrt{K_{Be,B} / D_{Be}} \tag{40b}$$

and

$$\psi_{Be} = R_0 \sqrt{K_{Be,B} / D_{Be}}; \tag{40c}$$

and

$$q_C = \frac{R_0}{R} \left[\frac{V}{\alpha_e^2 - \gamma_e^2} \sinh(\alpha_e R) + \frac{U}{\beta_e^2 - \gamma_e^2} \sinh(\beta_e R) \right.$$

$$\left. - \left\{ \frac{V}{\alpha_e^2 - \gamma_e^2} \sinh(\psi_{Ae}) + \frac{U}{\beta_e^2 - \gamma_e^2} \sinh(\psi_{Be}) - q_{C0} \right\} \frac{\sinh(\gamma_e R)}{\sinh(\psi_{Ce})} \right], \tag{41a}$$

where

$$\gamma_e = \sqrt{K_{Ce,C} / D_{Ce}}, \tag{41b}$$

$$\psi_{Ce} = R_0 \sqrt{K_{Ce,C} / D_{Ce}}, \tag{41c}$$

$$V = \frac{q_{A0}}{\sinh(\psi_{Ae})} \left\{ \frac{K_{Ae,C}}{D_{Ce}} - \left(\frac{K_{Ae,B}}{D_{Be}} \frac{K_{Be,C}}{D_{Ce}} \right) / \left(\alpha_e^2 - \beta_e^2 \right) \right\} \tag{41d}$$

and

$$U = \left(\frac{K_{Ae,B}}{D_{Be}} \frac{q_{A0}}{\alpha_e^2 - \beta_e^2} + q_{B0} \right) \left(\frac{K_{Be,C}}{D_{Ce}} \right) / \sinh(\psi_{Be}). \qquad (41e)$$

The overall rate of formation of component C is:

$$\bar{r}_{C,u} = -4\pi R_0^2 D_{Ce} \left(\frac{dq_C}{dR} \right)_{R=R_0}. \qquad (42)$$

Substituting Eq. (41) in Eq. (42) yields:

$$\bar{r}_{C,u} = -4\pi R_0 D_{Ce} \Big[W \left\{ \psi_{Ae} \cosh(\psi_{Ae}) - \sinh(\psi_{Ae}) \right\}$$

$$+ Y \left\{ \psi_{Be} \cosh(\psi_{Be}) - \sinh(\psi_{Be}) \right\}$$

$$+ Z \left\{ \psi_{Ce} \cosh(\psi_{Ce}) - \sinh(\psi_{Ce}) \right\} \Big], \qquad (43a)$$

where

$$W = V / \left(\alpha_e^2 - \gamma_e^2 \right), \qquad (43b)$$

$$Y = U / \left(\beta_e^2 - \gamma_e^2 \right) \qquad (43c)$$

and

$$Z = \left\{ q_{C0} - \frac{V}{\alpha_e^2 - \gamma_e^2} \sinh(\psi_{Ae}) - \frac{U}{\beta_e^2 - \gamma_e^2} \sinh(\psi_{Be}) \right\} / \sinh(\psi_{Ce}). \qquad (43d)$$

Combining the overall rate of consumption of component A defined by

$$-\bar{r}_{A,u} = 4\pi R_0^2 D_{Ae} \left(\frac{dq_A}{dR} \right)_{R=R_0}; \qquad (44)$$

with Eq. (39), yields:

$$-\bar{r}_{A,u} = 4\pi R_0 D_{Ae} q_{A0} \left\{ \psi_{Ae} \coth(\psi_{Ae}) - 1 \right\}. \qquad (45)$$

For a uniform distribution, the selectivity, defined by Eq. (36), is given by

$$S = - \frac{D_{Ce} \Big[W \left\{ \psi_{Ae} \cosh(\psi_{Ae}) - \sinh(\psi_{Ae}) \right\} + Y \left\{ \psi_{Be} \cosh(\psi_{Be}) - \sinh(\psi_{Be}) \right\} + Z \left\{ \psi_{Ce} \cosh(\psi_{Ce}) - \sinh(\psi_{Ce}) \right\} \Big]}{D_{Ae} q_{A0} \left\{ \psi_{Ae} \coth(\psi_{Ae}) - 1 \right\}}. \qquad (46)$$

2 A → C → D COMPOSITE-SURFACE DISTRIBUTION

In this reaction scheme, C is the desired product. For reasons already emphasized, the composite-surface distribution is the only one considered in this case. A similar scheme has been considered by Dadyburjor [7], with the difference that D represents in his case the coke, while in the present case it represents gaseous products. In contrast to the former case, in the latter the corresponding rate constant is greater in the silica–alumina than in zeolite. Again, first order reactions and isothermal conditions are assumed. The reaction-diffusion equations in the small spheres are given by Eqs. (6) and (7) of Table I. The solution of Eq. (7) for the boundary conditions (8) and (9) is:

$$p_C = \frac{r_0}{r}\left\{\left(p_{C0} + p_{A0}\frac{K_{Aa}}{D_{Ca}}\frac{1}{\left(\alpha^2 - \gamma^2\right)}\right)\frac{\sinh\left(\gamma r\right)}{\sinh\left(\psi_{Ca}\right)} - p_{A0}\frac{K_{Aa}}{D_{Ca}}\frac{1}{\left(\alpha^2 - \gamma^2\right)}\frac{\sinh\left(\alpha r\right)}{\sinh\left(\psi_{Aa}\right)}\right\}. \tag{47}$$

The reaction-diffusion process in the inner region of the large sphere containing only silica–alumina can be described by Eqs. (13) and (14) of Table I, while that in the outer composite region can be described by Eqs. (18) and (19). Combining Eq. (28) with (18) and Eq. (47) with (19) and taking into account the boundary conditions (31), one obtains:

$$\frac{d}{dR}\left(4\pi R^2 D_{Ae}\frac{dC_A}{dR}\right) = 4\pi R^2 K_{Ae,A}C_A, \tag{48}$$

where $K_{Ae,A}$ is defined by Eq. (32b), and

$$\frac{d}{dR}\left(4\pi R^2 D_{Ce}\frac{dC_C}{dR}\right) = 4\pi R^2\left(-K_{Ae,C}C_A + K_{Ce,C}C_C\right), \tag{49a}$$

where

$$K_{Ae,C} = K_{Ad}\varepsilon + 4\pi r_0 n K_{Aa}\left\{\psi_{Aa}\coth\left(\psi_{Aa}\right) - \psi_{Ca}\coth\left(\psi_{Ca}\right)\right\}/\left(\alpha^2 - \gamma^2\right) \tag{49b}$$

and

$$K_{Ce,C} = K_{Cd}\varepsilon + 4\pi r_0 n D_{Ca}\left\{\psi_{Ca}\coth\left(\psi_{Ca}\right) - 1\right\}. \tag{49c}$$

The boundary conditions and the matching conditions are given by Eqs. (20) through (22) and (24) of Table I.

The overall rate of formation of component C results from:

$$\bar{r}_C = -4\pi R_0^2 D_{Ce}\left(\frac{dC_C}{dR}\right)_{R=R_0} \tag{50}$$

and the overall rate of consumption of component A results from:

$$-\bar{r}_A = 4\pi R_0^2 D_{Ae}\left(\frac{dC_A}{dR}\right)_{R=R_0}. \tag{51}$$

The selectivity is defined by Eq. (36).

Expressions for C_A and C_C as a function of R are obtained by solving Eqs. (13), (14), (48) and (49) for the boundary conditions (20) through (22) and (24). In order to calculate $\bar{r}_C$ and S, these expressions are introduced in Eqs. (50) and (36). While Eqs. (13), (14), (48) and (49) are solved analytically, the coefficients in these solutions have been calculated numerically because of their complexity.

For the uniform distribution, an analytical expression can be easily obtained. The rate of consumption of component A for a uniform distribution is given by

$$-\bar{r}_{A,u} = 4\pi R_0 D_{Ae} C_{A0} \left\{ \psi_{Ae}\coth\left(\psi_{Ae}\right)-1 \right\}, \tag{52}$$

while the rate of formation of component C results from:

$$\bar{r}_{C,u} = -4\pi R_0 D_{Ce}\left[\left\{C_{A0}\left(K_{Ae,C}/D_{Ce}\right)/\left(\alpha_e^2-\gamma_e^2\right)+C_{C0}\right\}\left\{\psi_{Ce}\coth\left(\psi_{Ce}\right)-1\right\}\right.$$

$$\left.-C_{A0}\left(K_{Ae,C}/D_{Ce}\right)/\left(\alpha_e^2-\gamma_e^2\right)\left\{\psi_{Ae}\coth\left(\psi_{Ae}\right)-1\right\}\right]. \tag{53}$$

RESULTS AND DISCUSSION

As mentioned earlier, the calculations have been carried out for the catalytic cracking process which employs the zeolite/silica–alumina catalysts. The radius of the large sphere has been taken as 0.5 cm, which has the same order of magnitude as that of the commercial catalyst for a fixed or moving bed reactor [6]. The values of the diffusion coefficients used in the analysis are those suggested by Weisz [5]. Reaction rate constants whose values had the same order of magnitude as those recommended by Dadyburjor [7] were used. For the reaction $\overset{B\;\rightarrow\;C}{\underset{D}{\searrow\;\swarrow}}$, we have selected

$$\frac{K_{B1a}}{K_{Ba}} > \frac{K_{B1d}}{K_{Bd}} \text{ and } K_{Cd} > K_{Ca},$$

since the zeolite is more selective than silica–alumina for the formation of component C, and the secondary cracking ($C{\rightarrow}D$) occurs to a greater extent in the silica–alumina matrix than in zeolite. We also consider $K_{Ba} > K_{Bd}$, since the zeolite is more active than silica–alumina in the formation of compound C. In addition, K_{Ad} is assumed greater than K_{Aa}. For the reaction scheme $A{\rightarrow}C{\rightarrow}D$, we have chosen $K_{Aa} > K_{Ad}$ and $K_{Cd} > K_{Ca}$ for the same reasons as already mentioned above. Other values of the parameters, which do not satisfy the above inequalities, have also been used.

1 $\quad \overset{A\;\rightarrow\;B\;\rightarrow\;C}{\underset{D}{\searrow\;\swarrow}}$ **COMPOSITE-CENTER DISTRIBUTION**

It is convenient to scale the rate of formation of component C with respect to that for a uniform distribution (for which we take arbitrarily $\varepsilon = 0.26$). Let us denote this ratio by η_1. η_1 is plotted against ε, the volume fraction of silica–alumina in the core region, in Figure 2. The curves clearly show that the composite-center distribution and the uniform distribution have an optimum catalyst dilution. However, for the values of the parameters considered for catalytic cracking (dotted and dash-dotted curves) the overall rate of formation of component C in the catalyst pellet increases with λ at fixed values of ε. In contrast, for some values of the parameters which do not satisfy the

Optimum Design of Zeolite/Silica–Alumina Catalysts

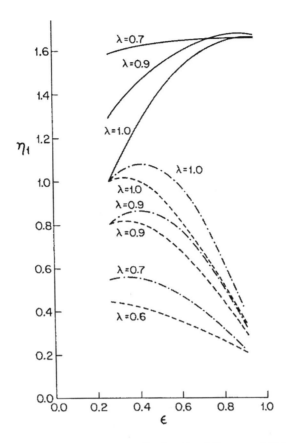

FIGURE 2 η_1 against ε for the composite-center distribution. $R_0 = 0.5$ cm, $r_0 = 0.005$ cm, $D_{Aa} = 0.1 \times 10^{-6}$ cm^2 sec^{-1}, $D_{Ba} = D_{Ca} = 0.1 \times 10^{-5}$ cm^2 sec^{-1}, $D_{Ad} = D_{Bd} = D_{Cd} = 0.1 \times 10^{-1}$ cm^2 sec^{-1}, $C_{A0} = 0.3 \times 10^{-4}$ mols cm^{-3} except for the solid curves for which $C_{A0} = 0.1 \times 10^{-3}$ mols cm^{-3}, $C_{B0} = C_{C0} = 0.1 \times 10^{-4}$ mols cm^{-3}.

	K_{Aa}(sec^{-1})	K_{B1a}(sec^{-1})	K_{B2a}(sec^{-1})	K_{Ca}(sec^{-1})
—·—	0.01	0.56	0.04	0.005
----	0.01	0.14	0.01	0.005
——	0.001	3.0	0.02	0.03
	K_{Ad}(sec^{-1})	K_{B1d}(sec^{-1})	K_{B2d}(sec^{-1})	K_{Cd}(sec^{-1})
—·—	0.5	0.012	0.006	0.007
----	0.5	0.012	0.006	0.007
——	10.0	0.48	0.002	0.001

inequalities noted above (the solid curves), the maximum value for the rate of formation of component C occurs for $\lambda = 0.9$ and $\varepsilon = 0.9$. In the latter case, the overall rate of formation of component C has a maximum as a function of both λ and the overall volume fraction of silica–alumina in the catalyst pellet, $\bar{\varepsilon}$. In Figure 3, the selectivity, scaled with respect to that for a uniform distribution having a volume fraction of 0.26 for the silica–alumina, is plotted against ε. The selectivity exhibits a weak maximum as a function of ε only for some values of λ and of the other parameters. One may note that for a uniform distribution ($\lambda = 1.0$), the selectivity decreases monotonically as ε increases. This happens since as ε increases, the rate of formation of component B as well as the fraction of component B which diffuses out to the external surface without further reaction increase. The latter effect is due to the small diffusivity in the small spheres.

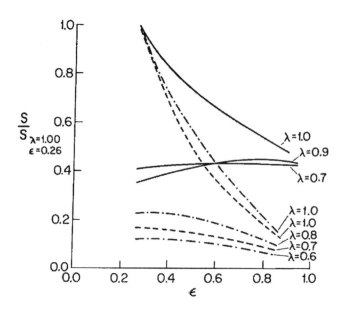

FIGURE 3 Selectivity, scaled with respect to that for a uniform distribution with $\varepsilon = 0.26$, against ε for the composite-center distribution. The parameter values for the curves are the same as those for the corresponding curves of Figure 2.

It is, however, more meaningful to compare pellets which contain the same amount of zeolite, although differently distributed. The ratio η_2 of the overall rate of formation of component C to that for a uniform distribution is plotted as a function of λ in Figure 4, for fixed values of the overall volume fraction of diluent (silica–alumina) in the catalyst pellet, $\bar{\varepsilon}$. The overall volume fraction of silica–alumina is related to its volume fraction in the core region, ε, as follows:

$$\varepsilon = \left(\lambda^3 - 1 + \bar{\varepsilon}\right)/\lambda^3. \tag{54}$$

The values of $\bar{\varepsilon}$ used in these calculations are in the range of 0.85–0.90, a range which is encountered in the commercial catalytic cracking processes [8].

As shown in Figure 4, λ has an optimum value for which the overall rate of formation of compound C exhibits a maximum. In comparison to the uniform distribution, the rate of formation of C increases by 28% for curve 1, but by only 2% for curve 5. The optimum value of λ, λ_{opt}, varies between 0.64 for curve 1 to 0.87 for curve 4. This maximum of the overall rate of formation of C as a function of λ can be explained as follows: An increase of the outer region of silica–alumina, by maintaining a fixed value of $\bar{\varepsilon}$, has the following consequences: It

(i) increases the amount of component B which is formed, thus increasing the driving force for the formation of component C;
(ii) increases the amount of component B which diffuses out to the external surface of the large sphere, without further penetration into the core (composite) region;
(iii) decreases the rate of diffusional mass transfer within the composite region;
(iv) increases the amount of component D which is formed from component C when it diffuses out to the surface of the large sphere.

The overall rate of formation of component C is increased by (i) but is decreased by (ii) through (iv). As a result, there exists a λ_{opt} which maximizes the overall rate of formation of C. As shown in

Optimum Design of Zeolite/Silica–Alumina Catalysts

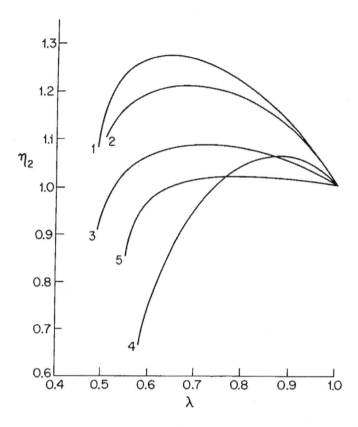

FIGURE 4 η_2 against the ratio, λ, of the inner radius to the radius of the pellet for fixed values of the average volume fraction of silica–alumina, for the composite-center distribution. $R_0 = 0.5$ cm, $r_0 = 0.005$ cm.

	$\bar{\varepsilon}$	K_{Aa}	K_{B1a}	K_{B2a}	K_{Ca}	K_{Ad}	K_{B1d}	K_{B2d}	K_{Cd}
1.	0.9	0.01	0.144	0.001	0.005	0.5	0.005	0.007	0.015
2.	0.9	0.1	0.13	0.005	0.003	1.5	0.005	0.008	0.015
3.	0.9	0.002	0.145	0.005	0.004	0.05	0.008	0.005	0.010
4.	0.85	0.01	1.4	0.05	0.02	0.5	0.012	0.009	0.060
5.	0.85	0.005	0.42	0.015	0.01	0.12	0.003	0.002	0.013

	$D_{Aa} \times 10^5$	$D_{Ba} \times 10^5$	$D_{Ca} \times 10^5$	D_{Ad}	D_{Bd}	D_{Cd}	$C_{A0} \times 10^3$	$C_{B0} \times 10^4$	$C_{C0} \times 10^4$
1.	0.01	0.1	0.1	0.01	0.01	0.01	0.1	0.1	0.1
2.	0.1	1.0	1.0	0.01	0.01	0.01	0.1	0.1	0.05
3.	0.01	0.1	0.1	0.01	0.01	0.01	0.1	0.1	0.1
4.	0.01	0.1	0.1	0.01	0.01	0.01	0.1	0.1	0.1
5.	0.01	0.05	0.1	0.01	0.05	0.1	0.09	0.2	0.1

The parameter units are the same as those in Figure 2.

Figure 5, the selectivity also presents a maximum as a function of λ, when the overall volume fraction $\bar{\varepsilon}$ of silica–alumina in the catalyst is fixed. Comparing Figures 4 and 5, one can note that the rate of formation of compound C and the selectivity exhibit a maximum almost at the same value of λ. This is a result of the fact that the rate of formation of C and the selectivity are affected by λ in the same manner.

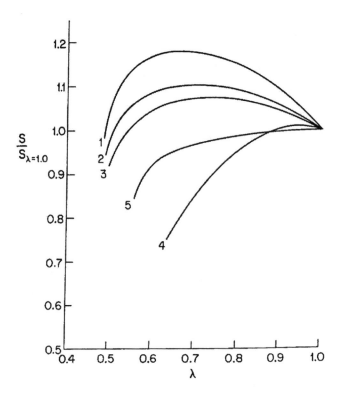

FIGURE 5 Selectivity, scaled with respect to that for the uniform distribution, as a function of λ for fixed values of, $\bar{\varepsilon}$ for the composite-center distribution. The parameter values for the curves are the same as those for the correspondingly numbered curves of Figure 4.

2 A→C→ D Composite-Surface Distribution

The ratio η_1 is plotted as a function of ε, the volume fraction of silica–alumina in the outer region, in Figure 6. An optimum catalyst dilution is noticed. The rate of formation of component C increases as λ decreases at fixed values of ε for the values of the parameters used for the dotted curves, whereas for the values of the parameters used for the dash-dotted curves (K_{Ca}> K_{Cd}), η_1 has a maximum for $\lambda = 0.6$ and $\varepsilon = 0.6$ ($\bar{\varepsilon} = 0.68$). This clearly indicates that the overall rate of formation of component C can exhibit for some values of the parameters a maximum with respect to both λ and $\bar{\varepsilon}$. In Figure 7, the selectivity, scaled with respect to that for a uniform distribution with $\varepsilon = 0.26$, is plotted against ε. Each of the solid curves of this figure presents a weak maximum. However, for the dotted curves, the selectivity decreases monotonically with ε, while, as shown in Figure 6, the rate of formation of component C has a maximum. This happens because an increase in ε increases the rate of formation of component D since K_{Cd}> K_{Ca} and this effect dominates when $K_{Cd} \gg K_{Ca}$. The selectivity increases monotonically with ε for the dash-dotted curves, since in those cases K_{Ca} is larger than K_{Cd}.

Figures 8 and 9 present plots of η_2 vs λ and of selectivity (scaled with respect to that for the uniform distribution) vs λ, respectively, for a fixed value of $\bar{\varepsilon}$. The relationship between ε and $\bar{\varepsilon}$. now has the form:

$$\varepsilon = \left(\bar{\varepsilon} - \lambda^3\right)/\left(1 - \lambda^3\right). \tag{55}$$

Optimum Design of Zeolite/Silica–Alumina Catalysts

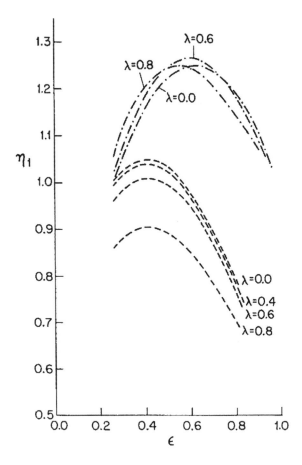

FIGURE 6 η_1 as a function of ε for the composite-surface distribution. $R_0 = 0.5$ cm, $r_0 = 0.005$ cm, $D_{Aa} = D_{Ca} = 0.1 \times 10^{-5}$ cm^2 sec^{-1}, $D_{Ad} = D_{Cd} = 0.1 \times 10^{-1}$ cm^2 sec^{-1}, $K_{Aa} = 1.5$ sec^{-1}, $K_{Ad} = 0.15$ sec^{-1}, $C_{A0} = 0.1 \times 10^{-3}$ mols cm^{-3}, $C_{C0} = 0.1 \times 10^{-4}$ mols cm^{-3}.

	K_{Ca}(sec^{-1})	K_{Cd}(sec^{-1})
--------	0.03	0.3
-·-·- --	0.3	0.1

The value of $\bar{\varepsilon}(=0.8)$ used in these calculations is somewhat smaller than those used for commercial cracking catalysts. As shown in Figure 8, η_2 can have a maximum as a function of λ for fixed value of $\bar{\varepsilon}$, which is observed to be 1.056 for curve 1, and is 1.250 for curve 3. The maximum of η_2 occurs because (i) the rate of formation of component C increases with (increasing) λ; (ii) the rate of formation of component D decreases with λ; (iii) the diffusional mass transfer in the outer composite region decreases with λ; and (iv) a part of the component C which is formed in the outer region penetrates through the silica–alumina region to form component D. As shown in Figure 9, the selectivity also has a maximum, which occurs for the same reasons as those noted above. The increase in the selectivity is, however, smaller than that in the rate of formation of the desired product since the consumption of component A also increases as the rate of formation of component C increases.

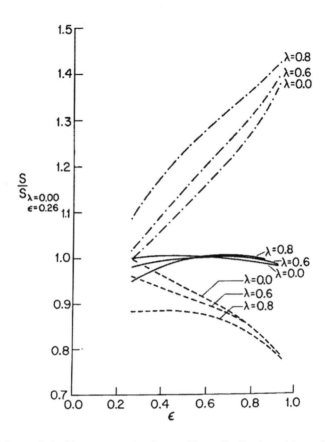

FIGURE 7 Selectivity, scaled with respect to that for a uniform distribution with $\varepsilon = 0.26$, as a function of ε for the composite-surface distribution. The parameter values for the dotted and dash-dotted curves are the same as those for the corresponding curves of Figure 6. The parameter values for the solid curves are the same as those for the other two except for $K_{Ca} = 0.1$ sec^{-1} and $K_{Cd} = 0.2$ sec^{-1}.

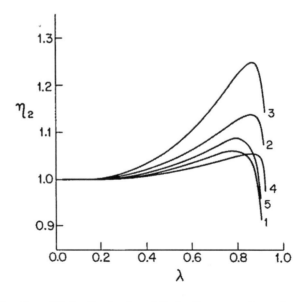

FIGURE 8 η_2 as a function of λ for fixed value of $\bar{\varepsilon}$, for the composite-surface distribution. $R_0 = 0.5$, $r_0 = 0.005$, $\bar{\varepsilon} = 0.8$, $D_{Aa} = D_{Ca} = 0.1 \times 10^{-5}$, $D_{Ad} = D_{Cd} = 0.1 \times 10^{-1}$, $C_{A0} = 0.1 \times 10^{-3}$, $C_{C0} = 0.1 \times 10^{-4}$.

Optimum Design of Zeolite/Silica–Alumina Catalysts

	K_{Aa}	K_{Ca}	K_{Ad}	K_{Cd}
1.	0.15	0.002	0.001	0.03
2.	1.5	0.002	0.001	0.03
3.	1.5	0.002	0.001	0.15
4.	1.5	0.03	0.15	0.3
5.	1.5	0.1	0.15	0.2

The parameter units are the same as those in Figure 6.

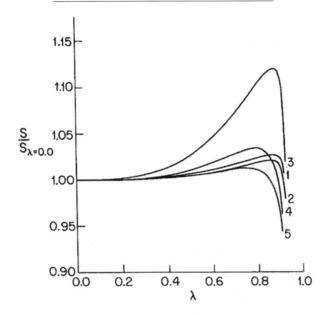

FIGURE 9 Selectivity, scaled with respect to that for a uniform distribution, as a function of λ for fixed value of $\bar{\varepsilon}$, for the composite-surface distribution. The parameter values for the curves are the same as those for the correspondingly numbered curves of Figure 8.

Dadyburjor showed that for the series reaction $(A \xrightarrow{K_A} C \xrightarrow{K_C} D$, where C is the desired product, and $K_{A,\,zeolite} > K_{A,\,matrix}$ and $K_{C,\,zeolite} > K_{C,\,matrix})$, the dilute active surface distribution (which is the same as the composite-surface distribution in this paper) exhibits a maximum for selectivity. In the present calculations, the composite-surface distribution also exhibits a maximum even though $K_{A,zeolite} > K_{A,\,matrix}$ and $K_{C,\,matrix} > K_{C,\,zeolite}$. One can, therefore, conclude that for the series reaction, the composite-surface distribution maximizes the selectivity if the rate constant for the desired product is greater in the zeolite than in the silica–alumina, regardless of the rate constant of the secondary reaction being larger or smaller in the zeolite.

CONCLUSION

An analysis has been carried out for two nonuniform distributions of composite catalysts, i.e., the composite-center and composite-surface distributions. The overall rate of formation of the desired product can exhibit, for some parameter values, a maximum with respect to both (1) the ratio of the radius of the core region to that of the catalyst pellet and (2) the overall volume fraction of silica–alumina.

For fixed values of the overall volume fraction of silica–alumina in the catalyst, the nonuniform distributions can lead to greater rates of formation of the desired product and the selectivity than the uniform ones.

NOMENCLATURE

C	concentration in the outer region of the large sphere, mol cm^{-3}
D	diffusion coefficient, cm^2 sec^{-1}
F	defined by Eq. (30d)
G	defined by Eq. (30e)
H	defined by Eq. (30f)
K	rate constant, sec^{-1}
$K_{Ae, A}$	effective rate constant of component A, in the composite region, in the reaction-diffusion equation for component A (defined by Eq. (32b))
$K_{Ae, B}$	effective rate constant for component A, in the composite region, in the reaction-diffusion equation for component B (defined by Eq. (33b))
$K_{Be, B}$	effective rate constant for component B, in the composite region, in the reaction-diffusion equation for component B (defined by Eq. (33c))
$K_{Ae, C}$	effective rate constant for component A, in the composite region, in the reaction-diffusion equation for component C (defined by: for composite-center, Eq. (34b); for composite-surface, Eq. (49b))
$K_{Be, C}$	effective rate constant for component B, in the composite region, in the reaction-diffusion equation for component C (defined by Eq. (34c))
$K_{Ce, C}$	effective rate constant for component C, in the composite region, in the reaction-diffusion equation for component C (defined by: for composite-center, Eq. (34d); for composite-surface, Eq. (49c))
n	number of the small spheres per unit volume in the composite region
p	concentration in the small spheres, mol cm^{-3}
q	concentration in the inner region of the large sphere, mol cm^{-3}
r	radial coordinate in the small sphere
r_0	radius of the small sphere
R	radial coordinate in the large sphere
R_0	radius of the large sphere
$\bar{r}$	overall rate of formation in the large sphere
S	selectivity defined by Eq. (36)
U	defined by Eq. (41e)
V	defined by Eq. (41d)
W	defined by Eq. (43b)
Y	defined by Eq. (43c)
Z	defined by Eq. (43d)

Greek Letters

α, β, γ	defined by Eqs. (28c), (29c) and (30c), respectively
$\alpha_e, \beta_e, \gamma_e$	defined by Eqs. (39b), (40b) and (41b), respectively
ε	the volume fraction of silica–alumina in the diluted region
$\bar{\varepsilon}$	overall volume fraction of silica–alumina in the catalyst pellet
η_1	ratio of the rate of formation of component C to that for a uniform distribution which has $\varepsilon = 0.26$
η_2	ratio of the rate of formation of component C to that for a uniform distribution when the amount of zeolite is fixed
λ	ratio of the radius of the inner region to the radius of the catalyst pellet

Optimum Design of Zeolite/Silica–Alumina Catalysts

$\psi_{Aa}, \psi_{Ba}, \psi_{Ca}$	Thiele Moduli in the small sphere defined by: for A, Eq. (28b); for B, Eq. (29b); for C, Eq. (30b)
$\psi_{Ae}, \psi_{Be}, \psi_{Ce}$	defined by: for A, Eq. (39c); for B, Eq. (40c); for C, Eq. (41c)

Subscript

0	surface value
A, B, C	component A, B and C
u	uniform distribution
a	active material (zeolite)
d	diluent (silica–alumina)
e	effective value
max	maximum value
opt	optimum value

REFERENCES

1. Ruckenstein, E., *A.I.Ch. E.J.,* **16**, 151 (1970).
2. Nandapurkar, P.J., and Ruckenstein, E., *Chem. Eng. Sci.,* **39**, 371 (1984).
3. Dadyburjor, D.B., *A.I.Ch. E.J.,* **28**, 720 (1982).
4. Nace, D.M., Volt, S.E., and Weekman, V.W., Jr., *Ind. Eng. Chem. Proc. Des. Dev.* **10**, 530 (1971).
5. Weisz, P.B., *Chem. Technol.,* 498 (1973).
6. Shankland, R.V., *Adv. Cat.* **6**, 271 (1954).
7. Dadyburjor, D.B., Ind. Eng. Chem. Fundam. **24**, 16 (1985).
8. Thomas, C.L., and Barmby, D.S., *J. Catal.,* **12**, 341 (1968).

3.9 Effect of the Strong Metal-Support Interactions on the Behavior of Model Nickel/Titania Catalysts[*]

Eli Ruckenstein and Sung H. Lee

Department of Chemical Engineering, State University
of New York, Buffalo, New York 14260

Received August 12, 1986; revised November 3, 1986

INTRODUCTION

The chemical and physical interactions between the metal particles and the support bring about changes in the morphology of the supported particles and consequently affect the activity and selectivity of the supported metal catalysts. The effect of such interactions on the morphology of the metal particles supported on irreducible oxides such as Al_2O_3 and SiO_2 has been studied extensively by using transmission electron microscopy (*1-8*). A detailed study in this direction has been carried out recently by Sushumna and Ruckenstein (*9-11*) who investigated the behavior of Fe/Al_2O_3 model catalysts in reducing and oxidizing atmospheres. TEM observations of the behavior of metal particles supported on reducible oxides also have been presented (*12-17*). In the case of Fe/TiO_2 samples heated in H_2, Tatarchuk and Dumesic (*14*) observed thin and flat particles and also an apparent decrease in the observable material on the surface of the substrate. They suggested that the flat morphology was a result of extension and the apparent decrease in observable material was due to a diffuse spreading of iron over the substrate surface and/or into the substrate. They suggested that in an oxygen atmosphere, iron partially diffuses back to the surface and forms large particles which are present as $FeTi_2O_5$ (*15*). Similar behavior was also observed with the Pt/TiO_2 system by Baker *et al.* (*12, 13*). Platinum particles had thin morphology upon heating in hydrogen, while they appeared to be thicker upon heating in oxygen. Simoens *et al.* (*16*) reported that the nickel particles were thin and flat at both 150 and 550°C upon heating in hydrogen. The average particle size was, however, somewhat greater at 550°C than at 150°C. On the other hand, heating at 700°C caused severe sintering. In a recent paper (*18*), Raupp and Dumesic reported that nickel particles on titania were relatively resistant to sintering in a reducing atmosphere at temperatures up to 650°C and that the particles had a flat shape, as also confirmed indirectly by CO temperature-programmed desorption. However, discussions on the role of the metal-support interactions on the morphology of the metal particles supported on reducible oxides have been

---

[*] *Journal of Catalysis* 104, 259–278 (1987). Republished with permission.

Effect of the Strong Metal-Support Interactions on Nickel/Titania Catalysts

provided by only a few authors. In addition, it is to be noted that in general the results arrived at in the literature with these systems were based on observations of short term behavior such as heating for up to only 1 h.

The aim of this paper is to examine in more detail the role of the strong metal-support interactions in affecting the shape of the particles and the behavior, in general, of crystallites supported on titania. For this purpose, Ni/TiO$_2$ model catalysts have been observed over extended periods of heating in reducing and oxidizing atmospheres. The major events observed in this study include an alternate change in the shape of the particles and formation of cavities in the substrate during heating in hydrogen; considerable extension of particles over the substrate and/or diffusion of Ni ions into the substrate during heating in oxygen; and restoration to large nickel particles upon subsequent heating in hydrogen, followed by their extension during additional heating in hydrogen. The latter observations are in contrast to those reported previously by various authors.

EXPERIMENTAL

Preparation of sample. Titania films (rutile form) of approximately 500-Å thickness, used as supports for the model catalysts, were prepared by heating titanium foils (99.98%, 0.025 mm thick, Alfa Products) in oxygen at 300-350°C for 1.5 h and stripping the oxide film off by dissolving the unoxidized titanium. The titania films were then rinsed in distilled water, picked up on gold electron microscope grids, and allowed to dry. Nickel films were deposited onto the titania films by vacuum evaporating nickel wire of 99% purity in an Edwards 306 vacuum evaporator, under a pressure of better than 10^{-6} Torr. *Heat treatment.* The samples were heated inside a quartz tube, placed in a furnace. After the sample was introduced into the tube, the tube was flushed with helium for at least 30 min. The temperature was then raised in a helium atmosphere from room temperature to the predetermined temperature, at which point helium was switched to the desired gas. After it was heated in the desired atmosphere for the predetermined duration of time, the tube was cooled down to room temperature again in a helium atmosphere. The flow rates of the gases during heating were maintained at about 150 cm^3/min. The gases used in the experiment were all ultrahigh-purity grade purchased from Linde Division, Union Carbide Corporation. Hydrogen and helium were both 99.999% pure, the former having less than 1 ppm O$_2$ and less than 2 ppin moisture and the latter having less than 3 ppm moisture. Oxygen, 99.99% pure, contained less than 3 ppm moisture. Except for the heat treatments in wet hydrogen, the ultrahigh-purity hydrogen was further purified by successive passage through a Deoxo (Engelhard Industries) unit, a silica gel column, and a bed of 15% MnO on SiO$_2$, and finally a 4A molecular sieve bed immersed in liquid nitrogen. For some experiments in wet hydrogen atmosphere, the as-supplied hydrogen was bubbled through distilled water. Helium was also further purified by passing it through a 4A molecular sieve bed immersed in liquid nitrogen. *TEM observation.* After each heat treatment, the samples were examined and the same regions of each sample were photographed using a JEOL 100U transmission electron microscope operated at 80 kV.

RESULTS

Several samples of 10-Å initial metal film thickness have been investigated in hydrogen and oxygen atmospheres at different temperatures. Table 1 summarizes the heat treatments for each sample and the corresponding micrograph numbers.

TABLE 1
Heat Treatments

Sample	Treatment	Figure
A	H_2 at 700°C for 23 h	1
	O_2 at 700°C for 3 h	5
	H_2 at 700°C for 4 h	5
B	H_2 at 700°C for 16 h	2
	H_2 at 700°C for 115 h	6
	O_2 at 700°C for 6 h	6
	H_2 at 700°C for 4 h	6
C	H_2 at 500°C for 23 h	3
D	H_2 at 250°C for 47 h	4

A. BEHAVIOR IN ADDITIONALLY PURIFIED HYDROGEN AT 700°C

The results on heating a specimen (sample A) at 700°C in additionally purified hydrogen are shown in Fig. 1. The major results observed were (1) alternate changes in the shape of the particles, (2) formation of cavities or channels in the substrate beneath the particles, and (3) severe sintering. These results are described separately.

Alternate changes in the shape of the particles. Following the deposition of the Ni film, the sample (sample A) was heated in H_2 at 250°C for 2 h to generate very small crystallites. Subsequently, the sample behavior was investigated at 700°C in H_2. Figure 1a shows the micrograph following heating at 700°C for 1 h. During the subsequent 4 h of heating, the crystallites agglomerated to form larger particles and maintained a circular shape (Fig. 1b). On heating for an additional 3 h, a change in the shape of the crystallites was observed (Fig. 1c). White (lighter in contrast) annular patches were formed adjacent to the periphery of (but not all around) the particles. The white patch around the particles disappeared after heating for an additional 3 h (Fig. 1d). Two more cycles of such changes were observed on further heating for a total of 23 h (Fig. 1). Another sample (sample B) of the same loading, whose micrographs are not shown, also exhibited similar alternations in the shape of the crystallites. Three cycles of shape alternations were observed over a period of 16 h, and, on further heating for up to 115 h, no further shape alternations were detected with this sample.

The electron diffraction patterns were also obtained after each heat treatment. The (d-spacing values for sample A are listed in Table 2. They show that titania remained as rutile TiO_2 throughout the entire heat treatment. (It is likely that in a hydrogen atmosphere, there is localized reduction of the substrate beneath and in the immediate vicinity of the particles, which is, however, not detected by electron diffraction in our case.) In addition, after heating for 1 h at 700°C, two additional rings appeared in the diffraction pattern. One of them corresponded to Ni, but the other did not correspond to any stoichiometric compound associated with nickel. During the subsequent 2 h of heating, the latter ring disappeared gradually, while another ring corresponding to Ni appeared. This indicates that the unknown compound mentioned above was gradually reduced to Ni and that the particles were finally present as mostly Ni. After a total of 11 h of heating, a new ring whose d value

Effect of the Strong Metal-Support Interactions on Nickel/Titania Catalysts

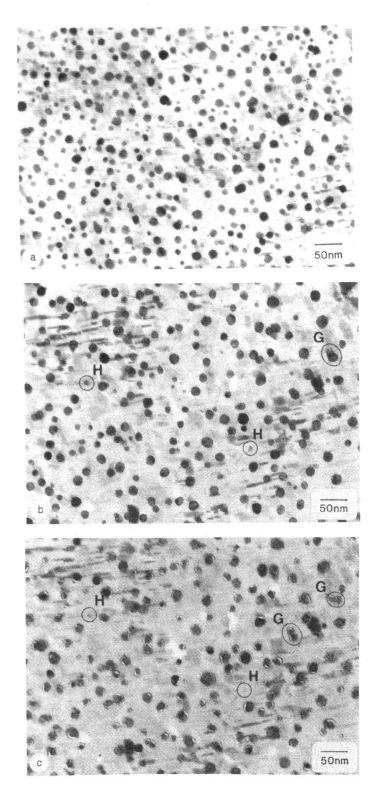

FIG. 1. Time sequence of the same region of a specimen (A) heated at 700°C in purified hydrogen, (a) 1 h, (b) 5 h, (c) 8 h. *(Continued)*

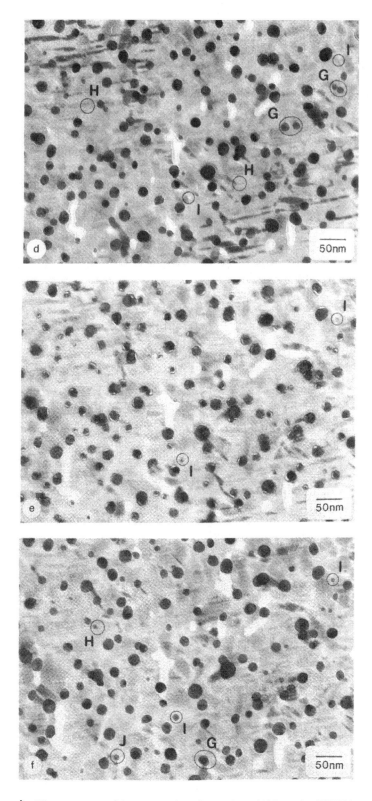

FIG. 1. (Continued) Time sequence of the same region of a specimen (A) heated at 700°C in purified hydrogen, (d) 11 h, (e) 12 h, (f) 13 h.   (*Continued*)

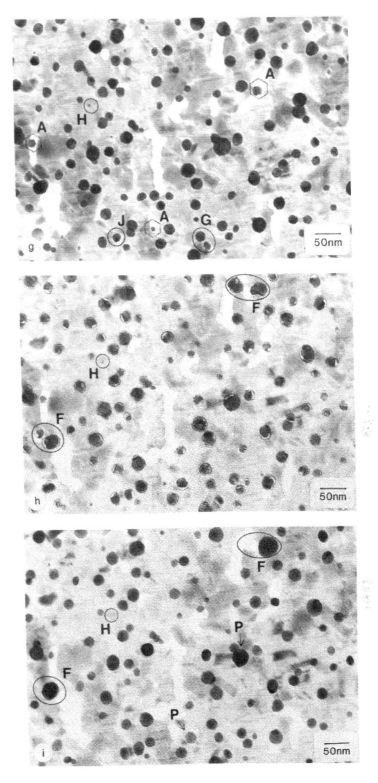

FIG. 1. (Continued) Time sequence of the same region of a specimen (A) heated at 700°C in purified hydrogen, (g) 16 h, (h) 20 h, (i) 23 h.

TABLE 2

Electron Diffraction Analysis for Sample A

d-Spacing from ASTM Card (Å)				d-Spacing from Diffraction Pattern (Å)			
TiO_2	Ni	NiO	Ni_3Ti	. 11 h, H_2	3h, O_2	1 h, H_2	4 h^a H_2
3.250	2.034	2.410	2.13	3.244	4.209	3.244	3.252
2.487	1.762	2.088	2.07	2.485	3.676	2.488	2.482
2.188	1.246	1.476	1.95	2.177	3.250	2.191	2.181
1.6874				2.045	2.707	2.084	2.058
				1.923	2.481	1.690	1.681
				1.769	2.181		
				1.683	1.829		
					1.687		

was close to the major d value of Ni_3Ti appeared. Following a total of 16 h of heating, the intensity of this ring appeared to be much stronger than before, and it increased further on subsequent heating while the intensity of the rings corresponding to Ni decreased gradually. These results indicate that nickel formed an intermetallic compound with titanium or TiO_{2-x} in hydrogen at 700°C.

The effect of moisture on the behavior of nickel crystallites was investigated by heating a different sample in wet hydrogen at 700°C. The presence of moisture, deliberately introduced in the gas stream by bubbling the hydrogen through distilled water, caused the cessation of the alternate changes in shape described before. In fact, the particles remained circular in shape throughout the entire heat treatment. This is in contrast to the behavior observed with Fe/Al_2O_3 (9) where impurities, such as moisture and/or O_2, in the as-supplied hydrogen caused changes in shape.

Formation of cavities or channels in the substrate beneath the particles. As shown in Fig. 1, the migration of a particle exposes a cavity in the substrate beneath the particle. Some particles generated cavities at their original location only, while others developed channels along their tracks over a fairly long distance. It is likely that there is substrate within the cavities and in the channels which is, however, too thin to be seen in the micrographs. There are particles located in the channels which indicate that there is indeed at least a thin layer of substrate underneath, to support the particles (Fig. 1g, A). Figure 2 shows micrographs of another sample (sample B) which exhibited similar cavity formation. In these micrographs, it is clear that the cavity region is thinner than the other regions of the substrate. Region B in Fig. 2 shows the cavity underneath the particle, exposed as a result of the migration of the particle out of the cavity. A different region, C, shows a particle which migrated completely out of the cavity but did not migrate far away from the cavity. In another region, D, a particle migrated away from the cavity without forming a channel along its track. Region E, on the other hand, shows a particle which formed a channel along its track. It should be noted that the cavity sizes are approximately the same as those of the corresponding particles.

Sintering. On heating sample A for 1 h at 700°C in additionally purified hydrogen, large particles were formed (Fig. 1a). The larger of these particles had diameters in the range 150-200 Å, while at 500°C, the larger particles in a different sample of the same loading had diameters of about 50 Å. even after 3 h of heating (Fig. 3). Considerable sintering occurs at 700°C while at temperatures of 500°C or less, relatively little sintering occurs.

Effect of the Strong Metal-Support Interactions on Nickel/Titania Catalysts

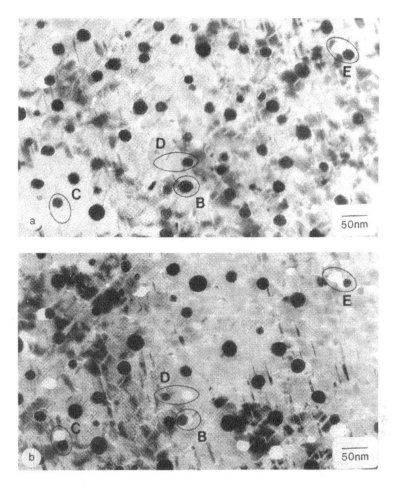

FIG. 2. Sequence of changes in a specimen (B) on heating at 700°C in purified hydrogen. The same region is shown after (a) to h and (b) 15 h.

At 700°C, sintering continued during a total of 23 h of heating and led to the growth of the particles. The larger particles now had diameters between 300 and 400 Å. During heating at 700°C, various events of sintering were observed (Fig. 1): migration and subsequent coalescence with neighboring particles (F), separation of two contacting or overlapping particles (G), disappearance of particles following a gradual decrease in size (H), appearance of new particles (I), and increase in the size of small particles without coalescence (J).

B. BEHAVIOR IN ADDITIONALLY PURIFIED HYDROGEN AT 500 AND 250°C

Figure 3 shows selected results for sample C heated at 500°C in additionally purified hydrogen. On heating for 3 h (Fig. 3a), the particles appeared to have a torus shape with a small remnant particle in the cavity. After an additional 8 h of heating, the cavities were filled in (Fig. 3b). A repeated alternation in the shapes of the particles between a torus and a continuous shape was observed on further heating (Figs. 3c-e). The rings in the selected-area electron diffraction patterns were identified to be from the rutile form of TiO_2. After a total of 19 h of heating, new rings attributed to Ni appeared and the intensity of these rings increased on further heating. One may note that during this period, the growth of the particles and the decrease in the number of particles occurred only to a small extent.

The results on heating at 250°C in additionally purified hydrogen were, however, different and are shown in Fig. 4. Initial particles were formed on heating sample D for 2 h at 250°C (Fig. 4a). On further heating for up to a total of 47 h there was very little change (Figs. 4b and c). However, some particles migrated and coalesced with nearby particles (K). One may also note that some other particles disappeared, probably due either to migration and subsequent coalescence with adjacent particle or to direct ripening (L).

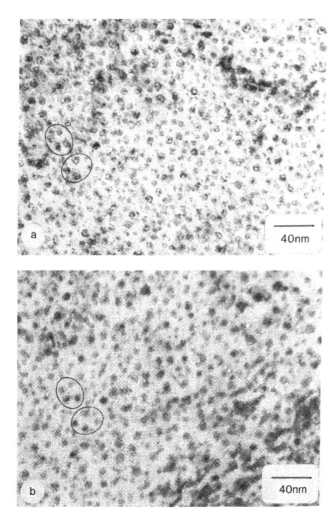

FIG. 3. Sequence of changes in the same region of a specimen (C) heated at 500°C in purified hydrogen. (a) 3 h, (b) 11 h. *(Continued)*

Effect of the Strong Metal-Support Interactions on Nickel/Titania Catalysts

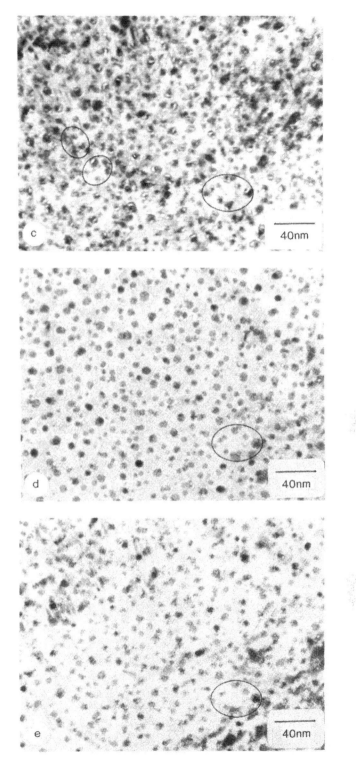

FIG. 3. (Continued) Sequence of changes in the same region of a specimen (C) heated at 5000e in purified hydrogen. (c) 15 h, (d) 19 h, (e) 23 h.

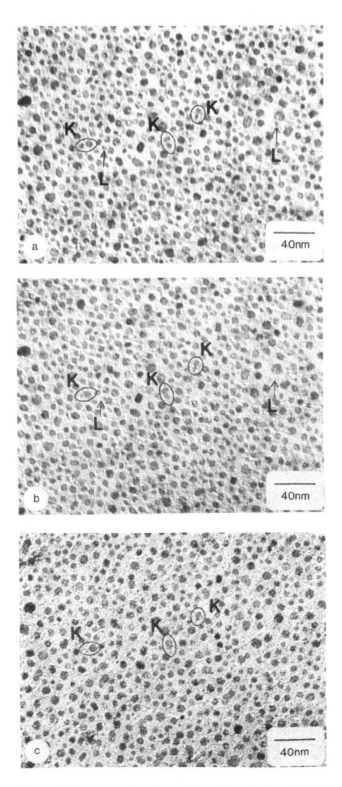

FIG. 4. Sequence of changes in the same region of a specimen (D) on heating at 250°C in purified hydrogen, (a) 2 h, (b) 4 h, (c) 47 h.

C. Behavior on Alternate Heating in Oxygen and Hydrogen

Sample A, which was heated in hydrogen at 700°C for 23 h before (Fig. 1i), was further heated at 700°C alternately in oxygen and additionally purified hydrogen. The results are shown in Fig. 5. Following 1 h of heating in oxygen (Fig. 5a), most of the particles decreased in size considerably or disappeared. This suggests that the particles probably spread over the substrate as thin films, which could not be seen in the micrographs, and/or diffused into the substrate. In fact, relatively large particles could be observed to have extended over the substrate, as marked in regions P in the Figs. 1i and 5a. In the electron diffraction pattern, several new rings, which had not been detected during prior heating in hydrogen, now appeared. These new rings, whose d values did not correspond to any stoichiometric compound, are perhaps from a non-stoichiometric compound of NiO and TiO_2. On heating for an additional 2 h in oxygen (Fig. 5b), many small particles further decreased in size (M) or disappeared (N) while the thick films around the large particles further extended over the

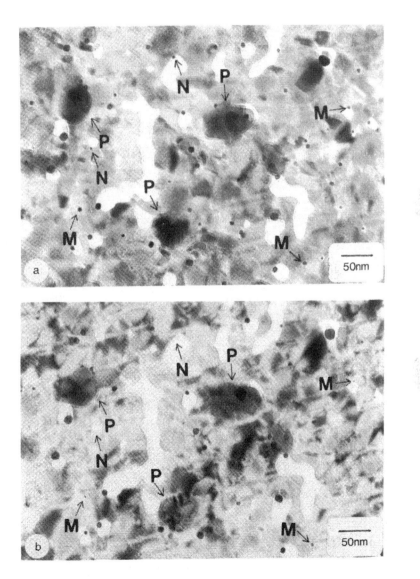

FIG. 5. Sequence of changes in sample A of Fig. 1 on subsequent alternate heating in oxygen and hydrogen at 700°C. The region shown here is the same as that of Fig. 1, after (a) 1 h. O_2, (b) 3 h, O_2. *(Continued)*

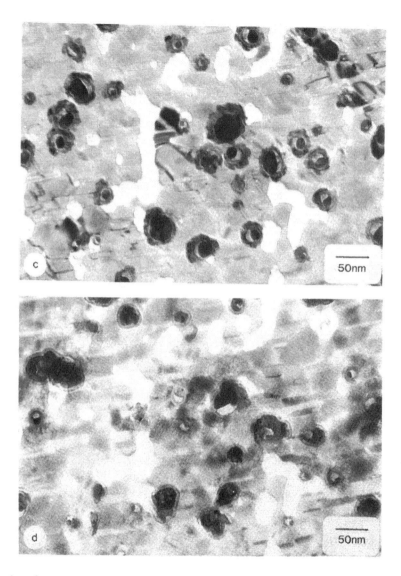

FIG. 5. (Continued) Sequence of changes in sample A of Fig. 1 on subsequent alternate heating in oxygen and hydrogen at 700°C. The region shown here is the same as that of (c) 1 h, H_2, (d) 4 h, H_2.

substrate (P). There was no change in the electron diffraction pattern. The sample was then heated in hydrogen for 1 h at the same temperature (Fig. 5c). A number of large particles with a core-and-ring structure now appeared on the substrate, indicating that during the previous heating in oxygen, the particles had indeed spread out on the substrate or had diffused into it and were not lost. The electron diffraction pattern indicated the presence of NiO. After 3 more hours of heating in hydrogen (Fig. 5d), some particles extended, coalesced, and/or rearranged, while some other particles disappeared completely probably due to extension. The electron diffraction pattern indicates that during the above heating, NiO was gradually reduced to Ni. Sample B, which was previously heated in hydrogen for 115 h at 700°C, was also heated alternately in oxygen and hydrogen at 700°C (Fig. 6). On heating in oxygen for 1 h, the particles extended considerably and formed thick patches with irregular contours (Fig. 6b) and the electron diffraction pattern indicated the presence of NiO. It is worth noting that the extension was perhaps more pronounced with sample A than with sample B, but it had to be inferred from the "loss of material" from the original crystallites as they extended

Effect of the Strong Metal-Support Interactions on Nickel/Titania Catalysts

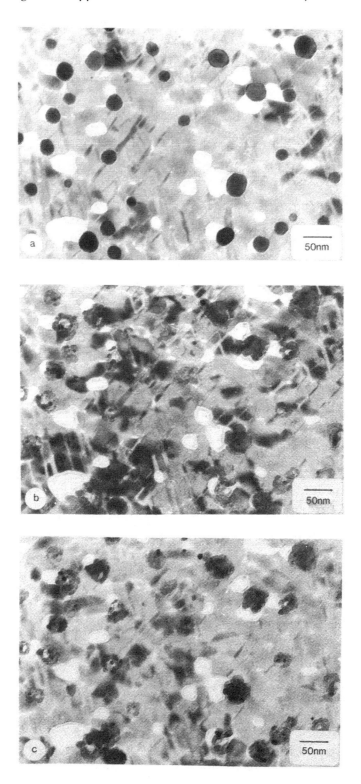

FIG. 6. Sequence of changes in sample B of Fig. 2 on subsequent alternate heating in oxygen and hydrogen at 700°C. The region shown here is the same as that of Fig. 2, after (a) 115 h, H_2, (b) 1 h O_2, (c) 3 h, O_2.
(Continued)

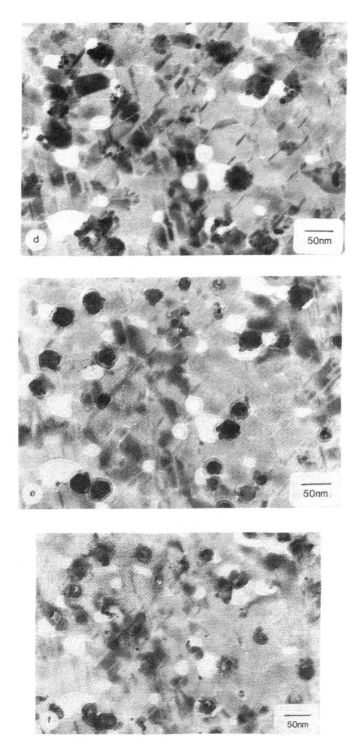

FIG. 6. (Continued) Sequence of changes in sample B of Fig. 2 on subsequent alternate heating in oxygen and hydrogen at 700°C. The region shown here is the same as that of Fig. 2, after (d) 6 h, O_2, (e) 1 h, H_2, (f) 4 h, H_2.

Effect of the Strong Metal-Support Interactions on Nickel/Titania Catalysts

and left behind very small remnant particles following heating in oxygen. On additional heating of sample B in oxygen, most of the particles split to form a few smaller particles (Figs. 6c and d). In the diffraction patterns, NiO rings and rings corresponding to probably some nonstoichiometric compound of NiO and titania (as in the case of sample A) were detected. However, NiO rings were more intense. Subsequent heating in hydrogen for 1 h at the same temperature brought about the contraction of the patches (Fig. 6e). The electron diffraction results indicated the presence of mostly NiO in the sample. On a further 3 h heating in hydrogen, the oxide was reduced to Ni and the particles were observed to have extended again, as a result of the concomitant reduction of titania and its subsequent strong interaction with the particles. Similar results of extension in oxygen, followed by contraction in hydrogen and again extension during additional heating in hydrogen, were observed also with another sample (whose micrographs are, however, not included here) when heated alternately in hydrogen and oxygen successively at 400, 500, 600, and 700°C.

DISCUSSION

The various phenomena, such as spreading, contraction, shape changes, which were observed during heating of model catalysts of nickel supported on titania in H_2 and O_2 environments were presented in the previous section. In this section, we employ the concept of wetting to explain the observed results. The wettability of a substrate by a crystallite is determined by the equilibrium of the interfacial free energies as expressed by Young's equation

$$\gamma_{sg} - \gamma_{cs} = \gamma_{cg} \cos\theta, \tag{1}$$

where γ_{sg} and γ_{cg} are the surface free energies per unit area of the substrate and crystallite, respectively, γ_{cs} is the interfacial free energy per unit area between crystallite and substrate, and θ is the equilibrium wetting angle. However, the interfacial free energy between crystallite and substrate is given by

$$\gamma_{cs} = \gamma_{cg} + \gamma_{sg} - \left(U_{int} - U_{str}\right) \tag{2}$$

$$\equiv \gamma_{cg} + \gamma_{sg} - U_{cs},$$

where U_{int} is the interaction energy per unit area between crystallite and substrate, and U_{str} is the strain energy per unit area due to the mismatch of the two lattices. When a chemical interaction takes place at the interface between crystallite and substrate, U_{cs} becomes very large and γ_{cs} thus decreases. This favors the extension of the crystallite over the substrate. This can happen in a reducing atmosphere, because of the strong interactions between TiO_{2-x} beneath the crystallite and metal, as well as in an oxygen atmosphere, because of the formation of a chemical compound between the oxidized metal and substrate. Increased values of γ_{sg} as well as smaller values of γ_{cg} also favor the extension of the crystallites.

The reduced form, TiO_{2-x}, as obtained in a hydrogen atmosphere, being nonstoichiometric, provides a higher value of γ_{sg} than the nonreduced form. In addition, the oxidized crystallite has a lower γ_{cg} than the metal and, in a reducing atmosphere, the migration of TiO_{2-x} moieties over the surface of the crystallites decreases the value of γ_{cg}.

SHAPE CHANGES IN ADDITIONALLY PURIFIED HYDROGEN

It is well known that the Group VIII metals supported on reducible metal oxides such as TiO_2, Nb_2O_5, V_2O_3, and Ta_2O_5 exhibit strong metal-support interactions resulting in the low-temperature suppression of hydrogen and carbon monoxide chemisorption after high-temperature reduction (ca. 500°C) (19-21). The migration of a monolayer or submonolayer of TiO_{2-x} over the surface of the crystallite because of its "strong interactions" with the metal appears to be responsible for this effect.

Interfacial reactions between nickel and titania have been observed to take place, though at 1500°C (22). It is likely that the high dispersion of metal in our system allows the reaction between nickel and titania to occur at much lower temperatures. Indeed, in the electron diffraction patterns a ring whose d value was close to the major d value of Ni_3Ti was observed on heating Ni/TiO_2 specimens in hydrogen at 700°C, suggesting the presence of a compound between Ni and Ti^{x+}. The presence of strong interactions at the crystallite-substrate interface decreases γ_{cs}. The substrate underneath and near the particles is most likely a reduced form of TiO_2, even though such a compound was not detected in our electron diffraction patterns. The reduction of TiO_2 to such a lower, nonstoichiometric oxide (TiO_{2-x}) leads to an increase in γ_{sg}, since a nonstoichiometric compound has a higher γ_{sg} than a stoichiometric compound. In addition, the possible migration of TiO_{2-x} moieties over the surface of the crystallite, as a monolayer or submonolayer, decreases the value of γ_{cg}, again because of their strong interactions with the metal. Such decreases in γ_{cs} and γ_{cg} as well as an increase in γ_{sg} will lead to an extension of the particles as can be seen from Eq. (1). Furthermore, the gradual increase in the intensity of the ring corresponding to the compound with a d value close to that of Ni_3Ti and the gradual decrease in the intensity of the rings corresponding to Ni suggest that there is a material transfer between substrate and particle. The observation of a cavity beneath the particle in the micrographs suggests that the direction of this transfer is most likely from the substrate to the particle. The cavity formation is discussed later in more detail. The migration of the TiO_{2-x}, layer away from the surface of the substrate beneath as well as near the particle leads to a contact between the particle and the unreduced TiO_2. Consequently, the particle contracts to a higher wetting angle both because TiO_2 has a lower surface free energy than a nonstoichiometric oxide (TiO_{2-x}) and because the interaction between the particle and the unreduced TiO_2 is weaker and hence γ_{cs} is larger. When the interactions between the particle and substrate are stronger over a short distance inward from the leading edge, this portion of the particle is strongly held to the substrate. Therefore, when the particle contracts, for reasons explained above, a part of the particle is detached from the main body most probably along the periphery of the cavity, but not all around the particle due to the heterogeneity of the substrate. Thus, a gap is formed between the main body and the detached portion. However, for some particles, the de-tached portion is not detected because it is too thin to be observed. In the micrographs, the gap appears to be brighter than the other regions of the substrate. This indicates that the substrate in the gap and probably underneath the particle is most likely thinner than in the other regions of the substrate, suggesting that there is a material transfer away from the substrate as already mentioned before.

On subsequent heating in H_2, the exposed TiO_2 is further reduced to a lower, nonstoichiometric oxide and extension of the particle will follow as a result of the accompanying reaction between the nickel particle and the reduced TiO_2. The migration of TiO_{2-x}, away from the cavity exposes another layer of the unreduced TiO_2 and, as explained above, leads to the contraction of the particle. Such alternate changes in extension and contraction will be observed until the rate of reaction between nickel and the reduced substrate and the migration of TiO_{2-x} onto or into the particle are considerably limited kinetically.

The alternate changes of the particles between a circular and a torus shape observed at 500°C in H_2 (Fig. 3) are also most likely a result of contractions and extensions. Extension leads to a torus shape, while contraction leads to a circular shape.

BEHAVIOR OF PARTICLES DURING ALTERNATE HEATING IN O_2 AND H_2

As shown in Figs. 5 and 6, alternate heating in O_2 and H_2 brought about dramatic changes. Baker *et al.* reported that platinum particles supported on titania exhibited an extended, pillbox morphology on heating in hydrogen at 550°C or higher, whereas the particles contracted to a globular morphology on heating in O_2 at 600°C (13). Similar results were obtained by Tatarchuk and Dumesic in the case of iron particles supported on titania (15). In contrast, in the present experiments at 700°C,

Effect of the Strong Metal-Support Interactions on Nickel/Titania Catalysts **309**

Ni particles supported on TiO_2 were observed to be in a more contracted morphology in a H_2 atmosphere and considerably extended in an O_2 atmosphere. Following heating in O_2 at 700°C, the drastic decrease in the number of particles and/or in their sizes (sample A, Fig. 5) suggests that either films spread out from around the particles onto the substrate or material diffused into the substrate. Diffraction patterns indicate that the particles were oxidized to NiO and then gradually changed to a nonstoichiometric compound of NiO and TiO_2. Such strong interactions between the particles and the support, leading to the formation of a compound, will lead to the diffusion of NiO into the substrate to enable compound formation in the bulk, to further decrease the free energy of the system. It will also decrease the interfacial free energy between the particle and substrate. In addition, γ_{cg} for the NiO is lower than that for the metal. The smaller values of γ_{cs} and γ_{cg} lead to the extension of the particles. When the driving force for spreading is sufficiently large, the particles, especially the small ones (N in Fig. 5), are likely to spread out as a thin film undetectable in the micrographs.

For sample B which was previously heated in H_2 for a longer time (115 h), the extension and diffusion of the particles during their heating in O_2 occurred to a smaller extent than for sample A, which was prereduced for shorter time. This may be due to the fact that the substrate in the sample prereduced for a long time (sample B) was reduced to a greater extent than the substrate in the sample prereduced for a short time (Sample A). Consequently, sample B is expected to form TiO_{2-x} in greater amounts than sample A. Therefore, on heating in O_2 sample B may have residual TiO_{2-x}, while sample A may be oxidized to TiO_2 completely. The small remnant particles which are shown in Figs. 5b and 6d will have a globular morphology since once the particle has interacted with the substrate and formed a compound at the interface (and most of the particle has extended out as a thin detectable or undetectable film), the remnant particle will be in contact not with TiO_2 but with the compound of NiO and TiO_2. Since U_{cs} between the particle (NiO) and the latter compound is no longer as large as that between NiO and TiO_2, the remnant particle may have a larger wetting angle and may therefore have a globular morphology. In a reducing atmosphere, the interaction compound is probably reduced, and NiO diffuses out to the surface and forms new particles on the substrate surface. After a sufficiently short time of heating in a reducing atmosphere, the latter is present as mostly TiO_2. Subsequently the particles are gradually reduced to Ni, and TiO_2 is reduced to TiO_{2-x}. The interactions between the Ni particles and the reduced TiO_2 and the higher γ_{sg} of TiO_{2-x} lead to the extension of the particles. However, the extension of the particles in a hydrogen atmosphere occurs to a smaller extent than that in an oxygen atmosphere. This suggests that the interaction between the particle and the substrate is greater in an O_2 atmosphere than in a H_2 atmosphere. While γ_{sg} is larger in the latter case than in the former, γ_{cs} is probably much smaller in an oxygen environment than in a reducing one. In addition, γ_{cg} could also be smaller in an oxygen atmosphere than in hydrogen. As a result, the extension in the oxygen atmosphere can be greater.

FORMATION OF CAVITIES

When Ni/TiO_2 specimens are heated in H_2 at 700°C, cavities are formed in the substrate via removal of a part of substrate material as a result of the formation of an intermetallic compound. The material which is removed is transferred most likely onto and/or into the metal particles. It may also migrate onto the substrate surface. Initially the particles were present as Ni. However, on further heating at 700°C in H_2, the electron diffraction patterns indicate that a compound whose d value is close to that of Ni_3Ti is formed, suggesting formation of a compound between the particles and the reduced Ti species. It was inferred from the electron diffraction patterns that on further heating, the above mentioned compound increased in amount gradually while Ni decreased. Therefore, it is likely that TiO_{2-x} diffuses into and/or onto the particles, and subsequently an intermetallic compound is formed in the particles. This leads to a thinning of the substrate beneath the particle and to the creation of a cavity. The migration of reduced titania (TiO_{2-x}) moieties onto the surface of metal particles supported on TiO_2 during reduction in hydrogen has been reported before (*16, 23*). Ko and Gorte (*24*), in addition, reported that the reduced titania probably diffused into the metal particles. The latter is probably true in the present case also as inferred from the compound between Ni and Ti species detected in the electron diffraction patterns. Recently, Dumesic *et al.* also reported the formation of cavities in a Ni/TiO_2 system (*25*).

A part of the cavity formed underneath the particle is exposed on one side of the particle in the electron micrographs, as a result of a random displacement of the particle away from the cavity (B in Fig. 2). When the particle actually migrates out of and away from the cavity, the entire cavity can be seen, as marked (C) in Fig. 2.

CONCLUSION

Various phenomena have been observed to occur during heating of Ni/TiO_2 model catalysts both in additionally purified hydrogen and in oxygen atmospheres. Alternate changes in crystallite shape were observed on heating in additionally purified hydrogen at 500 and 700°C. These alternations are suggested to be associated with the extension and contraction of the particles. Extension of the particles is a result of the strong interactions between nickel and reduced TiO_{2-x}, which decrease the values of γ_{cs} and γ_{cg}, and of the increased value of γ_{sg} for a reduced TiO_2 (as compared to a nonreduced one). The above-mentioned reduced species migrates into and/or onto the particles and probably also onto the substrate surface, from the substrate underneath the particles. Such a removal of the reduced TiO_{2-x} layer from underneath the particle exposes the unreduced TiO_2 and when the particle comes into contact with this TiO_2 layer, it contracts. Electron diffraction patterns indicate that a compound whose *d* value is close to the major *d* value of Ni_3Ti is formed during heating in H_2 at 700°C. Migration of the reduced substrate species away from the substrate surface underneath the particles leads to the formation of cavities in the substrate. Some specimens were heated in oxygen and hydrogen atmospheres alternately at 700°C. The results show that the particles considerably extend on, and/or diffuse into, the substrate due to the formation of a compound between NiO and TiO_2 in an oxidizing atmosphere. During subsequent heating in H_2, the material lost from the particles to the substrate during the previous heating in O_2 comes back onto the substrate and forms new particles. These particles are initially composed of NiO and then are reduced to Ni. On further heating in H_2, extension of the particles is observed as a result of the interactions between Ni and TiO_{2-x}, but the extension in this case occurs to a much smaller extent than in an oxygen atmosphere. These results seem to suggest that overall the interactions between the particle and substrate, in the case of Ni/TiO_2, are stronger in an oxygen atmosphere than in a hydrogen atmosphere.

The present results also show that severe sintering occurs at 700°C while sintering at temperatures of 500°C or less occurs to a much smaller extent.

It is emphasized that the extension of the crystallites over rutile in a hydrogen atmosphere is due to (1) the decrease of the interfacial free energy between support and crystallite, γ_{cs}, caused by the strong interactions between TiO_{2-x} and metal; (2) the decrease of the surface free energy between crystallite and atmosphere, γ_{cg}, caused by the migration of a monolayer or submonolayer of TiO_{2-x}, over the surface of the crystallite because of the strong interactions between the two; and finally (3) the increased surface free energy, γ_{sg}, of the reduced TiO_2 as compared to the nonreduced TiO_2. In contrast, the extension of the crystallites over rutile in an oxygen atmosphere is a result of the lower surface free energy of the oxide as compared to that of the metal and of the lower interfacial free energy between crystallite and substrate caused, particularly, by the reaction between the oxidized crystallite and substrate. While the surface free energy, γ_{sg}, of the present substrate is higher in a reducing atmosphere than in an oxidizing one, γ_{cs} is probably much lower in the latter atmosphere, thus ensuring a greater extension during heating in oxygen. In addition, γ_{cg} could also be lower in the oxygen atmosphere.

Effect of the Strong Metal-Support Interactions on Nickel/Titania Catalysts

ACKNOWLEDGMENT

We are indebted to Dr. I. Sushumna for his most useful comments and suggestions.

REFERENCES

1. Ruckenstein, E., and Chu, Y. F., *J. Catal.* **59,** 109 (1979) (Section 3.1 of this volume).
2. Ruckenstein, E., and Chen, J. J., *J. Catal.* **70,** 233 (1981).
3. Heinemann, K., Osaka, T., Poppa, H., and Ava-los-Borja, M., *J. Catal.* **83,** 61 (1983).
4. Wang, T., and Schmidt, L. D., *J. Catal.* **66,** 301 (1980).
5. Derouane, E. G., Chludzinski, J. J., and Baker, R. T. K., *J. Catal.* **85,** 187 (1984).
6. Ruckenstein, E., and Lee, S. H., *J. Catal.* **86,** 457 (1984).
7. Ruckenstein, E., and Hu, X. D., *Langmuir* **1,** 756 (1985).
8. Nakayama, T., Arai, M., and Nishiyama. Y., *J. Catal.* **79,** 497 (1983).
9. Sushumna, I., and Ruckenstein, E., *J. Catal.* **90,** 241 (1984).
10. Sushumna, I., and Ruckenstein, E., *J. Catal.* **94,** 239 (1985) (Section 3.4 of this volume).
11. Ruckenstein, E., and Sushumna, I., *J. Catal.* **97,** 1 (1986).
12. Baker, R. T. K., Prestridge, E. G., and Garten, R. L., *J. Catal.* **59,** 293 (1979).
13. Baker, R. T. K., *J. Catal.* **63,** 523 (1980).
14. Tatarchuk, B. J., and Dumesic, J. A., *J. Catal.* **70,** 308 (1981).
15. Tatarchuk, B. J., and Dumesic, J. A., *J. Catal.* **70,** 335 (1981).
16. Simoens, A. J., Baker, R. T. K., Dwyer, D. J., Lund, C. R. F., and Madon, R. J., *J. Catal.* **86,** 359 (1984).
17. Singh, A. K., Pande, N. K,, and Bell, A. T., *J. Catal.* **94,** 422 (1985).
18. Raupp, G. B., and Dumesic, J. A., *J. Catal.* **97,** 85 (1986).
19. Tauster, S. J., Fung, S. C., and Garten, R. L., *J. Amer. Chem. Soc.* **100,** 170 (1978).
20. Tauster, S. J., and Fung, S. C., *J. Catal.* **55,** 29 (1978).
21. Tauster, S. J., *in* "Strong Metal-Support Interactions," (R. T. K. Baker, S. J. Tauster. and J. A. Dumesic, Eds.), ACS Symp. Ser. 298, p. 1. Amer. Chem. Soc., Washington, DC, 1986.
22. Humenik, M., Jr., and Kingery, W. D., *J. Amer. Ceram. Soc.* **37,** 18 (1954).
23. Jiang, X-Z, Hayden, T. F., and Dumesic, J. A., *J. Catal.* **83,** 168 (1983).
24. Ko, C. S., and Gorte, R. J., *J. Catal.* **90,** 59 (1984).
25. Dumesic, J. A., Stevenson, S. A., Sherwood, R. D,, and Baker, R. T. K., *J. Catal.* **99,** 79 (1986).

3.10 Simulation of the Behavior of Supported Metal Catalysts in Real Reaction Atmospheres by Means of Model Catalysts[*]

Sung H. Lee and Eli Ruckenstein

Department of Chemical Engineering, State University
of New York, Buffalo, New York 14260

Received January 10, 1987; revised April 13, 1987

INTRODUCTION

The loss of activity and/or selectivity of industrial catalysts in the course of reaction can be attributed to a wide variety of causes, which may be grouped into sintering, coking, poisoning, and phase transformation. However, in real situations these processes coexist and influence each other, and the deactivation of the catalyst is a result of their cooperative action. Therefore, an investigation of deactivation by the cooperative action of various processes should be undertaken; this is the scope of the present paper.

A meaningful case for such studies is the steam-reforming reaction, in which steam reacts with natural gas, primarily methane, to form H_2, CO, CO_2. While supported nickel is employed as the commercial catalyst (*1*), cobalt and iron can also be used, though they are less active. The high temperature (about 700–1000°C) and the presence of steam, methane, and hydrogen set severe conditions on the catalyst. As is well known, when these compounds are alone, the heating in hydrogen stimulates sintering, the heating in methane causes coking, while the heating in the presence of steam results in a phase transformation of alumina (*2*) as well as in the formation of a solid solution between the oxidized metal and support (*3*). However, when the above gases are together these processes influence one another. For instance, when carbon deposition constitutes the dominant process, steam and hydrogen are expected to have an influence on carbon formation, on the morphologies of carbon deposits, and on the rate of carbon growth. Indeed, carbon deposits and/or coke precursors, which are formed by the dissociation of CO or decomposition of hydrocarbons, are removed as CH_4 or CO by the reaction with H_2 or H_2O (*4–6*). The presence

---

[*] *Journal of Catalysis* 107, 23–81 (1987). Republished with permission.

Simulation of the Behavior of Supported Metal Catalysts in Real Reaction Atmospheres 313

of sufficient amounts of H_2 or H_2O minimizes the formation of condensed hydrocarbons and amorphous carbon or graphite (7). Therefore, coking under such conditions differs from that in pure methane or carbon monoxide. On the other hand, carbon formation may enhance or deter sintering. Indeed, sintering is stimulated by carbon gasification, since the resulting gases enhance the mobility of the crystallites on the substrate; in contrast the deposition of carbon around the particles impedes their migration.

In the very extensive literature dealing with deactivation in the steam-reforming reaction, the investigators have studied only the individual processes leading to deactivation of the catalysts, such as coking, sintering, or poisoning. Coking has received the greatest attention and the results have been summarized in numerous reviews (5–10). Only a few investigators have, however, studied carbon formation in real reaction systems. Rostrup-Nielsen (5) reported that, during steam reforming of naphtha, the coking rate on the nickel surface at temperatures near 500°C depends on the steam-to-carbon ratio and other factors. No coking was observed to occur on the cobalt/alumina catalyst, which has a poor activity for the reaction, though in atmospheres without steam, coke was generated (from methane, carbon monoxide, or olefins) on cobalt as on nickel. The deactivation of nickel supported on alumina was explained as a result of the blockage of the pore mouth by the coke deposited mainly close to the external surface. This blockage impedes the adsorption of steam on the crystallites and therefore steam can no longer depress carbon formation. In contrast, the poor activity of cobalt for steam reforming was explained by the blockage of the pore mouth by the metal oxide which rapidly forms in steam, rather than by the blockage by carbon. Jackson *et al.* (11) reported different deactivation mechanisms for Ni catalysts in a real reaction atmosphere and in one without steam. He noted that the filamentous carbon, which is generated on the catalysts when the hydrocarbon is alone, is no longer present to any appreciable extent under real conditions.

Sintering is also considered to be an important cause for catalyst deactivation in the steam reforming reaction (12, 13). Williams *et al.* (14) suggested that the principal cause of sintering in steam reforming is the thermal instability of the alumina support, namely the transformation, due to the presence of steam, of the γ-alumina to the alpha phase, which triggers the sintering of the metal particles. These investigators observed a very sharp early decrease in the surface areas of both nickel and alumina with time in the presence of steam and hydrogen. As pointed out by Satterfield (12), the crystallite size is increased to about 1 μm, as soon as the catalyst is brought in contact with the steam-reforming mixture.

The formation of an inactive phase in the metal crystallites is also a problem of concern. Borowiecki *et al.* (15) suggested that the surface reaction as well as the deposition of carbon promote the formation of an inactive "oxidized" metal form at the boundary between metal and support.

The poisoning by the additional constituents of the hydrocarbon feed stock also contributes to the loss of catalyst activity (16). However, this problem is beyond the scope of the present paper.

There is a lack of comparative studies regarding the cooperative deactivation (due to sintering, coking, and phase transformation) of various catalysts in atmospheres similar to those encountered in steam-reforming reactions. In the present investigation, model catalysts of Ni/Al_2O_3, Co/Al_2O_3, and Fe/Al_2O_3 were heated in various chemical atmospheres, such as methane, carbon monoxide, steam plus methane, steam plus methane and hydrogen, and steam plus methane, hydrogen, and carbon monoxide. The composition of these mixtures simulates either the inlet or the outlet composition of the industrial primary reformer furnace. The effect of steam was already reported (17). The object of the present investigation is to examine the cooperative deactivation in simulated reaction atmospheres, and to identify the dominant event or events in various atmospheres and for various metals as well as the interrelationship between them.

EXPERIMENTAL

A. Preparation of the Specimen

Thin, electron transparent alumina films, about 30 nm thick, were prepared by anodization of high-purity aluminum foils (99.999% Alfa Products) and the dissolution of the remaining aluminum in mercury chloride solutions (18). The thin alumina films were then deposited on gold microscope grids and heated at 800°C for 40 h. Nickel (99%, Alfa Products), cobalt (99.8%, Alfa Products), and iron (99.998% Alfa Products) were then vacuum deposited on the alumina substrate to a thickness of 1.5 nm. The specimens were heated in flowing ultrahigh-purity H_2 (99.999% pure, but containing traces of O_2 (less than 1 ppm) and H_2O (less than 3 ppm); Linde Division, Union Carbide Co.) at 500°C for at least 5 h and then at 700°C for at least 5 additional hours in order to change the metal films to well-defined crystallites. In addition, the as-supplied hydrogen was further purified to eliminate the traces of O_2 and water by passing it through a Deoxo unit (Engelhard Industries) and then through a molecular sieve bed immersed in liquid nitrogen. This further-purified H_2 was always employed when heating in H_2 alone. In all the other cases, the as-supplied hydrogen was employed.

B. Heat Treatment

The heating of the above specimen was carried out at 1 atm and 700°C in a quartz tube located in a furnace. After the specimen was introduced in the tube, ultrahigh-purity helium (99.999% pure, Linde Division, Union Carbide Co.) was allowed to flow through the tube during the period in which the temperature was raised to 700°C. Helium was then replaced by the desired feed stream, such as methane (ultrahigh-purity, 99.97%, <10 ppm O_2 and <6 ppm moisture; Linde Division, Union Carbide Co.), carbon monoxide (ultrahigh-purity, 99.8%; Matheson Gas Products), and three sets of gas mixtures containing at least two components among hydrogen, steam, methane, carbon monoxide, and helium. The compositions of the gas mixtures are listed in Table 1. The amounts of the individual components in the gas mixtures were adjusted with individual flow meters and the total flow rate without steam was maintained at about 200 ml/min. Steam was introduced into the gas mixtures by means of a water saturator maintained at a temperature at which the water vapor pressure can provide the desired concentration of steam in the gas mixture. Before being taken out for observation, the samples were cooled down slowly to room temperature, in a helium atmosphere. The observations have been made in the same region of the sample after each heat treatment, by means of a JEOL 100U TEM.

TABLE 1
A List of the Composition of Gas Mixtures

Gas mixture	Volume (%)				
	CH$_4$	Steam	H$_2$	CO	He
Mixture 1	50	50	0	0	0
Mixture 2	5	43	40	0	12
Mixture 3	5	43	40	12	0

Simulation of the Behavior of Supported Metal Catalysts in Real Reaction Atmospheres 315

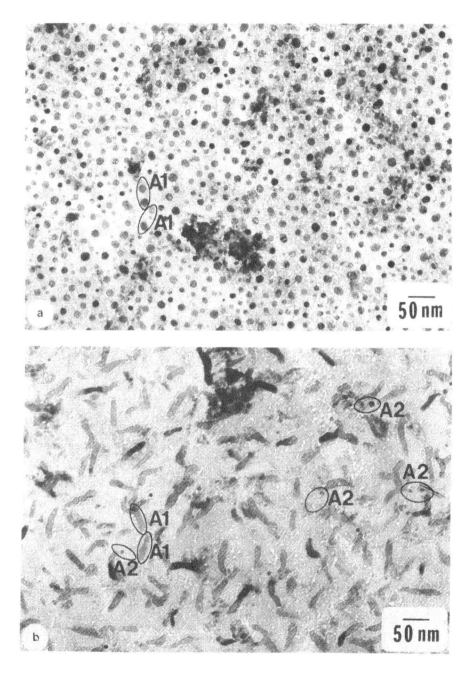

FIG. 1. Sequence of changes on heating a Ni/Al$_2$O$_3$ specimen in methane at 700°C after pretreatment in hydrogen. When the heating in a given atmosphere has taken place in several steps, the total time is indicated at each step. (a) 10 h (5 h at 500°C and 5 h at 700°C) H$_2$; (b) 1.5 h CH$_4$. *(Continued)*

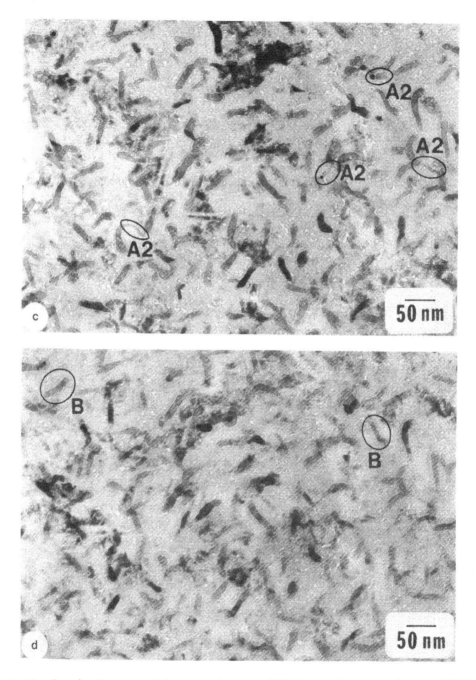

FIG. 1. (Continued) Sequence of changes on heating a Ni/Al$_2$O$_3$ specimen in methane at 700°C after pretreatment in hydrogen. When the heating in a given atmosphere has taken place in several steps, the total time is indicated at each step. (c) 3 h CH$_4$ (the specimen was heated for 1.5 more hours in CH$_4$ after the heat treatment (c)); (d) 2 h O$_2$ *(Continued)*

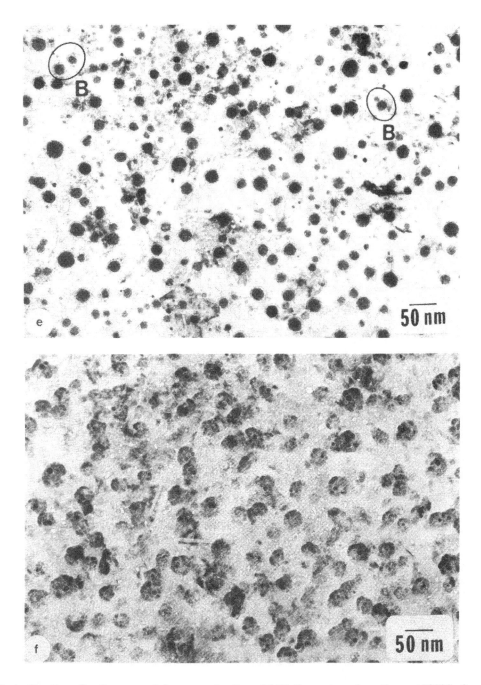

FIG. 1. (Continued) Sequence of changes on heating a Ni/Al$_2$O$_3$ specimen in methane at 700°C after pretreatment in hydrogen. When the heating in a given atmosphere has taken place in several steps, the total time is indicated at each step. (e) 3 h H$_2$; (f) 1.5 h CH$_4$.

RESULTS

1. HEATING IN METHANE

A. Ni/Al$_2$O$_3$

a. *Heating in CH$_4$ after pretreatment in H$_2$.* On heating in hydrogen for 10 h (5 h at 500°C and subsequently 5 h at 700°C), globular particles were formed (Fig. 1a). The electron diffraction pattern indicated the presence of Ni. Subsequently, the specimen was heated in CH$_4$. During heating in CH$_4$ for 1.5 h, filamentous (elongated) structures replaced the globular particles (Fig. 1b). Following elongation, most of the particles coalesced with their neighbors (A1 in Fig. 1b). At the tip of some filaments, dark particles were detected (A2 in Fig. 1b). On further heating for 1.5 additional hours, these darker particles moved over the substrate and decreased in size or disappeared, generating filaments along their trajectories (A2 in Fig. 1c). It is likely that the lost material from the particles was left behind in the filaments. After particles were heated for a total of 4.5 h, no significant change was observed. The electron diffraction patterns indicated the presence of NiO, probably because of the presence of moisture and oxygen in methane. Hence the major component of the filament was, most probably, NiO. It is, however, likely that carbon deposition and its penetration into the particles is in some way responsible for the filamentous structure. In some of the specimens, the particles extended, maintaining however the globular shape, and no filament formation was observed during heating in CH$_4$. In such cases, it is not clear whether carbon was deposited around the particles, because of the difficulty in discerning carbon deposition from extension of the particle. The specimen of Fig. 1c was subsequently heated in oxygen for 2 h. The filaments which contained mostly NiO remained almost unaffected (Fig. 1d), with only marginal loss of material, probably due to carbon gasification. There was no change in the diffraction pattern. The specimen was subsequently heated in H$_2$ for 3 h (Fig. 1e). During this heat treatment, the filaments transformed to globular shapes, as a result of contraction. In some regions (B in Figs. 1d and 1e), two or more particles were formed from a filament. The diffraction pattern indicated that NiO was reduced to Ni. The specimen was again heated in CH$_4$. After heating for 1.5 h, the particles extended and split into several interconnected subunits (Fig. 1f). Some small particles disappeared, most probably because of spreading out as thin undetectable films. One may note that most of the extended particles have a small darker particle supported on a very extended patch. On further heating in CH$_4$, for up to 15.5 h, the particles extended a little further and the darker remnant particles decreased in size or disappeared. However, filaments were not generated, indicating that the already coked specimens do not act in the same way as the fresh specimens do. However, on another sample, whose micrographs are not included in this paper, filaments which contained mostly NiO were generated in CH$_4$ even after heating the coked specimen in O$_2$ and subsequently in H$_2$, but to a much smaller extent than the first time.

It is inevitable to expose the specimen to air during repeated heating and subsequent observation by TEM. In order to check whether the exposure to air affects the morphology of the particles, a sample was continuously heated in CH$_4$ for 5 h at 700°C. The result is shown in Fig. 2 and is quite similar to that obtained by heating a specimen in several steps for almost the same total heating time (Fig. 1c). These results point out that the repeated exposure to air does not affect in an important way the morphology of the particles.

Simulation of the Behavior of Supported Metal Catalysts in Real Reaction Atmospheres

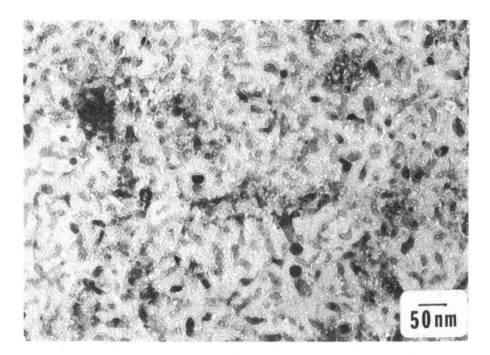

FIG. 2. Ni/Al$_2$O$_3$ specimen after heating in CH$_4$ for 5 h continuously at 700°C following heating in H$_2$ for 10 h (5 h at 500°C and 5 h at 700°C).

b. *Heating in CH$_4$ after pretreatment in steam.* Since particles were not generated on heating in steam at 700°C for up to 15 h, the specimen was first heated in H$_2$ for 10 h (5 h at 500°C and 5 h at 700°C) and subsequently in steam also at 700°C. Figure 3a shows the result of heating in steam for 5 h. Compared to the pretreatment in H$_2$ (Fig. 1a), the particles appear to be thinner. On heating the specimen of Fig. 3a in CH$_4$ for 4 h, dark patches, most likely carbon deposits, but no filaments appeared (C in Fig. 3b).

B. Co/Al$_2$O$_3$

After heating in H$_2$ for 10 h the particles acquired globular shapes (Fig. 4a). Electron diffraction indicated the presence of α-Co. However, some rings of Co$_2$O$_3$ and Co$_3$O$_4$ were also identified. On further heating in CH$_4$ for 30 min, most of the large particles increased in size (D in Fig. 4b), either because of their extension, or, most likely, because of carbon deposition around and on the extended particles. In some regions (E in Fig. 4b), some neighboring particles coalesced. In region F, two neighboring particles deformed and became interconnected, possibly by carbonaceous films. After heating for up to 15 h, in several steps, some of the particles extended marginally and some became so thin that one could see the structure of the substrate beneath them (Fig. 4c). Some particles decreased in size, probably because of the spreading out from them of thin films undetectable by electron microscopy. On subsequent heating in H$_2$ for 2 h at 700°C (Fig. 4d), numerous new small particles appeared in regions where no particles were observed before, indicating that indeed during heating in CH$_4$ some particles and part of other particles spread out over the substrate as undetectable films. These undetectable films ruptured during heating in H$_2$ and contracted to form small particles as suggested and explained in Ref. (*19*). A specimen which was heated continuously in CH$_4$ for 15 h (Fig. 5) provided results similar to those observed on heating in several steps for the same length of time (Fig. 4c).

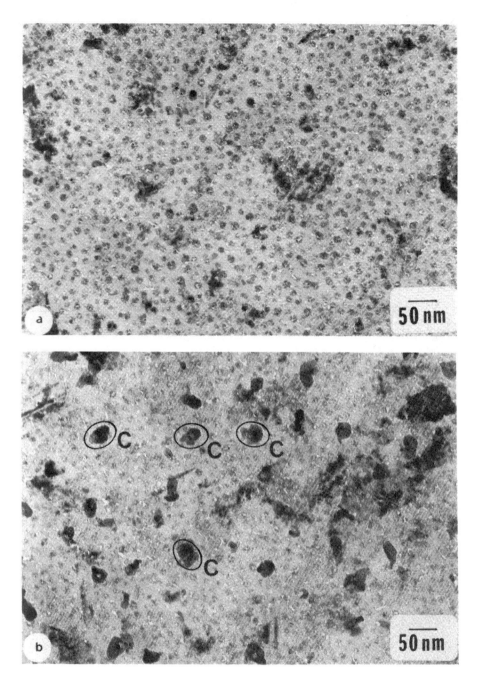

FIG. 3. Sequence of changes on heating a Ni/Al$_2$O$_3$ specimen in methane at 700°C after pretreatment in steam. (a) 10 h (5 h at 500°C and 5 h at 700°C) in H$_2$ and subsequently 5 h in steam at 700°C; (b) 4 h CH$_4$.

Simulation of the Behavior of Supported Metal Catalysts in Real Reaction Atmospheres 321

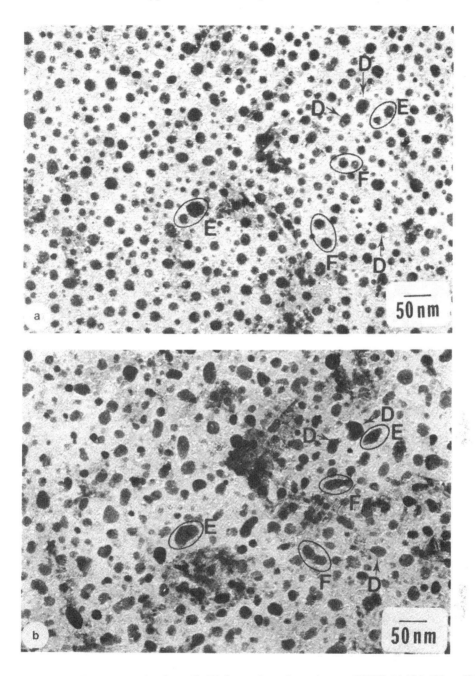

FIG. 4. Sequence of changes on heating a Co/Al$_2$O$_3$ specimen in methane at 700°C. (a) 10 h (5 h at 500°C and 5 h at 700°C) H$_2$; (b) 0.5 h CH$_4$. (*Continued*)

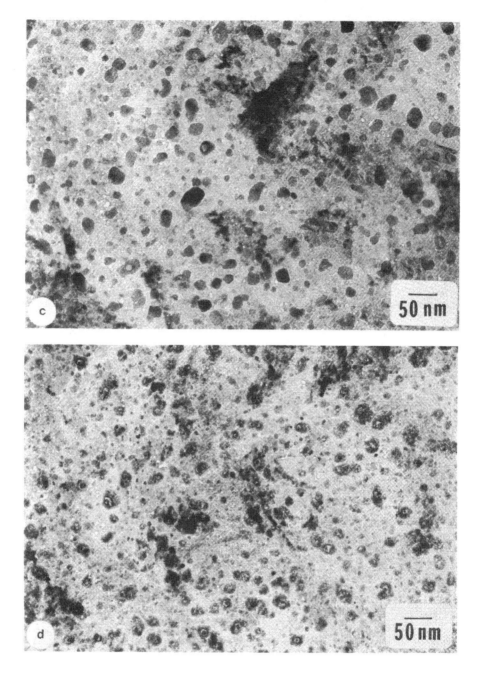

FIG. 4. (Continued) Sequence of changes on heating a Co/Al$_2$O$_3$ specimen in methane at 700°C. (c) 15 h CH$_4$; (d) 2 h H$_2$.

Simulation of the Behavior of Supported Metal Catalysts in Real Reaction Atmospheres 323

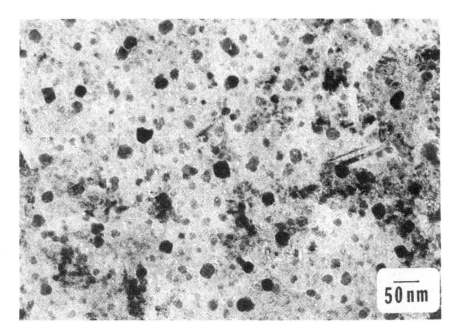

FIG. 5. Co/Al$_2$O$_3$ specimen after heating in CH$_4$ for 15 h continuously at 700°C following heating in H$_2$ for 10 h (5 h at 500°C and 5 h at 700°C).

C. Fe/Al$_2$O$_3$

Figures 6a–6e show the micrographs of a specimen which had a 7.5-Å-thick initial loading. The initial particles were generated on heating in H$_2$ for 5 h at 500°C and an additional 5 h at 700°C (Fig. 6a). α-Fe and γ-Fe$_2$O$_3$ (or Fe$_3$O$_4$) were identified in the electron diffraction pattern. (Since γ-Fe$_2$O$_3$ and Fe$_3$O$_4$ have the same lattice constant, one cannot differentiate between them in the electron diffraction pattern.) Upon heating in CH$_4$ for 20 min, some particles extended (Fig. 6b), and other particles disappeared, leaving sometimes small remnant particles, most likely as a result of their spreading out as thin films undetectable by electron microscopy. After a total of 2 h of heating in CH$_4$, most of the particles increased in size (Fig. 6c). They continued to grow, acquiring more elongated or disordered shapes, on additional heating in several steps, for up to 15 h (Fig. 6d). The growth of the particles was, most likely, a result of carbon deposition around and on the particles. The specimen was subsequently heated in FF. After 1 h of heating in H$_2$ (Fig. 6e), most of the particles decreased in size, probably because of the gasification of carbon. Figures 7a–7c provide information about a 15-Å loading specimen on heating in CH$_4$. Its behavior was quite similar to that of the 7.5-Å loading. Figure 7a shows the initial particles which were formed after heating in H$_2$ for 5 h at 500°C and 5 additional hours at 700°C. α-Fe and γ-Fe$_2$O$_3$ (or Fe$_3$O$_4$) were detected in the electron diffraction pattern. Upon heating in CH$_4$ for 1.5 h, large patches of irregular shape appeared (Fig. 7b). Coke was probably deposited around and on the particles, and/or carbon films interconnected neighboring particles, thus generating large patches. One may also note that some particles became elongated. In the electron diffraction pattern, γ-Fe$_2$O$_3$ (or Fe$_3$O$_4$) was identified, indicating that most of the particles were oxidized probably because of the impurities, such as O$_2$ and moisture, present in CH$_4$. The heating in CH$_4$ for a total of 14.5 h (Fig. 7c) led to the growth of carbonaceous films around the particle and also to the formation of filaments, most likely composed of γ-Fe$_2$O$_3$ (Fe$_3$O$_4$) and carbon deposits. It is worth noting that the filaments are straight and do not have a thick particle at their tip; in contrast, the filaments encountered in the case of Ni/Al$_2$O$_3$ were not straight and did have a thick particle at their leading tip. In Fig. 7c, the particles are probably entirely covered by carbonaceous films and therefore their activity is expected to be extremely reduced. Indeed, there was very little change on further heating for up to 19 h.

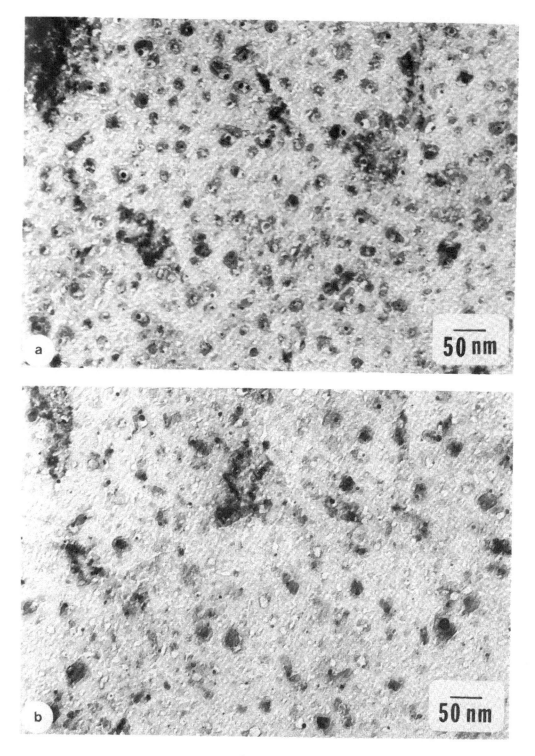

FIG. 6. Sequence of changes on heating a 7.5-Å loading Fe/Al$_2$O$_3$ specimen in methane at 700°C. (a) 10 h (5 h at 500°C and 5 h at 700°C) H$_2$; (b) 20 min CH$_4$. *(Continued)*

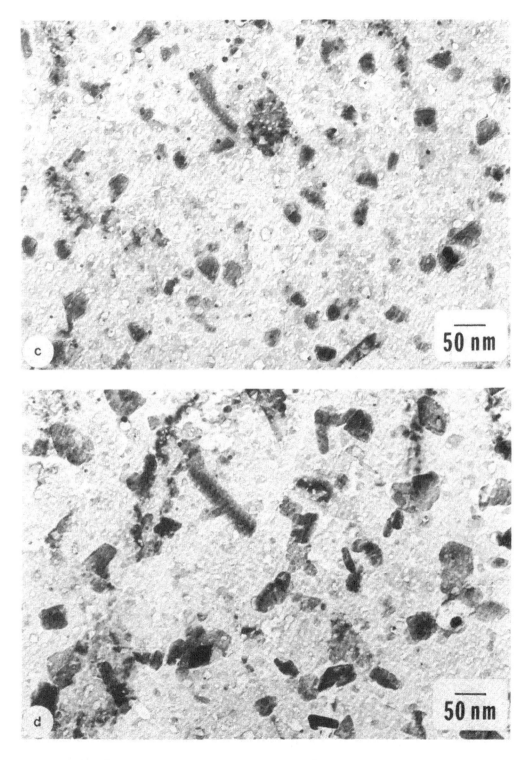

FIG. 6. (Continued) Sequence of changes on heating a 7.5-Å loading Fe/Al$_2$O$_3$ specimen in methane at 700°C. (c) 2 h CH$_4$; (d) 15 h CH$_4$. (*Continued*)

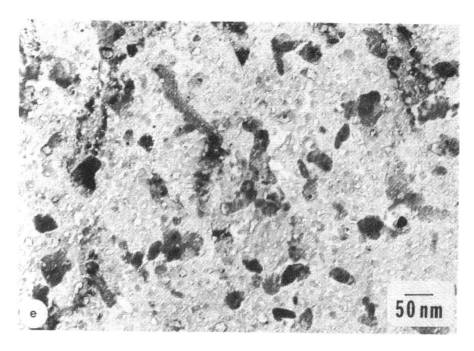

FIG. 6. (Continued) Sequence of changes on heating a 7.5-Å loading Fe/Al$_2$O$_3$ specimen in methane at 700°C. (e) 1 h H$_2$.

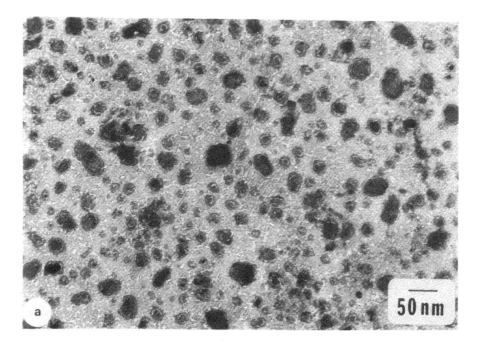

FIG. 7. Sequence of changes on heating a 15-Å loading Fe/Al$_2$O$_3$ specimen in methane at 700°C. (a) 10 h (5 h at 500°C and 5 h at 700°C) H$_2$. *(Continued)*

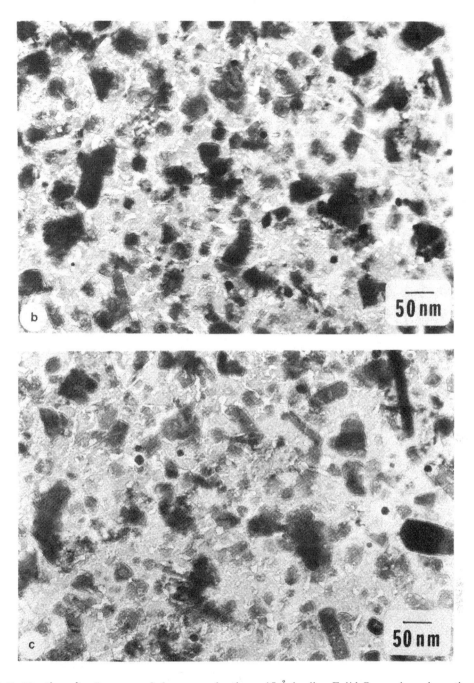

FIG. 7. (Continued) Sequence of changes on heating a 15-Å loading Fe/Al$_2$O$_3$ specimen in methane at 700°C. (b) 1.5 h CH$_4$; (c) 14.5 h CH$_4$.

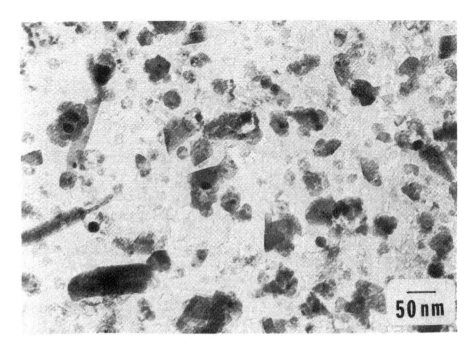

FIG. 8. Fe/Al$_2$O$_3$ specimen after heating in CH$_4$ for 20 h continuously at 700°C, following heating in H$_2$ for 10 h (5 h at 500°C and 5 h at 700°C).

Another 15-Å loading specimen was heated continuously in CH$_4$ for 20 h (Fig. 8). The obtained micrograph was very similar to that obtained after heating in several steps for the same length of time, indicating that repeated exposure to air had no significant effect on the morphology of the particles.

The specimen whose micrographs are given in Figs. 7 was further heated in various atmospheres. The micrographs given in Figs. 9a–9e represent a different area of the specimen. Figure 9a represents that area, after heating in CH$_4$ for 19 h, and shows elongated and irregular particles which, very likely, contain carbon in addition to crystallite material. Electron diffraction indicated the presence of γ-Fe$_2$O$_3$ (Fe$_3$O$_4$). On subsequent heating in H$_2$ for 6 h, the carbon was gasified, and the filaments recontracted to more globular shapes, leaving traces on the substrate (Fig. 9b). The contracted particles were in some cases located at one of the ends of the previous filaments (G) and in other cases in their middle (H). Some of the irregular particles contracted to form thick small particles having films beneath them (I). α-Fe was detected by electron diffraction. The bottom films were probably ungasified carbon and/or partially reduced or unreduced γ-Fe$_2$O$_3$ (Fe$_3$O$_4$). In addition, many small new particles appeared on the substrate, most probably because the undetectable γ-Fe$_2$O$_3$ (Fe$_3$O$_4$) films which spread out during heating in CH$_4$ ruptured and contracted to form particles after their reduction to α-Fe. In order to verify whether the bottom films and the remnant traces remaining after the filament contraction were carbonaceous deposits, the specimen was heated in steam for 2 h (Fig. 9c). As shown in the micrograph, the films and the remnant traces did not disappear, indicating that they were composed mostly of iron or iron oxides. One may note that the thick particles extended on the substrate. γ-Fe$_2$O$_3$ (Fe$_3$O$_4$) was detected in the electron diffraction pattern. On subsequent heating in H$_2$ for 4 h, the particles contracted (Fig. 9d). Further, the specimen was once again heated in CH$_4$. On heating for 1 h (Fig. 9e) the particles extended; it was however little change on additional heating for up to 5 h. One may note that the filamentous structures, which were identified on the fresh specimen, have no longer appeared.

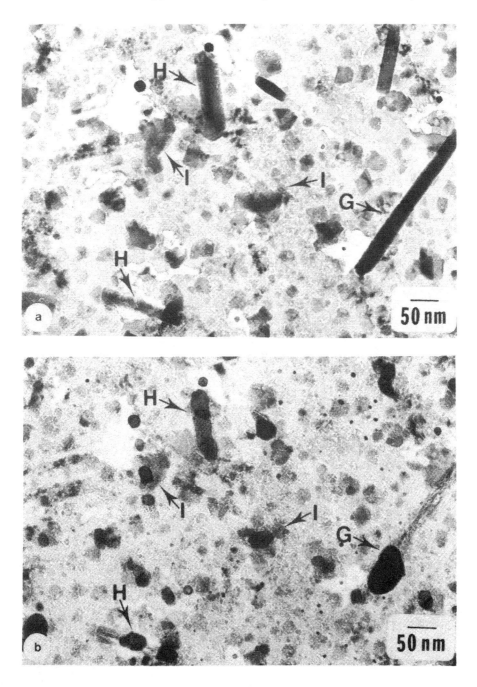

FIG. 9. Sequence of changes on heating the specimen of Fig. 7. The region shown here is different from that of Fig. 7. (a) 19 h CH$_4$; (b) 6 h H$_2$; *(Continued)*

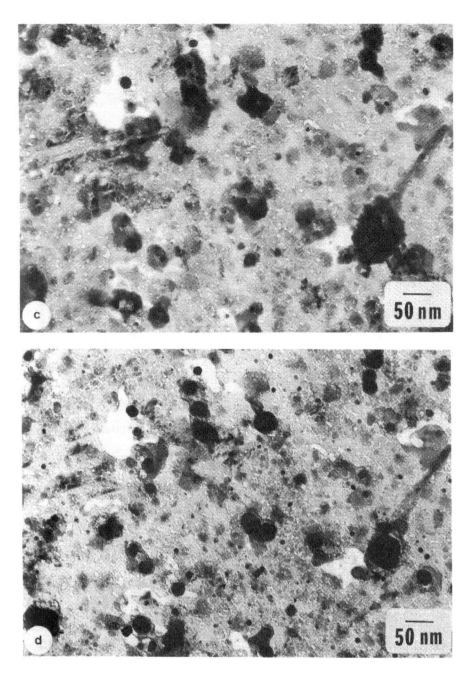

FIG. 9. (Continued) Sequence of changes on heating the specimen of Fig. 7. The region shown here is different from that of Fig. 7. (c) 2 h steam; (d) 4 h H_2. *(Continued)*

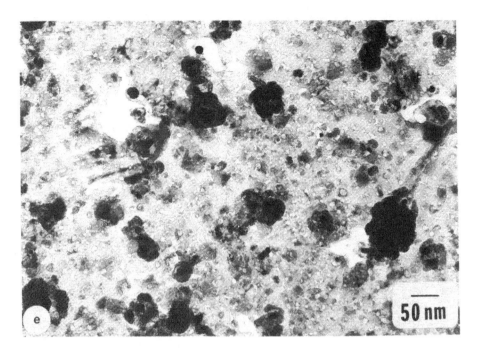

FIG. 9. (Continued) Sequence of changes on heating the specimen of Fig. 7. The region shown here is different from that of Fig. 7. (e) 1 h CH$_4$.

2. Heating in CH$_4$ + Steam

A. Ni/Al$_2$O$_3$

First, the specimen was heated in H$_2$ for 10 h, 5 h at 500°C and another 5 h at 700°C (Fig. 10a). The specimen was further heated in CH$_4$ + steam for 20 min at 700°C (Fig. 10b). The micrograph shows that most of the particles extended over the substrate as thin patches having small particles at one of the corners. Some particles acquired an elongated filamentous shape (J in Fig. 10b). NiO was detected by electron diffraction. On further heating in CH$_4$ + steam for 40 more min, most of the particles acquired short filamentous shapes (Fig. 10c). Some of the filaments had a dense particle at their tip, while others did not. This suggests that, as in the CH$_4$ atmosphere, the nickel particles transformed into filaments by depositing material behind them on the substrate during their migration. The electron diffraction pattern indicated NiO as the major component. Perhaps, the filaments contain also some carbon; this compound could not, however, be detected by electron diffraction. Compared to the filaments produced during heating in CH$_4$, the present ones are shorter. In addition, the number of filaments is smaller than the number of particles present before the last heat treatment (compare Figs. 10b and 10c). On further heating for 1 more hour, the filamentous particles changed their shape, becoming more irregular, and their number decreased (Fig. 10d). This indicates that the particles had relatively high mobilities on the substrate and coalesced. This probably happened because of the gasification by steam of the carbon deposits; the resulting gas lifted the particles, displacing them over the substrate. After the particles were heated for an additional 3 h, very little change occurred. After heating in CH$_4$ + H$_2$O, the specimen was heated in H$_2$ for 2 h at 700°C. The result is shown in Fig. 10e. The particles contracted and, in addition, new very small particles appeared. This is a result of the reduction of NiO to Ni,

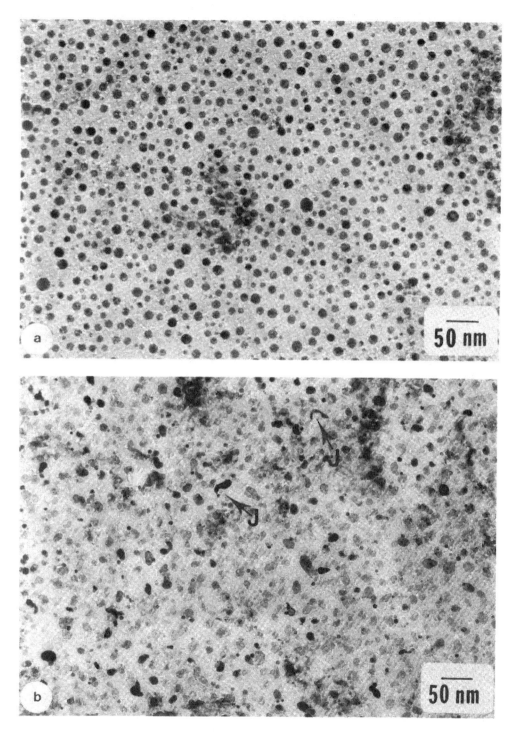

FIG. 10. Sequence of changes on heating a Ni/Al$_2$O$_3$ specimen in methane plus steam at 700°C. (a) 10 h (5 h at 500°C and 5 h at 700°C) H$_2$; (b) 20 min CH$_4$ + steam. *(Continued)*

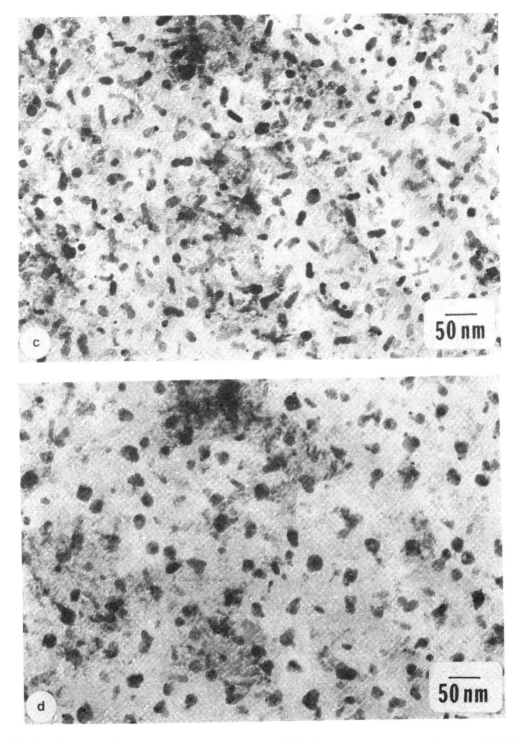

FIG. 10. (Continued) Sequence of changes on heating a Ni/Al$_2$O$_3$ specimen in methane plus steam at 700°C. (c) 1 h CH$_4$ + steam; (d) 2 h CH$_4$ + steam (the specimen was heated for additional 3 h in CH$_4$ + steam after the heat treatment (d)). *(Continued)*

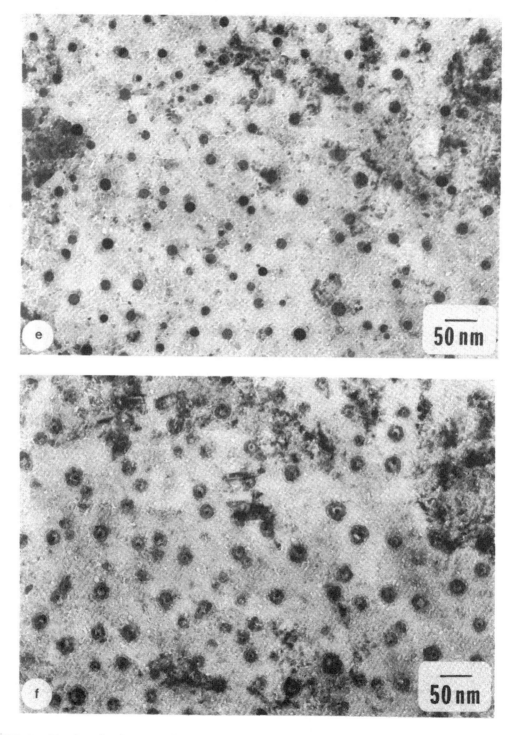

FIG. 10. (Continued) Sequence of changes on heating a Ni/Al$_2$O$_3$ specimen in methane plus steam at 700°C. (e) 2 h H$_2$; (f) 1 h CH$_4$ + steam.

which was detected by electron diffraction. The small particles were formed by the rupture of the undetectable films, which spread out over the substrate during heating in CH_4 + steam, and the subsequent contraction of the resulting patches. The specimen was heated again in CH_4 + steam, but the results were different from those of the previous heating in CH_4 + steam. After heating for 1 h, the shape of the particles changed to a torus which contained a core particle in the cavity (Fig. 10f). The outside diameter of the torus was larger than that of the particle before heating. This indicates that the change in shape is associated with the extension of the particles on the substrate. Electron diffraction indicated the presence of NiO. This (torus and core) structure is very similar to that observed upon heating Ni/Al_2O_3 in O_2 at 500°C (20). On heating for 3 additional hours in CH_4 + steam, little change occurred and no filaments were formed. This indicates that, in these circumstances, carbon was not deposited, probably because the catalyst has lost its catalytic activity for methane decomposition.

B. Co/Al$_2$O$_3$

Figure 11a represents the initial state after heating in H_2. Electron diffraction indicated the presence of α-Co. After heating in CH_4 + steam for 30 min, some particles extended on the substrate (L in Fig. 11b) and other smaller particles (M) decreased in size, most likely because of the spreading out from them of thin films undetectable by electron microscopy. Electron diffraction indicated that the particles contained mostly CoO. This suggests that the extension of the particles is associated with their oxidation. One may note that some particles are composed of one or several small and dark particles located on the top or on the edge of an extended patch (N1 in Fig. 11b). Some of these dark particles moved away from the main body (N2 in Fig. 11b). On further heating for up to 2 h, most of the particles decreased in size (Fig. 11c), becoming thinner, either because of the emission of thin undetectable films, or because of the loss of material to the substrate or to the gas stream. In addition, large patches (O) appeared in a few regions as a result of either extension and subsequent interconnection of neighboring particles and/or more likely because of carbon deposition. After heating for 1.5 more hours, a large amount of material was lost from the large patches (O in Fig. 11d), while most of the particles remained unchanged.

This suggests that the patches contained carbon deposits, which were subsequently gasified by steam. Further, the specimen was heated in H_2 for 1 h (Fig. 11e). Some particles contracted and formed small thick particles on the top of thin extended patches (P in Fig. 11e). The electron diffraction patterns indicated the presence of α-Co, Co_2O_3, and Co_3O_4. One may note again that many new particles appeared in regions where particles were not present before the heating in H_2. This is probably a result of the rupture of the undetectable films and the contraction of the resulting thin particles; it also indicates that most of the "lost material" from the particles remained on the substrate and was not permanently lost to the gas stream.

C. Fe/Al$_2$O$_3$

After heating for 30 min in CH_4 + steam, the shape of the particles of Fig. 12a changed to a torus with a small remnant particle in the cavity (Fig. 12b). The torus and the core particle were interconnected by a film. Only γ-Fe_2O_3 (or Fe_3O_4) was detected by electron diffraction. This indicates that the change in shape is associated with the oxidation of iron. Coke was also probably deposited around the particles, thus increasing their size, or the increase in size was a result of the extension of the particles. On further heating for up to 5 h, the cavities were filled, either because of the extension of the remnant particles and/or because of carbon deposition (Fig. 12c). After being heated in H_2 for 2 h, the specimen was again heated in CH_4 + steam. After heating for 1 h in CH_4 + steam, in some of the particles a part, the small remnant particle in most of the cases, moved away from the main body, generating filamentous traces (Q in Fig. 12d). On heating for 3 additional hours, the growth of the filaments continued (Q in Fig. 12e). The leading particle, which was located at the tip of the

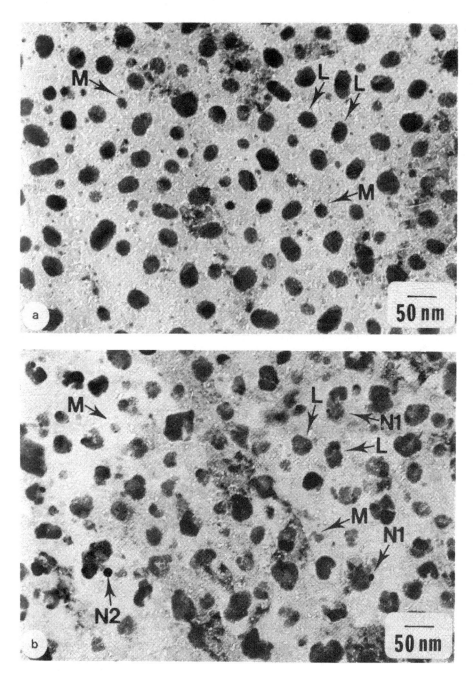

FIG. 11. Sequence of changes on heating a Co/Al$_2$O$_3$ specimen in methane plus steam at 700°C. (a) 10 h (5 h at 500°C and 5 h at 700°C) H$_2$; (b) 30 min CH$_4$ + steam. (*Continued*)

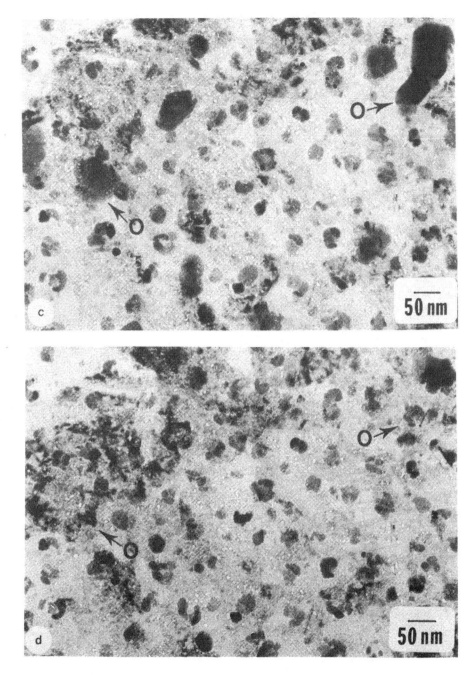

FIG. 11. (Continued) Sequence of changes on heating a Co/Al$_2$O$_3$ specimen in methane plus steam at 700°C. (c) 2 h CH$_4$ + steam; (d) 3.5 h CH$_4$ + steam. *(Continued)*

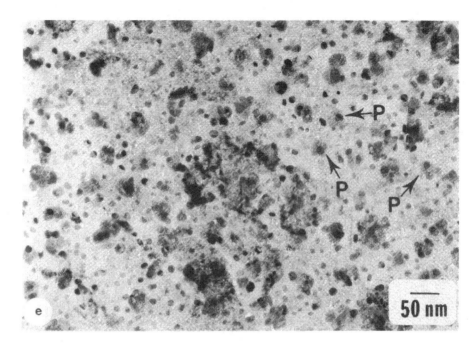

FIG. 11. (Continued) Sequence of changes on heating a Co/Al$_2$O$_3$ specimen in methane plus steam at 700°C. (e) 1 h H$_2$.

filament, did not decrease in size, indicating, very likely, that the filament was composed mostly of carbon. From this observation, one can infer that the heating of the coked specimen in H$_2$ is more effective for the regeneration of Fe/Al$_2$O$_3$ than of Ni/Al$_2$O$_3$ for CH$_4$ decomposition.

3. Heating in CH$_4$ + Steam + H$_2$

A. Ni/Al$_2$O$_3$

As is well known, if the sample is heated in the individual components of the above gas mixture, the processes of sintering, coking, and redispersion are relatively slow. For instance, to increase the size of the particles by sintering from 60 to 120 Å, heating in hydrogen for more than 18 h (20) is needed; to redisperse particles of 400 to 200 Å in size, heating in steam for 10 h (17) must be employed. For this reason, in those cases it is easy to follow the history of individual particles. In contrast, by heating the specimen in the above mixture of gases the changes occur so rapidly that it is impossible to follow, in some cases, the behavior of individual particles. The results obtained during heating of a specimen in CH$_4$ + steam + H$_2$ are shown in Figs. 13a–13e. After heating for only 10 min, the particles formed tails with small thick particles at the tip of the tail (Fig. 13a). In the electron diffraction pattern, NiO was the only compound detected, indicating that the tails were composed of NiO, and perhaps, amorphous carbon, as discussed later in the Discussion section. Further heating for up to 1.5 h caused severe sintering (Fig. 13b). While the particles became larger, their shape remained similar to that present before heating, with only an increased ratio of width to length of the tails. After heating for 2 more hours, the particles changed their shape to a torus with a core particle in the cavity (Fig. 13c). Possibly, because of the carbon deposits, the particles lost their catalytic activity for CH$_4$ decomposition and the gasification of the deposited carbon by H$_2$ and steam that followed caused the changes in shape. On further heating for 1.5 more hours, the shape of the particles became more irregular, and some of the particles became elongated (Fig. 13d). This suggests that after enough gasification of the deposited carbon, the particles regained their catalytic

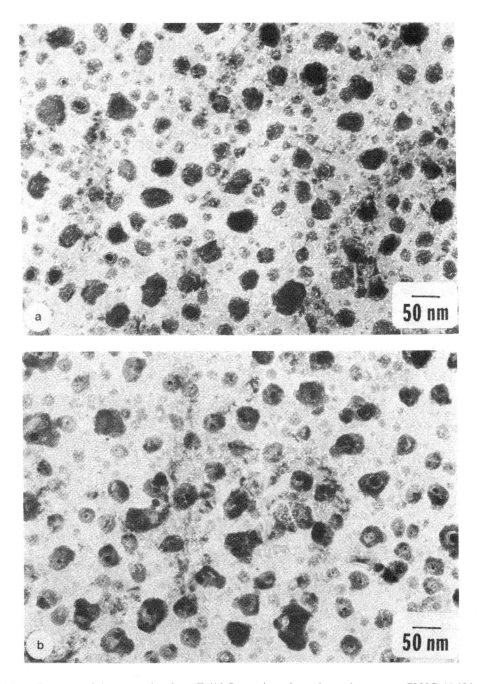

FIG. 12. Sequence of changes on heating a Fe/Al$_2$O$_3$ specimen in methane plus steam at 700°C. (a) 10 h (5 h at 500°C and 5 h at 700°C) H$_2$; (b) 30 min CH$_4$ + steam. *(Continued)*

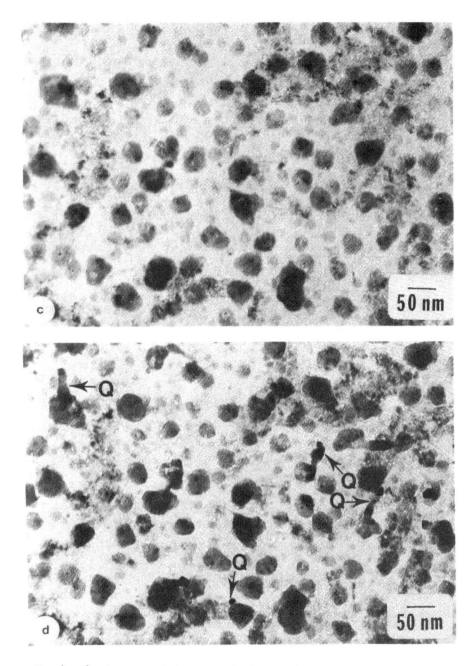

FIG. 12. (Continued) Sequence of changes on heating a Fe/Al$_2$O$_3$ specimen in methane plus steam at 700°C. (c) 5 h CH$_4$ + steam (the specimen was heated in H$_2$ for 2 h at 700°C between (c) and (d)); (d) 1 h CH$_4$ + steam. *(Continued)*

Simulation of the Behavior of Supported Metal Catalysts in Real Reaction Atmospheres 341

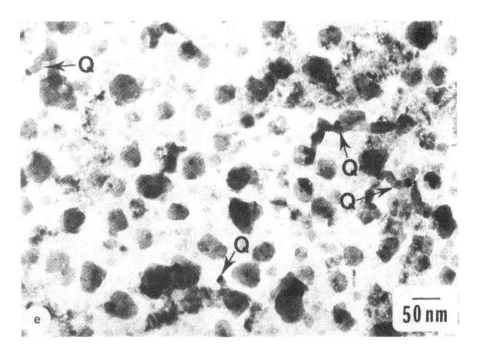

FIG. 12. (Continued) Sequence of changes on heating a Fe/Al$_2$O$_3$ specimen in methane plus steam at 700°C. (e) 4 h CH$_4$ + steam.

activity and could decompose CH$_4$ again. After heating in CH$_4$ + steam + H$_2$ for a total of 7 h, the specimen was heated in H$_2$ for 1 h. As shown in Fig. 13e, the particles acquired a spherical shape.

The main events that occurred during heating in CH$_4$ + steam + H$_2$ were somewhat similar to those observed during heating in CH$_4$ + steam. However, it is worth emphasizing the differences: (1) The particles formed tails more easily in CH$_4$ + steam + H$_2$ than in CH$_4$ + steam. (2) The particles lost their catalytic activity for decomposition of CH$_4$ more rapidly in the latter atmosphere. (3) The active catalyst had been better dispersed on the substrate, in the latter atmosphere. Indeed, for a specimen which was heat treated in CH$_4$ + steam, the subsequent heating in H$_2$ produced numerous new small crystallites in regions where no particles were present before heating; this was not, however, the case with the specimen which was heated in CH$_4$ + steam + H$_2$.

Figures 14a–14c show the micrographs of a specimen similar to those of Figs. 13a–13e, but belonging to a different batch, whose initial average particle size was larger. After 10 min of heating in CH$_4$ + steam + H$_2$, the particles changed their shape from globular to a torus with a small particle in the cavity (Fig. 14a). NiO was detected by electron diffraction. In contrast, the particles of the similar specimen of Figs. 13a–13e elongated after 10 min of heating in the same mixture. The heating for 20 more minutes brought about, however, a dramatic change very different from that of the similar specimen of Figs. 13a–13e (Fig. 14b). Hollow, very likely, carbonaceous filaments grew over the surface of the substrate and cut across each other, forming complex networks. Severe sintering of the particles also took place. Electron diffraction indicated that the particles were present mostly as Ni. Hence, the particles were reduced to metal nickel and sintered, (most probably) by migration followed by coalescence. It appears that the formation of the carbonaceous filaments is associated with the reduction of the particles. Indeed, in the specimen of Figs. 13a–13e, the particles were present as NiO throughout heating in CH$_4$ + steam + H$_2$ and they did not produce carbonaceous filaments. In addition, it is worth noting that the heating in methane or in methane + steam did not produce carbonaceous filaments either and the particles were also present as NiO. However, after additional heating for 30 more minutes in CH$_4$ + steam + H$_2$, the filaments disappeared and sintering

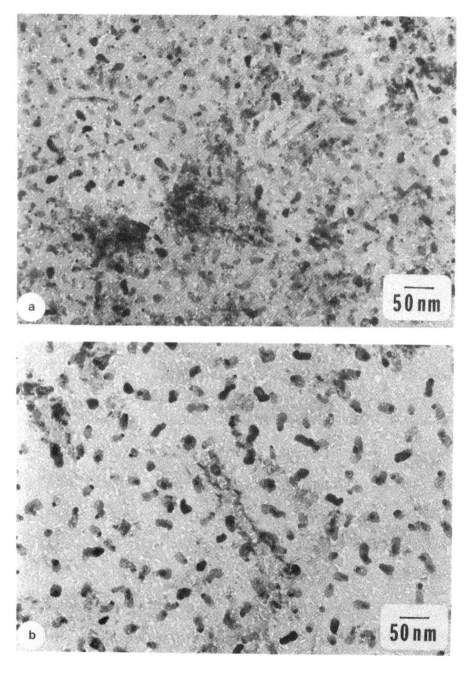

FIG. 13. Sequence of changes on heating a Ni/Al$_2$O$_3$ specimen in CH$_4$ + steam + H$_2$ at 700°C after pretreatment in H$_2$ for 10 h (5 h at 500°C and 5 h at 700°C). (a) 10 min CH$_4$ + steam + H$_2$; (b) 1.5 h CH$_4$ + steam + H$_2$. (*Continued*)

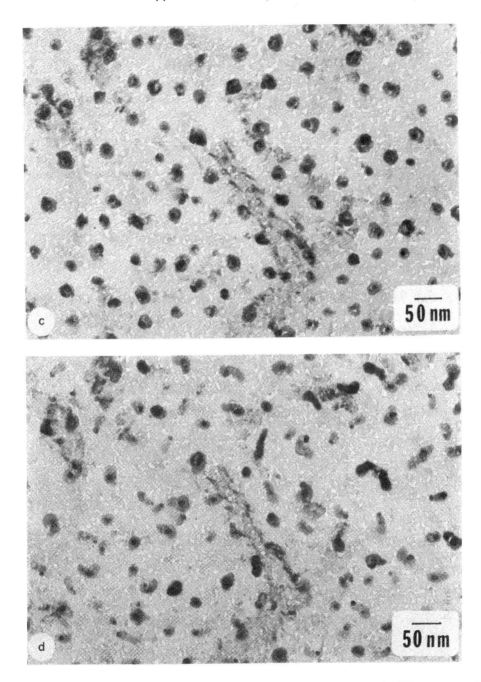

FIG. 13. (Continued) Sequence of changes on heating a Ni/Al$_2$O$_3$ specimen in CH$_4$ + steam + H$_2$ at 700°C after pretreatment in H$_2$ for 10 h (5 h at 500°C and 5 h at 700°C). (c) 3.5 h CH$_4$ + steam + H$_2$; (d) 5 h CH$_4$ + steam + H$_2$ (the specimen was heated for 2 more hours in CH$_4$ + steam + H$_2$ after the heat treatment (d)). *(Continued)*

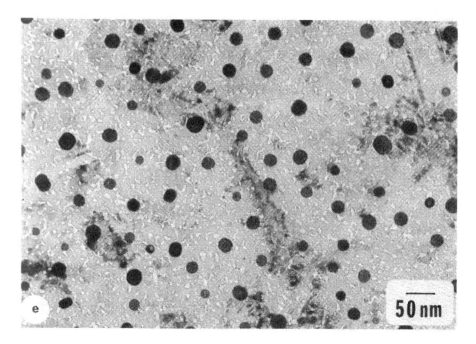

FIG. 13. (Continued) Sequence of changes on heating a Ni/Al$_2$O$_3$ specimen in CH$_4$ + steam + H$_2$ at 700°C after pretreatment in H$_2$ for 10 h (5 h at 500°C and 5 h at 700°C). (e) 1 h H$_2$.

continued to occur (Fig. 14c). Electron diffraction identified the presence of Ni. On heating for 1 more hour in CH$_4$ + steam + H$_2$ and subsequently in H$_2$ for 1 h, most of the particles remained without any significant change. Two similar specimens, from two different batches, behaved differently. The kinetics of oxidation and reduction of the crystallites, which is affected by the details of the preparation of the specimen, is probably responsible for the difference.

B. Co/Al$_2$O$_3$

On heating for 10 min in CH$_4$ + steam + H$_2$, the initial particles of Fig. 15a have extended over the substrate (Fig. 15b). The particles are now composed of an extended bottom that supports a darker small particle. One may also note the coalescence of neighboring particles (R1), as well as the dumbbell-shaped particles (R2). Heating for 20 more minutes caused the deformation of the particles, probably because of the penetration of deposited carbon (see Discussion) (Fig. 15c). It is worth noting that the dark particles moved over the extended bottom to its edge. Some of the dark particles moved out from the extended bottom and tails were formed while they were migrating (SI in Fig. 15c), perhaps because of the carbon formation. On further heating for up to 1.5 h, the dark particles changed their position within the extended bottom (S2 in Fig. 15d). In addition, a few dark particles moved further out from the extended bottom, forming tails which are probably composed of carbon and cobalt oxide (S3 in Fig. 15d). A few small particles disappeared, leaving some traces (S4 in Fig. 15d). After heating for a total of 5.5 h, the particles changed their shape, becoming more circular (Fig. 15e), and the dark particles moved from the edge of the extended bottom to their center. Now, the catalyst particles are no longer expected to be active for methane decomposition, because they are probably completely covered by carbon. However, heating for up to 10 h caused again tail formations, which were, however, shorter than before. This suggests that the coked particles were regenerated and able again to decompose methane, but with a lower activity than before. The specimen was subsequently

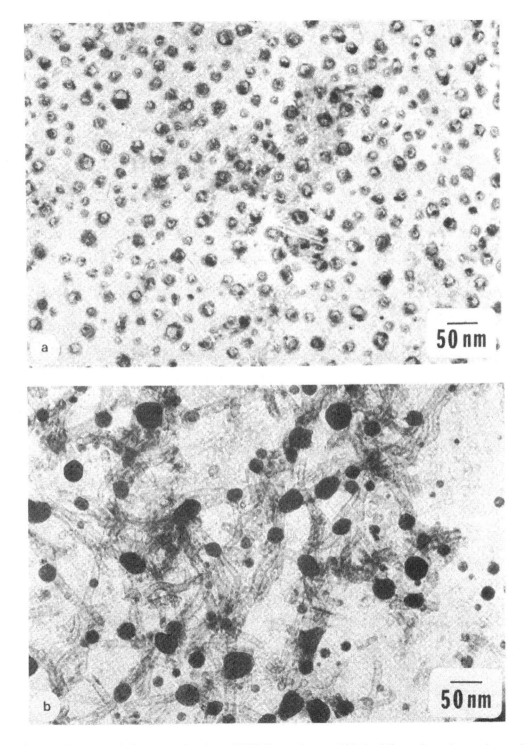

FIG. 14. Sequence of changes on heating a Ni/Al$_2$O$_3$ specimen, which is different from the specimen of Fig. 13, in CH$_4$ + steam + H$_2$, at 700°C after pretreatment in H$_2$ for 11.5 h (5.5 h at 500°C and 6 h at 700°C). (a) 10 min CH$_4$ + steam + H$_2$; (b) 30 min CH$_4$ + steam + H$_2$. *(Continued)*

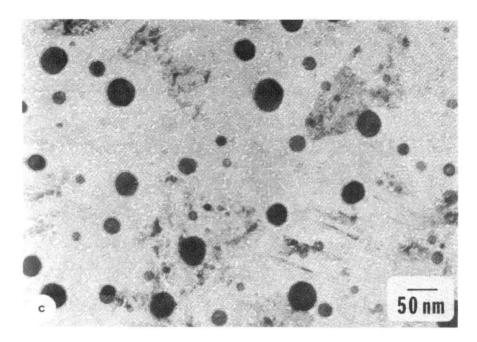

FIG. 14. (Continued) Sequence of changes on heating a Ni/Al$_2$O$_3$ specimen, which is different from the specimen of Fig. 13, in CH$_4$ + steam + H$_2$, at 700°C after pretreatment in H$_2$ for 11.5 h (5.5 h at 500°C and 6 h at 700°C). (c) 1 h CH$_4$ + steam + H$_2$.

heated in H$_2$ for 1 h (Fig. 15f). The particles decreased in size, because of contraction. The appearance of new small particles in regions where no particles were observed before indicates that during heating in CH$_4$ + steam + H$_2$ some particles and/or part of other particles spread over the substrate as undetectable films. The heating in H$_2$ caused the rupture of that film, and the contraction of the thin patches thus formed generated the small particles. Comparing Figs. 15a and 15f, it is plausible to conclude that some material was permanently lost to the gas stream. As mentioned earlier, in the case of Ni the heating in H$_2$ following heating in CH$_4$ + steam + H$_2$ did not produce new particles in regions where no particles were present before. Compared to nickel, the cobalt particles emit to a greater extent undetectable thin films over the surface of the Al$_2$O$_3$ substrate.

C. Fe/Al$_2$O$_3$

After heating the specimen of Fig. 16a in CH$_4$ + steam + H$_2$ for 10 min (Fig. 16b), one notes that most of the particles deformed and/or extended, and had one or several small darker particles supported on extended thin patches. In addition, a few particles elongated, acquiring a filamentous structure (T in Fig. 16b). Electron diffraction indicated the presence of γ-Fe$_2$O$_3$ (or Fe$_3$O$_4$). The filaments were most likely composed of carbon and some γ-Fe$_2$O$_3$ (or Fe$_3$O$_4$), and the extended thin patches contained probably deposited carbon. On heating for 20 more minutes (Fig. 16c), the particles became thinner and disintegrated in smaller, interconnected units, probably because of the formation of unstable carbides and their decomposition (see Discussion). It is likely that, because of disintegration, a part of the material was permanently lost to the gas stream. The only compound detected by electron diffraction was γ-alumina,

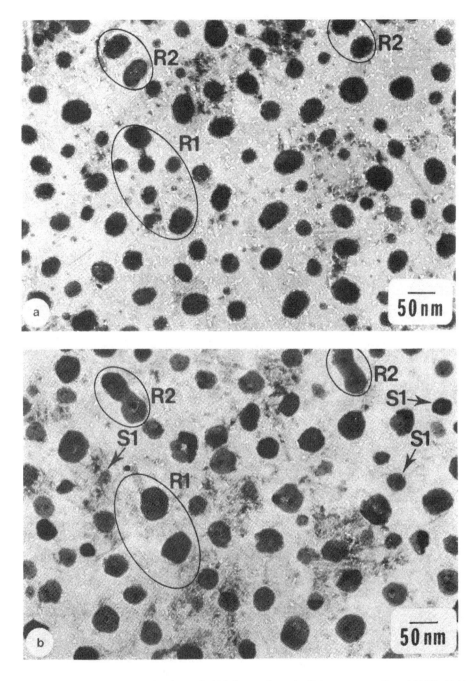

FIG. 15. Sequence of changes on heating a Co/Al$_2$O$_3$ specimen in CH$_4$ + steam + H$_2$ at 700°C. (a) 10 h (5 h at 500°C and 5 h at 700°C) H$_2$; (b) 10 min CH$_4$ + steam + H$_2$. *(Continued)*

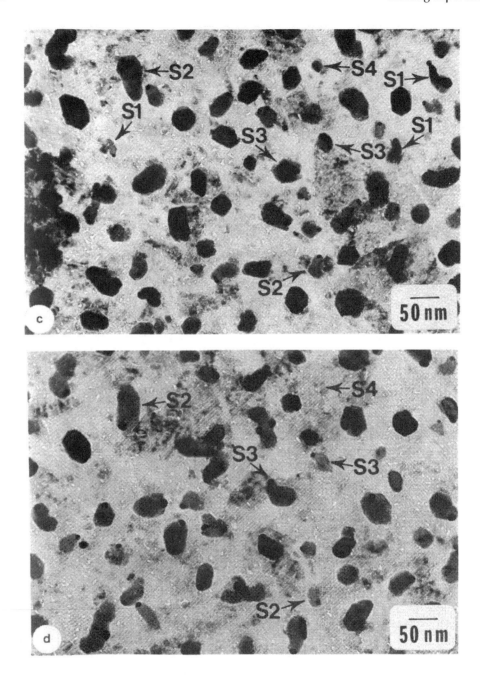

FIG. 15. (Continued) Sequence of changes on heating a Co/Al$_2$O$_3$ specimen in CH$_4$ + steam + H$_2$ at 700°C. (c) 30 min CH$_4$ + steam + H$_2$; (d) 1.5 h CH$_4$ + steam + H$_2$. *(Continued)*

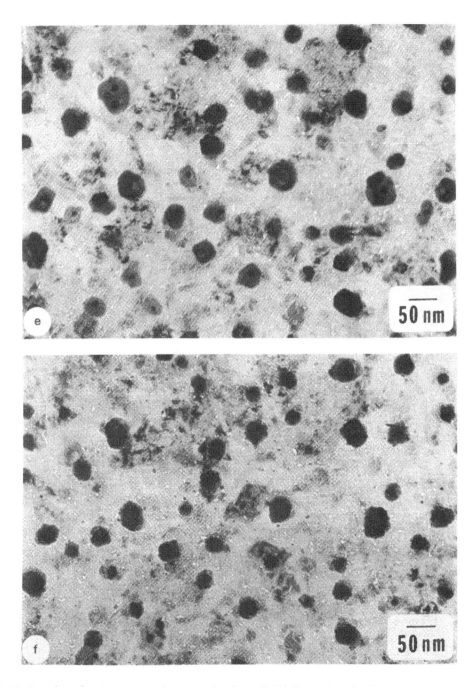

FIG. 15. (Continued) Sequence of changes on heating a Co/Al$_2$O$_3$ specimen in CH$_4$ + steam + H$_2$ at 700°C. (e) 5.5 h CH$_4$ + steam + H$_2$ (the specimen was heated for additional 4.5 h in CH$_4$ + steam + H$_2$ after the heat treatment (e)); (f) 1 h H$_2$.

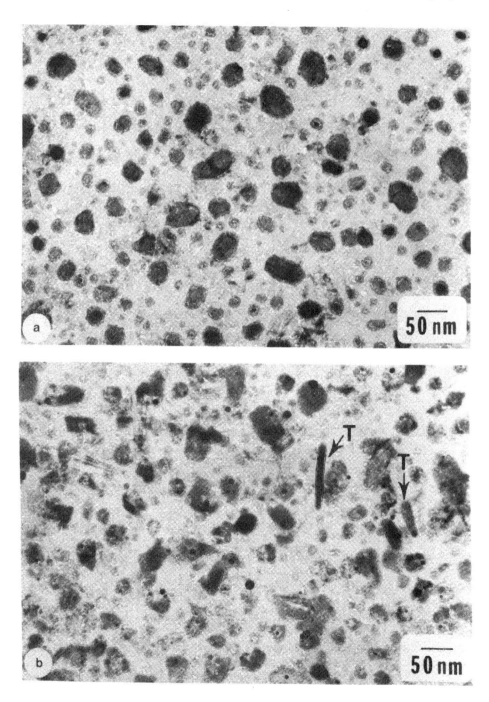

FIG. 16. Sequence of changes on heating a Fe/Al$_2$O$_3$ specimen in CH$_4$ + steam + H$_2$ at 700°C. (a) 10 h (5 h at 500°C and 5 h at 700°C) H$_2$; (b) 10 min CH$_4$ + steam + H$_2$. *(Continued)*

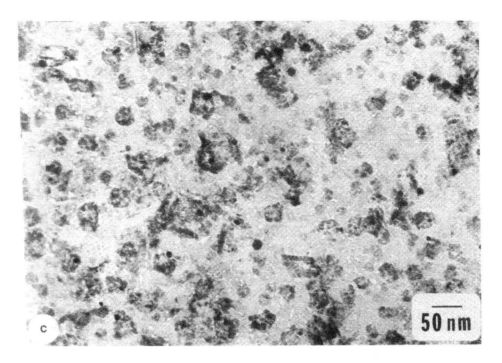

FIG. 16. (Continued) Sequence of changes on heating a Fe/Al$_2$O$_3$ specimen in CH$_4$ + steam + H$_2$ at 700°C. (c) 30 min CH$_4$ + steam + H$_2$.

probably because the particles were too thin. Further heating for up to 8.5 h did not cause any significant change. After particles were subsequently heated in H$_2$, no significant change was observed either. One may note that new particles did not appear in regions where no particles were observed before heating. This indicates that spreading as thin films, undetectable by electron microscopy, occurred in this case much less than for Co and that any "lost material" was permanently lost to the gas stream.

4. Heating of Ni/Al$_2$O$_3$ in CO

On heating the specimen of Fig. 17a in CO for 0.5 h, one notes that severe sintering occurred (Fig. 17b). Some filamentous carbon (Y1), as well as some large patches of carbon which covered several particles (Y2 in Fig. 17b), was observed. On further heating for a total of 2 h, the carbonaceous filaments in region Y1 disappeared, while new carbonaceous filaments appeared in other regions (Y3 in Fig. 17c). In addition, severe sintering continued to occur. The specimen was further heated for up to 4 h. Figures 17d$_1$ and 17d$_2$, which represent two different regions of the same specimen, show several types of carbon deposits:

  a. filamentous carbon (Y4 in Fig. 17 d$_1$),
  b. hollow filamentous carbon (Y5 in Fig. 17 d$_2$),

c. carbonaceous filament which wraps an elongated nickel particle (Y6 in Fig. 17 d$_1$),
d. thick large patches of carbon (Y7 in Fig. 17 d$_2$), and
e. films of carbon that cover a whole area of several neighboring particles (Y8 in Fig. 17 d$_1$).

However, in the electron diffraction pattern, no rings corresponding to graphite were detected, indicating that the deposited carbon was amorphous. With the exception of the elongated particles which were wrapped in the carbonaceous filaments, the catalyst particles maintained almost spherical shapes throughout the entire heating. Electron diffraction indicated that the crystallites were present as Ni. It is possible that a part of the metal was permanently lost to the gas stream as nickel carbonyl. It is, however, difficult to discriminate between the severe sintering that occurred and this loss of material.

Another specimen was heated continuously in CO for 4 h and its micrograph is shown in Fig. 18. All the types of carbon deposits, observed in the micrographs of the previous specimen which was heated in several steps for up to 4 h (Figs. 17a–17d$_2$), can be identified in Fig. 18. However, the number of filaments was greater in the latter case and most of them were hollow. In addition, some filaments swung very slowly during observation in the transmission electron microscope and were not sharp, while others were well focused. This means that the former filaments were not completely supported on the substrate.

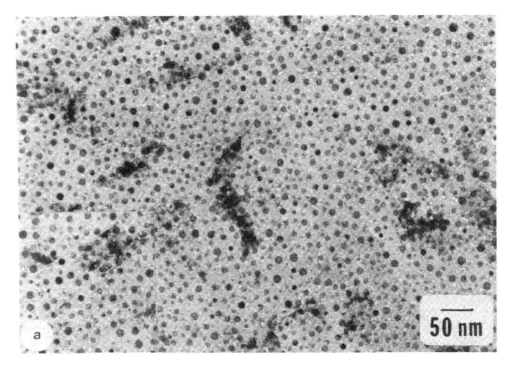

FIG. 17. Sequence of changes on heating a Ni/Al$_2$O$_3$ specimen in carbon monoxide at 700°C. (a) 10 h (5 h at 500°C and 5 h at 700°C) H$_2$. (*Continued*)

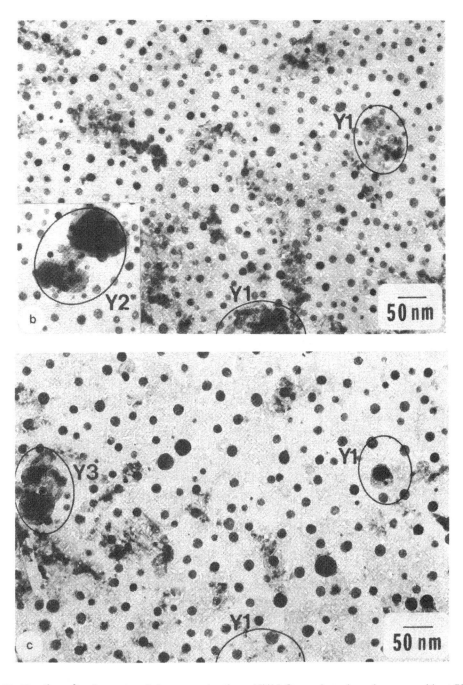

FIG. 17. (Continued) Sequence of changes on heating a Ni/Al$_2$O$_3$ specimen in carbon monoxide at 700°C. (b) 0.5 h CO; (c) 2 h CO. *(Continued)*

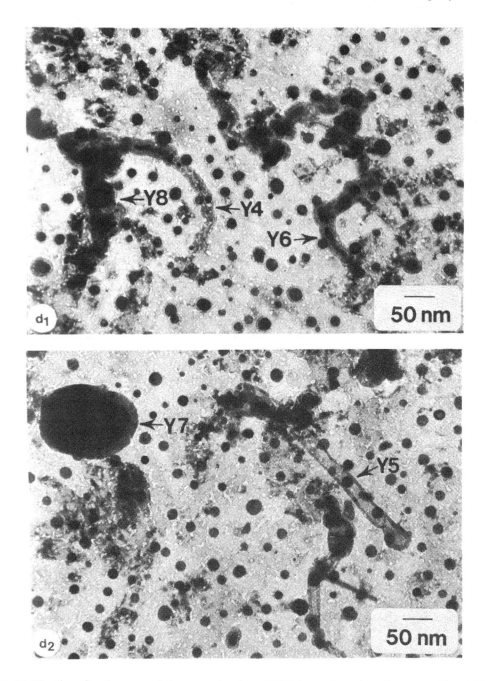

FIG. 17. (Continued) Sequence of changes on heating a Ni/Al$_2$O$_3$ specimen in carbon monoxide at 700°C. (d$_1$ and d$_2$) 4 h CO (d$_1$ and d$_2$ show two different regions of the same specimen).

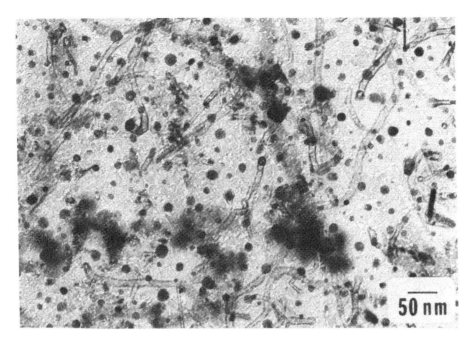

FIG. 18. Ni/Al$_2$O$_3$ specimen after heating in CO for 4 h continuously at 700°C following heating in H$_2$ for 10 h (5 h at 500°C and 5 h at 700°C). 5. HEATING IN CH$_4$ + STEAM + H$_2$ + CO.

5. HEATING IN CH$_4$ + STEAM + H$_2$ + CO

As mentioned earlier, different deactivation processes are dominant for the three systems during heating in CH$_4$ + steam + H$_2$, namely sintering and coking for Ni/Al$_2$O$_3$, spreading and coking for Co/Al$_2$O$_3$, and loss of material and coking for Fe/Al$_2$O$_3$. When CO was added to the ternary mixture, the results remained similar to those in the ternary mixture, but the changes occurred more rapidly.

A. Ni/Al$_2$O$_3$

After heating in the quaternary mixture for 5 min, most of the initial particles (Fig. 19a) became elongated (Fig. 19b) and the number of particles decreased, due to the mergence of the nearby particles (Z1 in Fig. 19b). In addition, some large patches, in which carbon interconnects and covers several particles, formed (Z2). NiO was detected by electron diffraction. On heating for 15 additional minutes, the particles contracted and acquired spherical shapes (Fig. 19c). Some neighboring particles coalesced (Z3), while some elongated particles which contacted one another split into several particles of spherical shape (Z1). However, most of the particles remained at their original locations with a change in shape only. Electron diffraction indicated that most of the particles were reduced to Ni, but NiO was also present. Comparing with the heating in the gas mixture without CO, one may note that NiO particles are more rapidly reduced to Ni in the CH_4 + steam + H_2 + CO mixture. After particles were further heated for up to 2 h, severe sintering occurred (Fig. 19d). As suggested in the case of heating in CO, it is possible that a part of the metal was permanently lost to the gas stream, because of carbonyl formation. Electron diffraction indicated that the particles were almost completely reduced to Ni after 1 h of heating. Another sample was heated continuously in the quaternary mixture for 2 h. The results were similar to those obtained with a specimen heated in several steps, for the same length of time.

B. Co/Al$_2$O$_3$

Many particles disappeared after heating the specimen of Fig. 20a in CH_4 + steam + H_2 + CO for only 5 min, and the remaining particles were deformed (Fig. 20b). The deformation is very likely associated with carbon deposition and its dissolution inside the particles (see Discussion). The electron diffraction pattern indicated that the particles were present mostly as CoO with traces of Co_2O_3. After heating for 15 more minutes, the particles acquired a more compact circular shape (Fig. 20c). One may also note that some particles split into two particles

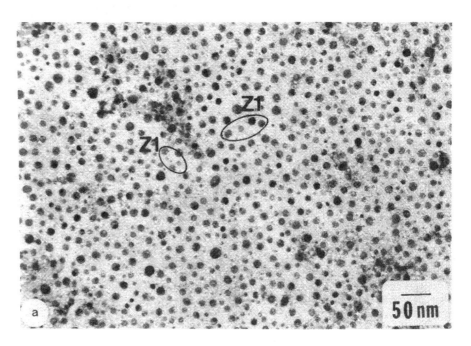

FIG. 19. Sequence of changes on heating a Ni/Al$_2$O$_3$ specimen in CH_4 + steam + H_2 + CO at 700°C. (a) 10 h (5 h at 500°C and 5 h at 700°C) H_2. *(Continued)*

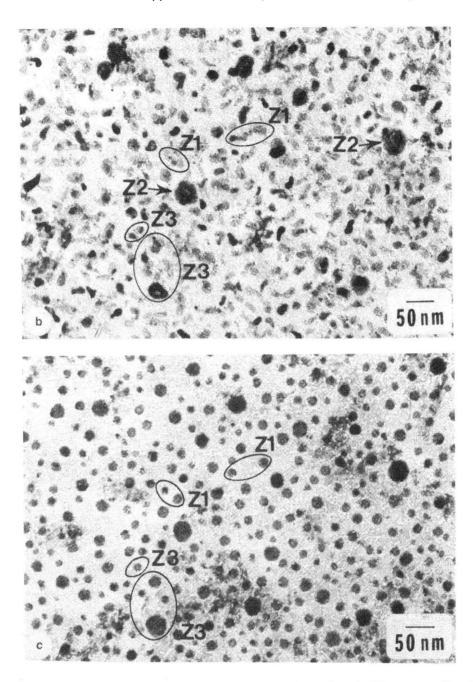

FIG. 19. (Continued) Sequence of changes on heating a Ni/Al$_2$O$_3$ specimen in CH$_4$ + steam + H$_2$ + CO at 700°C. (b) 5 min CH$_4$ + steam + H$_2$ + CO; (c) 20 min CH$_4$ + steam + H$_2$ + CO. *(Continued)*

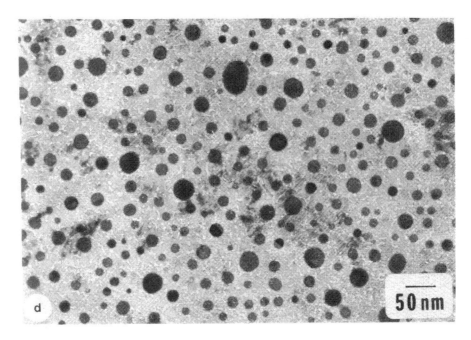

FIG. 19. (Continued) Sequence of changes on heating a Ni/Al$_2$O$_3$ specimen in CH$_4$ + steam + H$_2$ + CO at 700°C. (d) 2 h CH$_4$ + steam + H$_2$ + CO.

((a) in Fig. 20c). On further heating for up to 1 h (Fig. 20d) a few new particles appeared (b), while some particles decreased in size (c) and some particles extended (d). The specimen was subsequently heated in H$_2$ for 1 h (Fig. 20e). Numerous new particles appeared, indicating that at least a fraction of the disappeared material was not lost to the gas stream, but remained on the substrate as undetectable films.

C. Fe/Al$_2$O$_3$

After heating the specimen of Fig. 21a for only 5 min, one notes that most of the particles became smaller and thinner, indicating that a considerable amount of material was lost to the substrate and/or to the gas stream (Fig. 21b). On heating for up to 1 h, most of the particles became increasingly thinner (Fig. 21c). In addition, some of the dark particles (e) migrated over the substrate leaving white tracks. These tracks can be a result of the catalytic gasification of the carbon deposited on the substrate during the migration of the crystallites, migration which is enhanced by the gasification process. The specimen was subsequently heated in H$_2$ for 1 h (Fig. 21d). As shown in the micrograph, a few new particles appeared, while most of the thin particles remained unchanged. Comparing with Co/Al$_2$O$_3$, one may note that the amount of material lost permanently to the gas stream during heating in CH$_4$ + steam + H$_2$ + CO is greater for Fe/Al$_2$O$_3$.

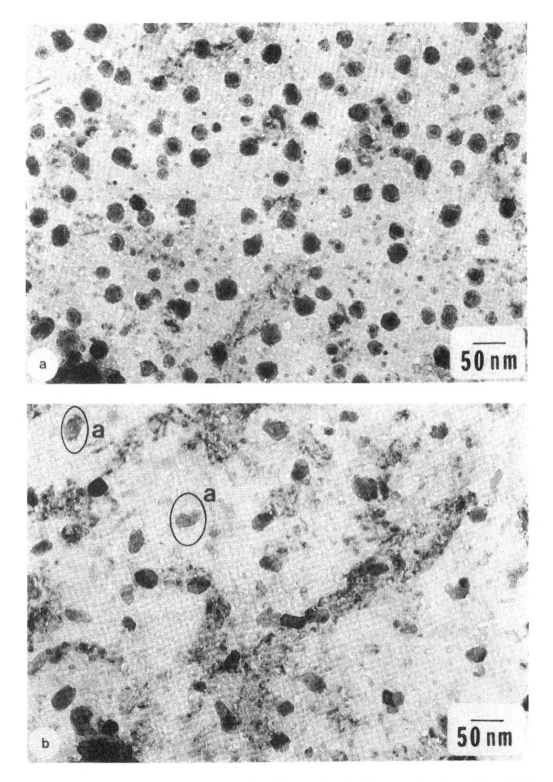

FIG. 20. Sequence of changes on heating a Co/Al$_2$O$_3$ specimen in CH$_4$ + steam + H$_2$ + CO at 700°C. (a) 10 h (5 h at 500°C and 5 h at 700°C) H$_2$; (b) 5 min CH$_4$ + steam + H$_2$ + CO. *(Continued)*

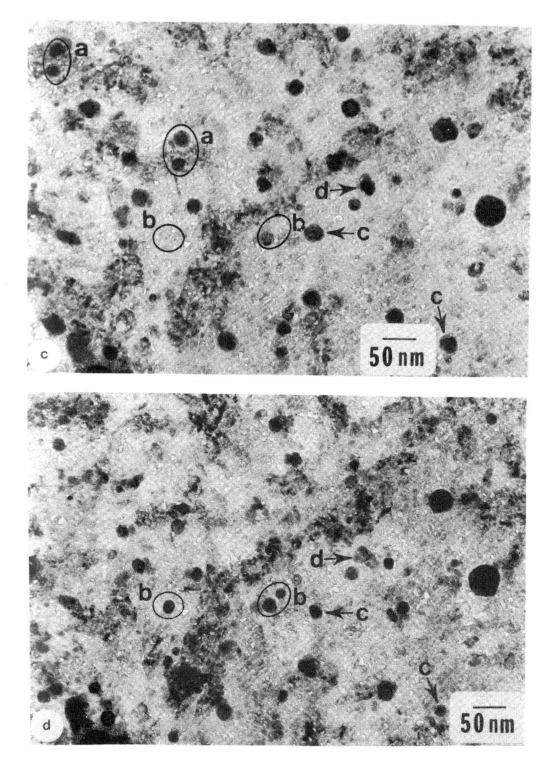

FIG. 20. (Continued) Sequence of changes on heating a Co/Al$_2$O$_3$ specimen in CH$_4$ + steam + H$_2$ + CO at 700°C. (c) 20 min CH$_4$ + steam + H$_2$ + CO; (d) 1 h CH$_4$ + steam + H$_2$ + CO. (*Continued*)

Simulation of the Behavior of Supported Metal Catalysts in Real Reaction Atmospheres 361

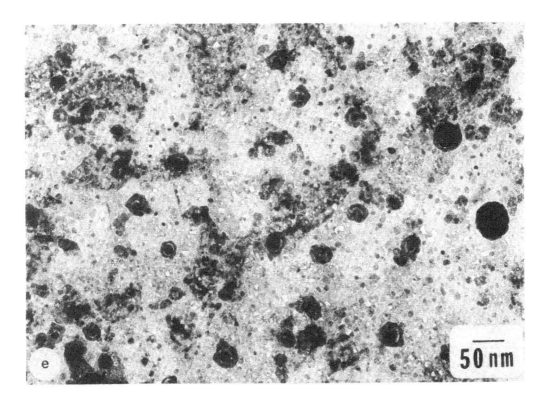

FIG. 20. (Continued) Sequence of changes on heating a Co/Al$_2$O$_3$ specimen in CH$_4$ + steam + H$_2$ + CO at 700°C. (e) 1 h H$_2$.

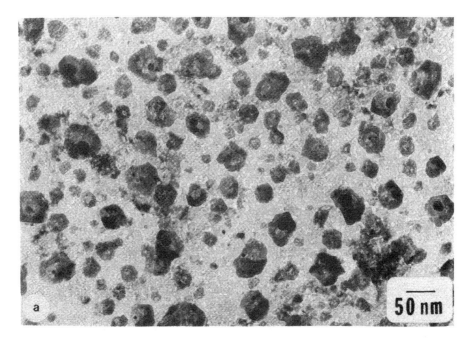

FIG. 21. Sequence of changes on heating a Fe/Al$_2$O$_3$ specimen in CH$_4$ + steam + H$_2$ + CO at 700°C. (a) 10 h (5 h at 500°C and 5 h at 700°C) H$_2$. *(Continued)*

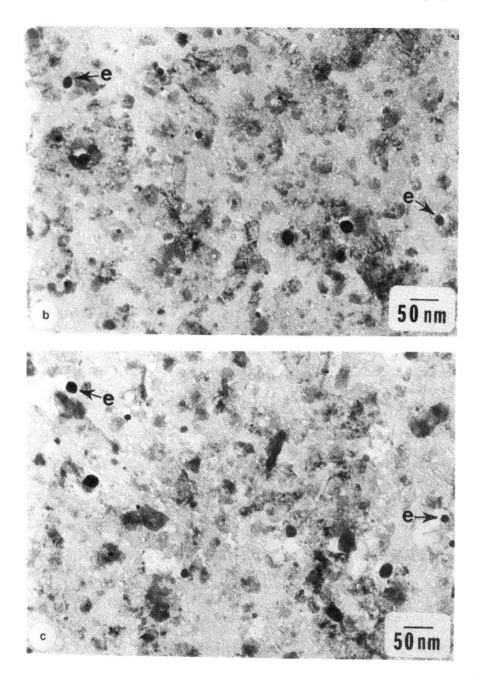

FIG. 21. (Continued) Sequence of changes on heating a Fe/Al$_2$O$_3$ specimen in CH$_4$ + steam + H$_2$ + CO at 700°C. (b) 5 min CH$_4$ + steam + H$_2$ + CO; (c) 1 h CH$_4$ + steam + H$_2$ + CO. (*Continued*)

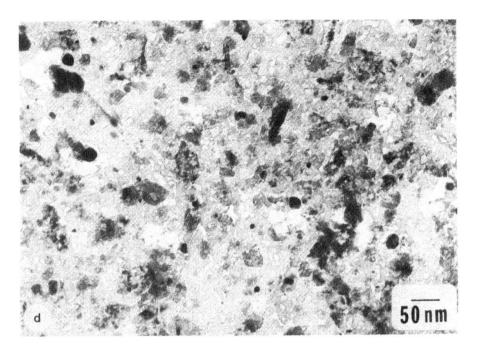

FIG. 21. (Continued) Sequence of changes on heating a Fe/Al$_2$O$_3$ specimen in CH$_4$ + steam + H$_2$ + CO at 700°C. (d) 1 h H$_2$.

DISCUSSION

1. Overall View of the Behavior of the Model Catalysts in Various Atmospheres

1a. Individual Components

Before the phenomena that occur in the simulated reaction atmospheres are examined, a brief review of the effects of the individual components on the model catalysts is useful.

It is well known that the presence of H$_2$ at elevated temperatures enhances sintering in the case of nickel on alumina catalysts. Although there is less information on the sintering of cobalt, the effect of H$_2$ on cobalt should be similar to that on nickel. The sintering of Fe/Al$_2$O$_3$ in H$_2$ was examined in some detail by Sushumna and Ruckenstein (21). Since Fe is extremely reactive, only traces of O$_2$ or H$_2$O are sufficient for the formation of iron oxide. The chemical interactions between this oxide and the alumina substrate appreciably diminish the rate of sintering.

The results reported by Ruckenstein and Hu (17) have shown that by heating in steam, the particles of Ni/Al$_2$O$_3$, Co/Al$_2$O$_3$, and Fe/Al$_2$O$_3$ extended and spread over the surface of the substrate, with a rate in the order Co > Ni > Fe.

All three metals are active and efficient catalysts for the decomposition of methane or disproportionation of carbon monoxide (10), and various types of carbon deposits have been reported (22–25).

The results presented in this paper show that heating in CH$_4$ changes the shape of the particles; the change is, however, different for the three metals. In the case of Ni, the particles became elongated and transformed to filaments, by depositing their material behind migrating leading particles, which decreased in size. The presence of small amounts of carbon was inferred from the subsequent heating of the specimen in O$_2$. The elongation of the particles is most likely a result of carbon deposition on the particles and its diffusion inside. In the case of Co, carbon was deposited around and on the particles, and, on further heating, the particles decreased in size, most probably because of spreading out from them of thin films, undetectable by electron microscopy. The presence of films could be inferred from the observation of new particles after the subsequent heating of the specimen in H_2. In the case of Fe, the particles transformed to extended patches and a few filaments formed, very likely as a result of carbon deposition. Carbonaceous filaments were not observed for any of the three metals. Baker *et al.* (26) also reported that carbonaceous filaments were not generated when heating in high-purity methane.

Upon heating Ni/Al_2O_3 in CH_4, following its pretreatment in steam, one notes that carbon was deposited around and/or on the particles. In this case, the NiO formed during pretreatment interacted more strongly than Ni with alumina. As a result, the particles did not migrate over the substrate, and carbon could not penetrate at the interface between particles and substrate to form filaments, for reasons discussed later in the paper.

The present experiments indicate that the main difference between heating in single gas components and their mixtures is the rapid change of the morphologies of the metal particles in the latter case. This may explain the rapid decline, noted by many investigators (*3, 14, 27, 28*), in the activity of the catalysts after they were subjected to reaction conditions. The previous investigators attributed the early deactivation to either sintering or coking. In contrast, we conclude that this deactivation is due to the coupling of sintering and coking. In other words, the rapid deactivation reflects the cooperative effect of the components of the gas mixture on the supported metal catalysts.

1b. CH_4 + Steam

Let us examine how this cooperative effect acts for the three metals. In the case of Ni/Al_2O_3, on heating the specimen in methane plus steam, one notes that the nickel particles are first oxidized by steam to nickel oxide, which is believed to constitute both the active catalytic species for the decomposition of methane (*29*) and the species responsible for the extension and spreading of the particles (*17*). However, the carbon deposited as a result of methane decomposition is subsequently gasified by steam. The gases thus resulting lift the particles and displace them over the surface of the substrate. This appears to be the main cause for the severe sintering that occurs on heating for 2 h the nickel model catalyst in the binary mixture of methane and steam.

In the case of Fe/Al_2O_3, steam has a much smaller effect because even traces of O_2 or water are enough for its oxidation, and therefore iron is already oxidized before the heating in methane plus steam and has already reacted with the substrate. Therefore, the particles did not migrate and remained at their initial locations, in spite of the gasification of carbon by steam.

In the case of Co/Al_2O_3, the cobalt particles acquired deformed shapes and undetectable films spread out from them over the substrate, indicating that cobalt spreads more easily than Ni or Fe over alumina. The undetectable films were inferred from the new particles which were formed after subsequent heating in H_2.

1c. CH_4 + Steam + H_2

In a mixture of H_2 + CH_4 + steam two different behaviors were observed for similar specimens of Ni/Al_2O_3, belonging to two different batches. In one specimen, the particles became elongated, forming tails behind dark leading particles. Electron diffraction indicated the presence of NiO, but carbon was also probably present. The behavior was similar to that observed on heating in CH_4 + steam, except that sintering occurred to a greater extent, because of the presence of H_2. This enhances gasification of carbon deposits and, in addition, ensures more contracted forms of the crystallites; both effects enhance sintering. In the other specimen, the particles acquired toroidal shapes with a small particle in the cavity, after 10 min of heating. NiO was identified by electron diffraction. Upon heating for 20 more minutes, the particles were reduced to Ni and sintered, and long hollow interlinked carbonaceous filaments appeared. However, on further heating, the carbonaceous filaments disappeared and sintering continued to occur. The disappearance of the carbonaceous filaments was due to the carbon gasification by H_2 and H_2O. The severe sintering was caused by the formation of Ni, which has a higher mobility than NiO, and by the gases produced by gasification. Note that in the former specimen, the particles were present as NiO, while in the latter, they were present as NiO for the first 10 min of heating and were reduced to Ni on subsequent heating. The different behavior of the two similar specimens appears to be of kinetic origin, being probably related to the reducibility of the NiO. These results indicate that heating in CH_4 + steam + H_2 oxidizes the Ni particles during the initial period. On subsequent heating, the particles can be reduced or not, and then they will have one or the other kind of behavior.

In the case of Co/Al_2O_2, the particles deformed during heating in CH_4 + steam + H_2, most likely (as discussed later) because of the penetration of the deposited carbon inside the particles. Compared to the heating in CH_4 + steam, less spreading of the particles occurred, because the particles were less oxidized. However, since on subsequent heating in H_2 only a few new particles appeared, it is clear that a large part of the material was permanently lost to the gas stream.

In the case of Fe/Al_2O_3, the particles were deformed throughout their heating in CH_4 + steam + H_2 and part of the material disappeared, being permanently lost to the gas stream. This was inferred from the subsequent heating in H_2, and happened, as discussed later in the paper, because of the disintegration of the particles.

1d. CO

The effect of heating in CO will be given only for Ni/Al_2O_3. Various types of carbon deposits, such as filamentous carbon, carbonaceous films around the particles, and large patches of carbon, were obtained. Compared with the heating in CH_4, the amount of carbon deposited was greater, indicating that on Ni particles the disproportionation of CO occurs more easily than the decomposition of CH_4. Electron diffraction indicated the presence of Ni during the entire heating in CO, as is in fact expected for such a strong reducing agent. As discussed later, the formation of filamentous carbon is associated with the reduced state of the particles.

1e. CH_4 + Steam + H_2 + CO

For Ni/Al_2O_3, on heating for 5 min, the particles became elongated, as throughout their heating in CH_4, CH_4 + steam, or CH_4 + steam + H_2, and carbon deposits formed. Electron diffraction indicated the presence of NiO. This indicates that the oxidizing effect of steam is stronger during the initial heating. On further heating, the particles acquired spherical shapes and gradually were reduced to Ni, indicating that the reducing effect of H_2 and CO gradually increased to become, finally, the stronger one. The absence of carbon deposits throughout the latter heat treatment was probably a result of a faster rate of gasification by H_2 and steam than rate of carbon formation.

For Co/Al_2O_3, on heating for 5 min in the quaternary mixture numerous particles disappeared and the remaining particles became deformed, probably because of the penetration of the deposited carbon inside the particles. The particles disappeared, either because material was transferred to the substrate as undetectable films or because it was permanently lost to the gas stream. On subsequent heating in H_2, new small particles appeared, indicating that at least a fraction of the disappeared material remained on the substrate as undetectable films.

For Fe/Al_2O_3, numerous particles disappeared on heating in the quaternary atmosphere. In contrast to Co/Al_2O_3, the material was permanently lost, as indicated by the fact that it was not restored as new particles on subsequent heating in H_2.

2. THE GROWTH OF CARBONACEOUS FILAMENTS AND THE METAL SUPPORT

Interactions

For Ni/Al_2O_3, carbonaceous filaments appeared in CO, and, in some specimens, in CH_4 + steam + H_2, but as a transient phenomenon. In CH_4 and CH_4 + steam, carbonaceous filaments were

366   Wetting Experiments

not observed, although elongated particles with filamentous structure did form. The pretreatment of a specimen in steam suppressed the formation of filamentous structures in CH_4.

The mechanism which was proposed for the carbon filament growth (*10, 30, 31*) involved the diffusion of carbon through the catalyst particles, driven either by a concentration difference (*10*) or by a temperature difference (*31*), and its precipitation between the particles and the support. The termination of the growth was attributed to the formation of a carbonaceous "skin" at the exposed (free) particle surface.

An additional important element should be, however, included. The initiation of carbon filaments, which have leading metal particles, implies that carbon can penetrate between crystallites and substrate. This, however, can happen only when the interfacial free energy between carbon and substrate plus the interfacial free energy between carbon and crystallite becomes smaller than the interfacial free energy between crystallite and substrate. This condition is satisfied in some cases. For instance, on heating metal foils in carbon containing atmospheres, one notes that filaments form with all three metals: Ni, Co, and Fe. This probably happens because a thin surface layer, with a composition different from that of the bulk, exists, and the interfacial tension between the metal foil and the thin surface layer is higher than the sum of the interfacial tensions between carbon and the metal foil and between carbon and the thin layer. As a result, carbon penetrates between metal and thin layer.

In the case of supported metal catalysts, however, strong interactions between crystallites and support will lower their interfacial free energy (*21*), and the above condition may no longer be satisfied. Therefore, the strength of the above-mentioned interactions is critical for filament formation on supported metal catalysts. If the above condition is not satisfied, only changes in the particle shape and accumulation of carbon around and on the particles can occur.

When Ni/Al_2O_3 specimen is heated in CH_4 + steam, Ni is easily oxidized by steam. The NiO which is thus formed interacts strongly with Al_2O_3 (*20*) and the interfacial free energy between crystallites and support decreases. The absence of carbonaceous filaments during heating in CH_4 + steam indicates that the low interfacial free energy between NiO particles and substrate, which arises because of the strong interactions between them, does not satisfy the above thermodynamic inequality for filament formation. In contrast, the formation of carbonaceous filaments during heating in CO shows that the interfacial free energy between Ni particles and substrate is high enough for the above thermodynamic condition to be satisfied. When Ni/Al_2O_3 specimens were heated in CH_4 + steam + H_2, different behaviors occurred, depending on the chemical state of the crystallites. When the crystallites were present as NiO throughout the entire heat treatment, no carbonaceous filaments were generated. But, in a very similar sample of a different batch, carbonaceous filaments were formed as a transient process after an induction period. The crystallites were present as NiO throughout the induction period, but they were reduced to Ni when filaments were formed. The filaments disappeared after subsequent heating, because of their gasification.

Since Fe is easily oxidized even by traces of O_2 and/or H_2O, and the oxide thus formed interacts strongly with Al_2O_3 (for details see Ref. (*21*)), these interactions are indeed much stronger for Fe/Al_2O_3 than for Ni/Al_2O_3. As a result, filaments should be absent in the former system. Indeed, carbonaceous filaments have not been observed on heating Fe/Al_2O_3 catalyst in an atmosphere containing CH_4 and/or CO. Co had a behavior similar to that of Fe, since it is also easily oxidized.

Once the particles are detached from the substrate, filaments will grow, by the further displacement of the particles. This can happen, however, only if the bulk diffusion of carbon through the particles and/or the surface diffusion of carbon over the surface of the particle are sufficiently rapid for a part of the metal to remain exposed to the chemical atmosphere, thus allowing the chemical decomposition of CH_4 or CO to continue on the surface of the particle. The carbon which diffuses through the particles forms transient carbide phases (*32*), with a higher content of carbon at the particle surface and a lower one at the interface between particle and substrate. This gradient in carbon content probably constitutes the driving force for diffusion through the particle. The increase in the percentage of carbon inside the crystallites lowers their melting point, as indicated by the phase diagram, and the likelihood of defect formation is thus increased. This in turn increases

Simulation of the Behavior of Supported Metal Catalysts in Real Reaction Atmospheres **367**

the rate of diffusion of carbon through the crystallites and accelerates the process of precipitation of carbon between particle and substrate.

3. DEFORMATION OF THE PARTICLES

For Ni, the particles became elongated during heating in CH_4, or in mixtures containing CH_4 and/or CO. For Co, the particles deformed on heating in mixtures containing CH_4 and/or CO. For Fe, the deformation of the particles was observed in CH_4 + steam + H_2. The elongation and deformation of the particles are probably associated with carbide formation. This is because carbon dissolves into the metal and forms carbides. The heating temperature being higher than the decomposition temperature of the stable carbides of Ni, Co, and Fe, only transient unstable phases form, from which some carbon can precipitate inside the particles. This generates internal stresses, leading to the deformation of the particles. Indeed, the carbides of Ni, Co, and Fe are not stable at the temperature of 700°C at which the specimens were heated. NiC decomposes in the temperature range of 300–400°C (33); Fe_2C and Fe_3C are not stable over 450°C, and decompose into iron and carbon (34); Co_3C decomposes in the temperature range of 300–350°C (35).

In addition, as the carbon content inside the particles increases, the melting point of the particles decreases, as indicated by the phase diagram, and thus the particles become softer. This may contribute to the deformation of the particles. In the case of Ni/Al_2O_3, the softened particles acquire high mobility and migrate over the substrate by liquid-like motion, depositing material behind the leading particle.

4. THE LOSS OF ACTIVE METAL

During heating of Fe/Al_2O_3 in the mixture of CH_4 + steam + H_2, a fraction of the particle material disappeared. This can be explained via unstable carbide formation, as follows: When small amounts of carbon penetrate and precipitate inside the particles, the latter will deform to release the internal stresses. This can disintegrate the particles, and at least a fraction of the resulting metal powder can be lost permanently to the gas stream. In the case of Co/Al_2O_3, a smaller amount of material was lost than in the case of Fe/Al_2O_3. This suggests that in the latter case, the particles disintegrate more easily.

On heating Fe/Al_2O_3 in CH_4 + steam + H_2 + CO, one notes that the loss of material was even greater than on heating in the gas mixture without CO. This is most likely due to the formation of volatile carbonyl. Blackmond and Ko (36) reported a nearly 30% weight loss of Ni supported on SiO_2 after CO chemisorption at 273 K, due to the formation of $Ni(CO)_4$. For Co/Al_2O_3, the loss of material during heating in CO containing mixture was smaller than that for Fe/Al_2O_3. For Ni/Al_2O_3, it is not clear whether the material was permanently lost during the heat treatment in the gas mixtures containing CH_4 and/or CO, because of the sintering of the particles.

CONCLUSION

In order to investigate the effect of the gas atmosphere on the physical and chemical changes of the particles in the steam reforming reaction, Ni/Al_2O_3, Co/Al_2O_3, and Fe/Al_2O_3 model catalysts were heated at 700°C in various atmospheres, such as CH_4, CO, and various gas mixtures containing CH_4 and/or CO.

During heating in CH_4, the above metals behaved differently. For Ni/Al_2O_3, the particles became elongated, acquiring filamentous shapes, with leading migrating particles leaving their material along their trajectory until they disappeared. The elongated particles were composed mostly of NiO and probably some carbon. For Co/ Al$_2$O$_3$, carbon deposits were formed around and on the particles, while for Fe/Al$_2$O$_3$, the particles initially extended on the substrate, and on further heating, carbon deposits were formed around the particles, as well as straight filaments, most likely composed of γ-Fe$_2$O$_3$ (or Fe$_3$O$_4$) and some carbon deposits.

On heating Ni/Al$_2$O$_3$ in CH$_4$ + steam, one notes that the particles extended initially on the substrate, because of their oxidation by steam. On further heating, the particles became elongated, but much less than in CH$_4$, and sintering occurred, because the gasification of deposited carbon enhanced the mobility of the particles on the substrate. For Co/Al$_2$O$_3$, the particles deformed, carbon deposits were formed around and on the particles, and undetectable films spread out from the particles over the substrate. For Fe/Al$_2$O$_3$, the particles extended on the substrate but less than for Co/Al$_2$O$_3$, because they were already partially oxidized before heating in CH$_4$ + steam and had therefore already reacted with the substrate.

On heating Ni/Al$_2$O$_3$ specimens in the gas mixture of CH$_4$, steam, and H$_2$, one notes that severe sintering occurred. In addition, hollow carbonaceous filaments were formed when the crystallites were present as Ni, and did not form when the crystallites were present as NiO. Carbonaceous filaments have also appeared when Ni/Al$_2$O$_3$ specimens were heated in CO. In this case, the crystallites were present as Ni. Hence, in the case of Ni/Al$_2$O$_3$, carbonaceous filaments can form if the particles are present as Ni and not as NiO.

A more general conclusion is that carbon filaments appear in those particular catalysts for which the interactions between crystallites and substrate are sufficiently weak. A thermodynamic criterion, based on wetting, is proposed for filament formation. The interfacial free energy between metal and substrate must be larger than the sum of the interfacial free energies between carbon and substrate, and carbon and metal. If this condition is satisfied, the crystallite can be detached from the substrate by a layer of carbon. The rate of growth of the filament is affected by the diffusion of carbon through the particles, diffusion which is probably facilitated by the formation of transient carbides.

In addition to filamentous carbon, other types of carbon deposits have also been identified. When steam-pretreated Ni/Al$_2$O$_3$ was heated in CH$_4$, carbon was deposited around and on the particles. Carbonaceous filaments were not generated because NiO interacts strongly with the substrate. Therefore, the interfacial free energy between substrate and crystallite is small, and the thermodynamic condition for filament formation is not satisfied.

Deposits of carbonaceous films around and/or on the particles were also observed in other cases, such as Co/Al$_2$O$_3$ and Fe/Al$_2$O$_3$ in CH$_4$ and in gas mixtures containing CH$_4$, and Ni/Al$_2$O$_3$ in CO. Carbonaceous filaments wrapping elongated particles, as well as thick large patches of carbon deposits, were observed throughout heating of Ni/Al$_2$O$_3$ in CO.

Upon heating Co/Al$_2$O$_3$ and Fe/Al$_2$O$_3$ in CH$_4$ + steam + H$_2$, one notes that the particles deformed, and some material was permanently lost to the gas stream, because of the disintegration caused by the precipitation of carbon inside the particles.

When Ni/Al$_2$O$_3$ was heated in CH$_4$ + steam + H$_2$ + CO, severe sintering occurred because the particles were reduced after a short time to Ni which sinters more easily, and the gasification of the deposited carbon increased the mobility of the particles. For Co/Al$_2$O$_3$, spreading of the particles and permanent loss of material were observed. For Fe/Al$_2$O$_3$, a considerable amount of material was permanently lost.

Unstable carbides are suggested to be responsible for the observed deformation of the particles. Carbide formation inside the particle, followed by carbon separation, generates internal stresses, which, in turn, lead to the deformation of the particles.

The investigation also indicates another aspect of catalyst deactivation, namely, the loss of active metal crystallites to the gas stream. This process occurs in different atmospheres for different reasons. On heating in the mixtures containing methane, the precipitation of carbon inside the metal disintegrates the crystallites, a part of which is then carried away by the gas stream. On heating in the mixtures containing carbon monoxide, the metal carbonyl that possibly forms may be in part responsible for the loss of particles.

ACKNOWLEDGMENT

Preliminary experiments regarding the present paper have been carried out by X. D. Hu (37).

REFERENCES

1. Satterfield, C. N., *in* "Heterogeneous Catalysis in Practice," Chemical Engineering Series, p. 286. McGraw–Hill, New York.
2. Schaper, H., Amesz, D. J., Doesburg, E. B. M., and Van Reijen, L. L., *Appl. Catal.* **9**, 129 (1984).
3. Takemura, Y., Yamamoto, K., and Morita, Y., *Int. Chem. Eng.* **7**(4), 737 (1967).
4. Rostrup-Nielsen, J. R., "Symposium on the Science of Catalysis and its Application in Industry, FPDIL, Sindri, February 22–24, 1979." Paper No. 39.
5. Rostrup-Nielsen, J. R., *J. Catal.* **33**, 184 (1974).
6. Gardner, D. C., and Bartholomew, C. H., *Ind. Eng. Chem. Prod. Res. Dev.* **20**, 80 (1981).
7. Bartholomew, C. H., *Catal. Rev. Sci. Eng.* **24**, 67 (1982).
8. Trimm, D. L., *Catal. Rev. Sci. Eng.* **16**, 155 (1977).
9. Baker, R. T. K., and Harris, P. S., *in* "Chemistry and Physics of Carbon" (P. L. Walker, Jr., Ed.), Vol. 14, p. 83. Dekker, New York, 1979.
10. Rostrup-Nielsen, J. R., and Trimm, D. L., *J. Catal.* **48**, 155 (1977).
11. Jackson, S. D., Thomson, S. J., and Webb, G., *J. Catal.* **70**, 249 (1981).
12. Satterfield, C. N., *in* "Heterogeneous Catalysis in Practice," p. 288. McGraw–Hill, New York, 1980.
13. James, P., and Van Hook, O., *Catal. Rev. Sci. Eng.* **21**, 1 (1980).
14. Williams, A., Butler, G. A., and Hammons, J., *J. Catal.* **24**, 352 (1972).
15. Borowiecki, T., Denis, A., Nazimek, B., Grzegorczyk. W., and Barcicki, J., *Chemia Stosowana* **27**(3), 229 (1983).
16. Phillips, T. R., Yarwood, T. A., Mulhall, J., and Turner, G. E., *J. Catal.* **17**, 28 (1970).
17. Ruckenstein, E., and Hu, X. D., *J. Catal.* **100**, 1 (1986).
18. Ruckenstein, E., and Malhotra, L., *J. Catal.* **41**, 303 (1976).
19. Ruckenstein, E., and Sushumna, I., *J. Catal.* **97**, 1 (1986).
20. Ruckenstein, E., and Lee, S. H., *J. Catal.* **86**, 457 (1984).
21. Sushumna, I., and Ruckenstein, E., *J. Catal.* **94**, 239 (1985) (Section 3.4 of this volume).
22. Baird, T., Fryer, J. R., and Grant, B., *Nature (London)* **233**, 329 (1971).
23. Evans, E. L., Thomas, J. M., Thrower, P. A., and Walker, P. L., *Carbon* **11**, 441 (1973).
24. Boehm, H. P., *Carbon* **11**, 583 (1973).
25. Hofer, L. J. E., Sterling, E., and McCartney, J. T., *J. Phys. Chem.* **59**, 1153 (1955).
26. Baker, R. T. K., Harris, P. S., Henderson, J., and Thomas, R. B., *Carbon* **13**, 17 (1975).
27. Hayer, R. E., Thomas, W. J., and Hayer, K. E., *J. Caral.* **92**, 312 (1985).
28. Jackson, S. D., Thomson, S. J., and Webb, G., *J. Catal.* **70**, 249 (1981).
29. Renshaw, G. D., Roscoe, C., and Walker, P. L., Jr., *J. Catal.* **18**, 164 (1970).
30. Lobo, L. S., Trimm, D. L., and Figreiredo, J. L., "Proc. 5th Int. Congr. Catal. 1972," p. 1125. 1973.
31. Baker, R. T. K., Barber, M. A., Harris, P. S., Feates, F. S., and Waite, R. J., *J. Catal.* **26**, 52 (1972).
32. Renshaw, G. D., Roscoe, C., and Walker, P. L., Jr., *J. Catal.* **22**, 394 (1971).
33. Hofer, L. J. E., Cohn, E. M., and Peebles, W. C., *J. Phys. Colloid. Chem.* **54**, 1161 (1950).
34. Jack, K. H., *Proc. R. Soc. London, Ser. A* **195**, 56 (1948).
35. Kosolapova, T. Ya., *in* "Carbides," p. 177. Plenum, New York, 1971.
36. Blackmond, D. G., and Ko, E. I., *J. Catal.* **94**, 343 (1985).
37. Hu, X. D., MS. thesis, State University of New York at Buffalo, 1985.

Index

Note: Page numbers followed by f and t refer to figures and tables respectively.

A

Adhesion, work of, 11, 12, 32–33, 97
Ag/Al_2O_3 catalyst, 244, 251f, 257f
Ag/Al_2O_3 catalyst/various chemical environments, in. *See also* Wetting by catalysts
  about, 244–245
  conclusion, 261
  discussion
    sintering/various atmospheres, 259–261
  experimental
    heat treatment, 245
    sample, preparation of, 245
  results
    heating in C_2H_4, 254–255
    heating in $C_2H_4+O_2$, 255–258
    silver crystallites, heating in H_2/O_2, 245–251
    silver particles/oxidation of ethylene, 251–254
Alumina substrate, 131, 156, 185, 229, 314
Aqueous environment, 25, 37, 44t, 62t, 73t

B

Biology related experiments
  blood compatibility of foreign surfaces, surface energetic criterion of
    about, 8–9
    biomaterial surfaces, characterization of, 15–16
    blood-biomaterial interactions, 9–11
    blood-biomaterial interfacial tension, 11–14
    conclusions, 22–23
    energetic criterion, discussion of, 16–17
    interfacial instabilities, low interfacial tensions, 22–23
    liquid surface tension, 23
    nomenclature, 23
    nonpolar solids, blood compatibility of, 14
    polymers, surface modification of, 18–22
    thermodynamic/kinetic considerations, 18
  deposits on surfaces, nondestructive approach to
    about, 2–3
    characterization of, 4–6
    discussion, 6–7
    germanium surfaces, hydroxyapatite on, 3–4
  solids in aqueous environment, surface characterization of, 25–27
  thin film surface coating
    about, 28–29
    blood-biomaterial interaction, 29–33
    conclusions, 55–56
    polymer surfaces, characterization of, 36–47
    preparation, 33–36, 39f
    surface modification for enhancing biocompatibility, 47–55
Biomaterial surfaces, characterization of, 14, 17
Blood, 10–14

Blood compatibility of foreign surfaces, surface energetic criterion of. *See also* Biology related experiments
  about, 8–9
  biomaterial surfaces, characterization of, 15–16
  blood-biomaterial interactions, 9–11
  blood-biomaterial interfacial tension, 11–14
  conclusions, 22–23
  energetic criterion, discussion of, 16–17
  interfacial instabilities, low interfacial tensions, 22–23
  liquid surface tension, 23
  nomenclature, 23
  nonpolar solids, blood compatibility of, 14
  polymers, surface modification of, 18–22
  thermodynamic/kinetic considerations, 18
Blood plasma, 10–11, 14, 18, 24, 30
Boltzmann constant, 141, 240
Boundary conditions, 272t, 277
Bovine serum albumin (BSA), 115, 117, 118f, 119f, 120f, 121f, 122f
BSA. *See* Bovine serum albumin (BSA)

C

C_2H_4, heating in, 254–255
Co/Al_2O_3 catalyst, 319, 335, 344, 354
Contact angle, 16, 25–27
Cyclohexane. *See* Hexane/cyclohexane

D

Deposits on surfaces, nondestructive approach to. *See also* Biology related experiments
  about, 2–3
  characterization of, 4–6
  discussion, 6–7
  germanium surfaces, hydroxyapatite on, 3–4
Desorption, 117, 119, 125
Diffusion
  coefficient of, 261, 269, 277
  mass transfer due to, 284, 287
Dispersion interactions, 23, 39, 96, 102, 265
Dissolution pressure, 240
Driving force, thermodynamic, 10, 11, 30–36, 123–126, 227

E

Electron microscopy, 4–5, 69, 150, 163
Energetic criterion, surface, 9, 16–17, 30
Entropy, 122, 265
Environment
  octane. *See* Octane environment
  polar/nonpolar, 70, 96, 99, 101t, 111t

372   Index

F

Fe/Al$_2$O$_3$, 323, 335, 346, 358
Free energy, interfacial, 30–33, 41, 44t, 49f, 50f, 51t

G

γ-Al$_2$O$_3$ substrate, 134t
Germanium surfaces, 3–4
Glass/siliconized glass. *See* Solid surface

H

Hamaker constant, 124, 125t
Heat treatment, 131, 242t, 245, 294t, 314
Hexane/cyclohexane, 95, 99, 100f, 101t, 104
Hydrogel. *See* Solid surface
Hydroxyapatite, 3, 4f, 5t, 6

I

Instability, thermal, 313
Internal reflection spectroscopy, 2–7
Ionic strength, 114, 116–117, 121, 126
Iron on alumina/role of physical and chemical interactions.
        See also Wetting by catalysts
   about, 184–185
   conclusion, 230–231
   discussion
     sintering behavior, 225
     splitting behavior, 229–230
     strong-metal-support-interaction (SMSI), 229
     wetting behavior, 226–229
   experimental
     active metal, deposition of, 185
     alumina supports, preparation of, 185
     heat treatments, 185–186
   results and discussion
     heating in hydrogen/purified hydrogen, 186–195,
       214–221
     heating in oxygen, 196–214, 221–224
     moisture/residual oxygen, effect of, 224
Irradiation, ultraviolet, 18, 19f, 49f, 87, 89t

L

Lennard-Jones potential, 264
Line tension, 141, 240

M

Marangoni effect, 238
Metal catalysts, behavior of/simulation of. *See also*
        Wetting by catalysts
   about, 312–313
   conclusion, 367–368
   discussion
     active metal, loss of, 367
     carbonaceous filaments, growth of, 365–367
     deformation of particles, 367
     various atmospheres, in, 363–365
   experimental
     heat treatment, 314–317
     specimen preparation, 314

   results
     CH$_4$ + steam + H$_2$ + CO, heating in, 355–363
     CH$_4$ + steam + H$_2$, heating in, 338–351
     CH$_4$ + steam, heating in, 331–338
     CO, heating in, 351–355
     methane, heating in, 318–331
Microstress, 260
Mobility, 14, 54, 84, 260, 367

N

Ni/Al$_2$O$_3$ catalyst, 318, 331, 338, 351, 355
Nickel/titania catalysts/strong metal-support interactions,
        effect of. *See also* Wetting by catalysts
   about, 292–293
   conclusion, 310
   discussion
     cavities formation, 309–310
     heating in O$_2$/H$_2$, 308–309
     shape changes, 307–308
   experimental
     heat treatment, 293
     sample preparation, 293
     TEM observation, 293
   results
     cavities/channels formation, 298
     oxygen/hydrogen, heating in, 303–307
     purified hydrogen, behavior in, 294–302
     sintering, 298–299
Ni/TiO$_2$ catalyst, 293, 308–310

O

Octane environment, 25–27, 39f, 41–42, 44t, 47t
Ostwald ripening, 162, 187t, 225, 239, 266

P

Pd/γ-Al$_2$O$_3$, 234–235
PdO, 238–239, 242. *See also* Supported metal crystallite,
       wetting during alternating heating
Photooxidation, 87, 89, 91, 93
Platelet adsorption, 10, 30
Platinum crystallite supported on alumina, redispersion of.
        See also Wetting by catalysts
   about, 130–131
   conclusions, 140–141
   discussion
     fracture, redispersion by, 137–139
     spreading, redispersion by, 139–140
   dissolution pressure/critical radius, derivation of, 141
   results
     O$_2$/H$_2$, effect of, 132–134
     Pt crystallite, redispesion/sintering of, 135
     wet H$_2$/wet N$_2$, effect of, 135–137
PMMA. *See* Polymethylmethacrylate (PMMA)
Poly(2-hydroxyethyl methacrylate) (poly(HEMA)), 98, 105
Poly(HEMA). *See* Poly(2-hydroxyethyl methacrylate)
       (poly(HEMA))
Polymeric solids, surface restructuring of/polymer-water
       interface, stability of. *See also* Wetting of
       polymers, experiments on
   about, 77–78
   characterization of/contact angle, 79–80

Index

conclusions, 84–85
preparation/characterization, aqueous
  environment, 78–79
results and discussions, 80–84
Polymeric surfaces, adsorption of proteins onto/bovine
  serum albumin. *See also* Wetting of polymers,
    experiments on
  about, 113–115
  adsorption experiments, 117
  characterization, 116–117
  conclusions, 126
  discussion
    desorption, 125
    ionic strength, effect of, 126
    isotherms, 122–125
    pH, effect of, 125–126
    rearrangement, 125
materials, 115–116
results
    adsorption isotherms, 117–118
    adsorption kinetics, 120–121
    desorption, 119
    ionic strength, effect of, 121–122
    pH, effect of, 121
Polymeric surfaces, stability of/ultraviolet irradiation.
    See also Wetting of polymers, experiments on
  about, 87
  conclusion, 93
  discussion, 90–93
  experimental/methodology, 87–89
  results, 89–90
Polymers, surface restructuring of. *See also* Wetting of
    polymers, experiments on
  about, 104
  conclusion, 111
  discussion, 108–111
  dynamic contact angle, 105–106
  materials, 104–105
  results, 107
Polymethylmethacrylate (PMMA), 87, 89f, 90f, 91t, 92f
Potential energy, film-substrate, 267
Pt/Al_2O_3, 143, 162, 166t, 167f, 173f
$PtAl_2O_3$ catalyst, crystallite migration in. *See also* Wetting
    by catalysts
  about, 162–164
  conclusion, 181
  discussion, 178–181
  experimental
    alumina support preparation, 164–165
  results
    changes in film thickness, 167f–180f
    events on $PtAl_2O_3$ catalyst, 165–166
Pt/alumina, redispersion of/via film formation. *See also*
    Wetting by catalysts
  about, 143
  conclusion, 160
  discussion, 157–160
  experimental, 144
  results
    film thickness changes, $PtAl_2O_3$, 145f, 146f,
      147f, 148f, 149f, 151f, 152f, 153f, 154f, 155f,
      156f, 157f
Pt crystallite, 130, 132, 134t, 135–137
PtO_2, 145, 150, 153, 159

R

Reaction-diffusion equation, 271, 272t
Redispersion, 130, 132, 134f, 139, 263
Relaxation time, 37–38, 62–63
Restructuring of polymer surfaces, environmentally
    induced/aqueous environment. *See also*
      Wetting of polymers, experiments on
  about, 62–63
  conclusion, 74–75
  nomenclature, 75
  results/discussion
    contact angle measurement, 70–74
    polymeric surfaces, preparation/characterization, 68–70
  surface free energy/relaxation time
    about, 63–64
    solid-water-octane contact angle, 65–66
    surface tensions, equations for, 66–68
Restructuring/polymer surfaces, 37–38, 41, 47, 56
Restructuring solid surfaces, free energy components.
    See also Wetting of polymers, experiments on
  about, 94
  conclusion, 103
  contact angle, 99
  discussion, 102–103
  estimation, 95–97
  materials, 97–99
  results, 99–102

S

Saliva, 3, 6f, 32
Silicone elastomer. *See* Solid surface
Sintering, 134, 225, 259, 263, 298
SMSI. *See* Strong-metal-support-interaction (SMSI)
Solids in aqueous environment, surface characterization
    of. *See* Biology related experiments
Solid surface
  glass/siliconized glass, 97, 98t, 99, 100f, 105t
  hydrogel, 13, 67, 98t, 100f, 105t
  hydrophilic, 15, 37, 98t, 100f, 105t
  hydrophobic, 14, 48, 98t, 102, 115t
  silicone elastomer, 97, 98t, 99, 100f, 105t
Splitting, 141, 166t, 212f, 229
Spreading, critical radius for, 141
Strong-metal-support-interaction (SMSI), 150, 160, 229
Stress. *See also* Microstress
  spreading generated, 229, 230, 239, 243
  internal, 137, 139, 367
Supported metal crystallites/wetting in sintering and
    redispersion in. *See also* Wetting by catalysts
  about, 263
  discussion, 266
  film formation, free energy of, 267
  thick film, wetting by, 264
  thin film, wetting by, 264–265
Supported metal crystallite, wetting during alternating
    heating. *See also* Wetting by catalysts
  about, 232
  conclusion, 242–243
  discussion
    H_2, heating in, 240–242
    O_2, heating in, 238–239
    PdO, emission of, 239–240

374 Index

Supported metal crystallite, wetting during alternating heating. *See also* Wetting by catalysts (*Continued*)
    experimental procedures, 233
    results, 233–238
Surface tension, 8, 30, 47t, 66, 139

T

Tearing, 207f, 229
Thin film surface coating. *See also* Biology related experiments
    about, 28–29
    blood-biomaterial interaction
        interfacial free energy, 32–33
        surface interactions, 30–32
        conclusions, 55–56
    polymer surfaces, characterization of
        about, 36–37
        free energy components/relaxation time, 38–43
        restructuring of, environmentally induced, 37–38
        results and discussion, 43–47
    preparation
        about, 33–35
        radio frequency sputter deposition, 35–36
    surface modification for enhancing biocompatibility, 47–55

W

Wetting by catalysts
    Ag/Al$_2$O$_3$ catalyst/various chemical environments, in
        about, 244–245
        conclusion, 261
        discussion, 259–261
        experimental, 245
        results, 245–258
    iron on alumina/role of physical and chemical interactions
        about, 184–185
        conclusion, 230–231
        discussion, 225–230
        experimental, 185–186
        results, 186–224
    metal catalysts, behavior of/simulation of
        about, 312–313
        conclusion, 363–368
        discussion, 363–367
        experimental, 314–317
        results, 318–363
    nickel/titania catalysts/strong metal-support interactions, effect of
        about, 292–293
        conclusion, 310
        discussion, 307–310
        experimental, 293
        results, 293–307
    platinum crystallite supported on alumina, redispersion of
        about, 130–131
        conclusions, 140–141
        discussion, 137–140
        dissolution pressure/critical radius, derivation of, 141
        results, 132–137

PtAl$_2$O$_3$ catalyst, crystallite migration in
    about, 162–164
    conclusion, 181
    discussion, 178–181
    experimental, 164–165
    results, 165–178
Pt/alumina, redispersion of/via film formation
    about, 143
    conclusion, 160
    discussion, 157–160
    experimental, 144
    results, 144–157
supported metal crystallites/wetting in sintering and redispersion in
    about, 263
    discussion, 266
    film formation, free energy of, 267
    thick film, wetting by, 264
    thin film, wetting by, 264–265
supported metal crystallite, wetting during alternating heating
    about, 232
    conclusion, 242–243
    discussion, 238–242
    experimental procedures, 233
    results, 233–238
zeolite/silica–alumina catalysts, optimum design of
    about, 269–271
    analysis, 271–282
    conclusion, 289
    nomenclature, 290–291
    results/discussion, 282–289
Wetting of polymers, experiments on
    polymeric solids, surface restructuring of/polymer-water interface, stability of
        about, 77–78
        characterization of/contact angle, 79–80
        conclusions, 84–85
        nomenclature, 85
        preparation/characterization, aqueous environment, 78–79
        results and discussions, 80–84
    polymeric surfaces, adsorption of proteins onto/bovine serum albumin
        about, 113–115
        adsorption experiments, 117
        characterization, 116–117
        conclusions, 126
        discussion, 122–126
        materials, 115–116
        results, 117–122
    polymeric surfaces, stability of/ultraviolet irradiation
        about, 87
        conclusion, 93
        discussion, 90–93
        experimental/methodology, 87–89
        results, 89–90
    polymers, surface restructuring of
        about, 104
        conclusion, 111
        discussion, 108–111
        dynamic contact angle, 105–106
        materials, 104–105
        results, 107

Index

restructuring of polymer surfaces, environmentally
induced/aqueous environment
about, 62–63
conclusion, 74–75
nomenclature, 75
results/discussion, 68–74
surface free energy/relaxation time, 63–68
restructuring solid surfaces, free energy components
about, 94
conclusion, 103
contact angle, 99
discussion, 102–103
estimation, 95–97
materials, 97–99
results, 99–102
Wetting, partial, 266

Y

Young equation, 38, 64, 139, 226, 239
Young modulus, 23, 230

Z

Zeolite/silica–alumina catalysts, optimum design of.
See also Wetting by catalysts
about, 269–271
analysis
composite-center distribution, 271–280, 282–286
composite-surface distribution, 281–282, 286–289
conclusion, 289
nomenclature, 290–291
results/discussion, 282–289